Forest Products Biotechnology

Forest Products Biotechnology

edited by

DR ALAN BRUCE

and

DR JOHN W. PALFREYMAN

both of the Scottish Institute for Wood Technology, University of Abertay Dundee, Scotland, UK

Taylor & Francis
Publishers since 1798

UK Taylor & Francis Ltd., 1 Gunpowder Square, London ECA 3DE
USA Taylor & Francis Inc., 1900 Frost Road, Suite 101, Bristol PA 199007

First Indian Reprint 2003

British Library Cataloguing in Publication Data

A catalogue record for this book is available from the British Library

ISBN 0-7484-0415-5

Library of Congress Cataloguing Publication data are available

Cover design by Jim Wilkie

Typeset in Times 10/12pt by Santype International Ltd., Salisbury, UK.
Printed at Chennai Micro Print Pvt. Ltd. (Export Division) 100% EOU
No. 34, Nelson Manickam Road, Aminjikarai, Chennai 600 029 (India).

For sale in india, Nepal and Bangladesh only

Contents

Contents

List of Contributors

DR J. J. MORRELL &
DR B. L. GARTNER
Department of Forest Products
Oregon State University
Forest Research Laboratory
Corvallis
Oregon OR 97331-5709
USA

DR T. L. HIGHLEY &
DR W. V. DASHEK
US Department of Agriculture
Forest Service
Forest Products Laboratory
One Gifford Pinchot Drive
Madison
Wisconsin WI 53705-2398
USA

DR G. DANIEL &
DR T. NILSSON
Department of Forest Products
The Swedish University of Agricultural
Sciences
Box 7008
S-750-07 Uppsala
Sweden

PROF. DR K. MESSNER
Technische Universitat Wien
Institut fur Biochemische Technologie
und Mikrobiologie
Getreidemarkt 9/172
A-1031 Wien
Austria

PROF. L. VIIKARI, DR J. BUCHERT &
DR A. SUURNÄKKI
VTT
Technical Research Centre of Finland
Biotechnology and Food Research
PO Box 1500
FIN-02044 VTT
Espoo
Finland

DR S. R. GUIOT &
MR J.-C. FRIGON
Biotechnology Research Institute
National Research Council
Montreal
Quebec
Canada

DR A. BORAZJANI &
DR S. V. DIEHL
Mississippi Forest Products
Laboratory
Mississippi State University
Box 9820
Mississippi State MS 39762
USA

DR A. MAJCHERCZYK & DR A. HÜTTERMANN
Institute of Forest Botany
University of Göttingen
Busgenweg 2
37077 Göttingen
Germany

DR A. KHARAZIPOUR
Forschung und Entwicklung
Pfleiderer Industrie GmbH & Co. KG
Duropal-Werk Arnsberg
Westring 19–21
D-59759 Arnsberg
Germany

DR J. R. OBST
US Department of Agriculture
Forest Service
Forest Products Laboratory
One Gifford Pinchot Drive
Madison
Wisconsin WI 53705-2398
USA

PROF. A. PIZZI
Ecole Nationale Superieure Des
Technologies et Industries Du Bois
(ENSTIB)
University of Nancy
27 Rue du Merle Blanc-BP 1041
88051 Epinal Cedex 9
France

PROF. J. N. SADDLER &
MR D. J. GREGG
Forest Products Biotechnology
Faculty of Forestry
University of British Columbia
270-2357 Main Mall
Vancouver
British Columbia
Canada

DR F. C. MILLER
140 Holl Road
Cabot
Pennsylvania PA 16023
USA

DR A. A. KADIR
Institut Penyelidikan Perhutanan
Malaysia
Forest Research Institute Malaysia
Kepong
52109 Kuala Lumpur
Malaysia

DR B. GOODELL &
DR J. JELLISON
Respectively: Department of Forest
Management and Department of Plant
Biology and Pathology
University of Maine
5722 Deering Hall
Orono
Maine 04469-5722
USA

DR A. BRUCE &
DR J. W. PALFREYMAN
Scottish Institute for Wood
Technology
School of Molecular & Life Sciences
University of Abertay Dundee
Bell Street
Dundee DD1 1HG
Scotland
UK

DR S. R. PALLI &
DR A. RETNAKARAN
Great Lakes Forestry Center
Canadian Forest Service
Sault Ste Marie
Ontario
P6A 2E5
Canada

DR A. SÉGUIN &
DR G. LAPOINTE
Natural Resources
Canadian Forest Service
Laurentian Forestry Centre
1055 du PEPS
PO Box 3800
Sainte-Foy
Quebec
G1V 4C7
Canada

DR P. J. CHAREST
Biotechnology Coordinator
Technology Transfer and
Commercialisation
Natural Resources Canada
Canadian Forest Service
Science Branch
580 Booth Street
7th Floor
Ottawa
Ontario
K1A 0E4
Canada

Wood as a Material

JEFFREY J. MORRELL AND BARBARA L. GARTNER

1.1 Introduction

One of the most important features of wood structure is that it is variable. Wood is a composite of various components (at spatial scales of microns to metres) that themselves are highly ordered. The relative abundance of these components, however, varies with environment and genetics. After discussing the importance of wood from several viewpoints, we overview the general characteristics of wood at its various scales, emphasizing features of importance to biotic degradation.

We often consider wood from our own narrow interests, but this resource must be viewed on a broader basis. Foremost, wood tissue serves to conduct moisture and nutrients from the roots to the foliage of the living tree. In this same context, it also serves to support the canopy structurally permitting trees to extend above other plants to capture additional sunlight. In its conductive role, wood functions by allowing passive movement in the axial direction of liquids through the lumens of its dead cells. About 5–40 per cent of the wood volume is made of parenchyma cells that are living during their functional lifespan (Panshin and de Zeeuw, 1980). These cells, oriented axially or radially, may be important storage sites for organic compounds, and as such are functionally related to the phloem (inner bark) where most organic compounds are transported. In addition, these living cells interact physiologically with the conducting cells (Sauter, 1972; Van Bel, 1995). These functions require that all cells be interconnected, which is accomplished with intercellular pits. Insects, fungi, bacteria and marine borers have all evolved unique strategies for utilizing all or part of the lignocellulosic matrix of the cell walls and the contents of the living cells (Zabel and Morrell, 1992). The activities of these agents represent the field of forest products pathology.

In its simplest application, the energy stored in wood is used to produce heat for cooking or for industrial process. Worldwide utilization of wood is primarily for fuel. Wood serves as an important structural resource, providing a renewable material with high strength per unit weight. In many countries, wood is the primary construction material.

Wood is also a major potential chemical feedstock for synthesis of more complex materials. Although currently under-utilized in developed countries in this regard, periods of energy shortages have encouraged the use of wood as chemical feedstock. As our supplies of non-renewable energy resources such as oil decline, we will increasingly move towards forest and agricultural resources for various chemical feedstocks.

Wood can represent a large volume of the waste material generated by industrialized societies. Cellulose and lignin are the two most abundant polymers on earth. Lignin is particularly problematic because of its resistance to degradation, and the development of methods for efficient degradation of this resource into more readily utilizable materials has received extensive study (Ericksson *et al.*, 1990; Higuchi, 1985, 1990).

Finally, it is important to consider that trees represent far more than wood products. For generations forests have been used, but we are increasingly concerned about non-utilization issues in forestry. Items related to riparian zones, species diversity and management of ecosystems have emerged as crucial issues in the past decade and are increasingly influencing forest management decisions. The trends towards reduced harvests on many lands and softer, less intensive forestry practices on others will continue as we become more knowledgeable about our forests and their many functions. As a result, it will be critical that the materials that are harvested be used with the greatest efficiency.

While specialists often view wood from their particular perspective, a broader understanding of this unique material will provide new insights into the potential problems and opportunities for its utilization.

1.2 Wood Polymers

The polymers cellulose, hemicellulose and lignin comprise 90–98 per cent of the wood mass (Table 1.1). Wood extractives including phenolic compounds, lipids, proteins, and other materials comprise the remainder. Each polymer serves a specific function in the living tree.

Cellulose contains repeating units of β 1-4 linked D-glucose. Individual chains may contain 1500 to 2000 glucose units and be 2.5–5.0 mm long. Cellulose provides the strength to the wood cell wall. The cellulose chains are oriented into crystalline units termed microfibrils. The highly ordered nature of these microfibrils gives wood its high tensile strength. This property is particularly important for trees, which

Table 1.1 Relative amounts of lignin, cellulose and hemicellulose in representative conifers and hardwoods

Wood species	Component level (%)		
	Cellulose	Lignin	Hemicellulose
Acer rubrum L.	45	24	29
Fagus grandifolia Ehrh	45	22	29
Picea glauca (Moench) Voss	41	27	31
Pinus strobus (L.) Carr	41	29	27

Data from Kollmann and Côté (1968) (Reproduced with kind permission of Springer-Verlag, Berlin, Germany)

must support a canopy subjected to stresses caused by phenomena such as wind, snow loads or crown asymmetries.

Hemicelluloses (MW 18 000 to 100 000) are a heterogeneous class of polymers containing glucose, galactose, mannose, xylose and/or other sugars. Hemicellulose lacks the crystallinity and microfibrillar structure of cellulose and thus does not contribute substantially to the structural properties of wood. For many years, the role of hemicellulose was poorly understood, but recent investigations suggest that it is an integral component of the lignocellulosic matrix and may play a role in the resistance of wood to impact or sudden loading (Timell, 1986). It is thought to link covalently to the lignin, and through hydrogen bonds to the cellulose (Whistler and Chen, 1991). Some researchers suggest that it acts as a coupling agent between the hydrophilic cellulose microfibrils and the hydrophobic lignin matrix. In the absence of this agent, water films could develop along the microfibril/lignin interface, decreasing the wood's strength considerably. A number of biodeterioration studies have shown that hemicelluloses are among the first polymers to be degraded, suggesting that their utilization may represent a first key step in the decay process (Winandy and Morrell, 1993). Selectively blocking access to this nutritional source could represent a more targeted strategy for protecting wood from biodeterioration although methods for accomplishing this task remain elusive.

Lignins are amorphous high-molecular weight and highly branched polymers composed of phenyl propane units with numerous types of linkages between individual units (Saka and Goring, 1985). A few lignin structures have been elucidated (e.g. Nimz, 1973) but in general their structures remain unknown, although the nature of the repeating units and many of the linkages have been studied intensively. Lignins provide rigidity to wood by encasing the cellulose, and allow wood to creep through their viscoelastic nature. Lignin is believed to help improve the durability of wood against microbial attack (Vance *et al.*, 1980) by coating and protecting the cellulose microfibrils. While many organisms have evolved the ability to utilize hemicellulose or cellulose, relatively few can decompose lignin effectively (Zabel and Morrell, 1992; Eaton and Hale, 1993). The lignified cell wall is thus capable of resisting deterioration for the many decades of a tree's lifespan. The nature of the lignocellulosic matrix remains the subject of considerable debate owing to the difficulty of *in situ* observation.

1.3 Extractives

Materials that can be removed via various soaking procedures are termed extractives and may include sugars, phenolics, lipids, fatty acids, proteins, waxes and a host of other materials. The sapwood (see below) has much lower extractive content than does the heartwood, and its extractives are located in the parenchyma cells (ray and/or longitudinal). In the heartwood, these materials are present in the parenchyma cells as well as in the conducting cells (tracheids, vessels) into which they have been deposited (extruded). Extractives in sapwood generally exhibit little toxicity to potential wood-invading organisms; in fact, there is an increasing body of evidence indicating that these minor components may play a critical role in sustaining microorganisms at the early stages of colonization (Merrill and Cowling, 1966; Abraham *et al.*, 1993). Attempts to exploit this need by limiting access to specific nutrients, for example destroying all thiamine in the wood to limit colonization by fungi (which require exogenous thiamine), have generally failed, probably because of the low levels of these nutrients required for normal metabolic functions and the adaptations of the invading organisms which enable them to subsist on low levels of these nutrients (Highley, 1970). Extractives in the heartwood vary widely in their toxicity. Heartwood of some species, such as cedar and redwood, is characterized by the presence of potent extractives which render the wood highly resistant to microbial or insect attack (Scheffer and Cowling, 1966). These compounds are synthesized largely from carbohydrates present in the ray cells as they die to form the heartwood.

The toxicity, type, quantity and locations of the extractives are related to the taxonomy of the plant species as well as the individual's history and the part of the plant under consideration (Hillis, 1987). Thus, the property of natural durability must be exploited with some care. One opportunity for exploiting natural durability is the use of either genetic selection or silvicultural practice to enhance the decay resistance of certain wood species. For example, it may be possible to select specific clones that produce more durable heartwood or to identify silvicultural practices that optimize heartwood production or heartwood decay resistance. At present, however, the rewards for such selection are limited and a change in philosophy concerning the merits of naturally durable wood would be necessary for implementation of such a strategy.

1.4 Wood Cell Wall

The cell wall is a multi-layered system composed of the primary wall and up to three secondary cell wall layers. The middle lamella is the region of attachment between the primary cell walls of adjacent cells. The cell wall layers differ markedly in terms of the relative ratios of lignin, cellulose and hemicellulose as well as in the orientation of the cellulose microfibrils (Table 1.2). These differences account for the variations in material properties as well as the resistance of individual cell wall layers to microbial attack.

Table 1.2 Relative amounts of cellulose, hemicellulose and lignin by cell wall layer in a theoretical hardwood and conifer

Wood type	Cell layer	Component level (%)		
		Cellulose	Hemicellulose	Lignin
Hardwood	P/ML	0–20	10–30	50–90
	S1	50–55	30–35	15–20
	S2	40–55	20–30	20–30
	S3	20–40	20–30	25–55
Conifer	P/ML	5–15	20–25	60–75
	S1	15–35	20–30	30–60
	S2	35–55	30–35	15–30
	S3	50–55	30–35	15

Data from Tsuomis (1991) and Panshin and de Zeeuw (1980) (Reproduced with kind permission of Chapman & Hall, London, UK and McGraw-Hill, Inc., New York, US)

The primary cell wall and middle lamella are both highly lignified regions where the cellulose microfibrils are less uniformly ordered (Table 1.2). These layers are generally considered together and often represent a small fraction of the overall volume of the wood cell wall.

The secondary cell wall has three regions: the S1 (closest to the primary cell wall), S2 and S3 (lining the cell interior). Each of the cell wall regions is itself made up of layers of material, but the layers within a region are more similar to one another than the layers of adjacent regions. The S1 and S3 cell wall layers have microfibrils oriented in nearly flat helices (equatorial). The S2 cell wall layer has microfibrils that are oriented in a steep helix nearly parallel to the longitudinal axis of the cell. The S2 layer generally is responsible for much of the tensile strength of wood because it generally contains a higher percentage of cellulose, is thicker, and has more axially-oriented microfibrils than do the other layers. This zone is also a preferred zone of attack for many soft rot fungi, whose Type 1 soft rot cavities tend to align closely with the microfibrillar angle (Bailey and Vestal, 1937).

In some plant species, there is also a warty layer on the lumen inside the S3 cell wall layer. This material is presumed to be either additional materials deposited on the S3 or accumulations of protoplasmic debris left after cell death.

1.5 Cell Types

An additional factor affecting wood properties and utilization is the composition and arrangement of cells of different types (Bodig and Jayne, 1982; Gartner, 1995). The sapwood of the living tree has conducting cells (tracheids and vessels) that are dead at maturity (lacking cytoplasm and nucleii), cells that are alive at maturity (ray and axial parenchyma, epithelial cells that surround resin canals), and fibre cells that may remain alive or dead. In woody plants from environments with distinct seasons, the wood is produced in annual increments, each made up of earlywood and latewood. Earlywood often differs greatly from latewood in its composition of cell types, dimensions of its cells, and density. Earlywood is generally less dense, has wider lumens and narrower cell walls than latewood.

Softwoods (gymnosperms) have only one type of conducting cell, the tracheid, a closed-ended member of one to several millimetres length and 10–60 μm diameter. These cells remain in the rank and file in which they were produced, giving the wood an ordered appearance. Softwoods seldom have much axial parenchyma, and only certain taxonomic groups of the softwoods have resin canals with their surrounding epithelial cells. Tracheids perform the dual function of longitudinal fluid conduction and structural support. The radial system of softwoods always contains ray parenchyma (cells involved in storage of phloem products and a myriad of other physiological and defence tasks), and may also contain ray tracheids and resin canals.

In contrast, hardwoods (angiosperms) are usually characterized by wood of more heterogeneous structure. The conducting members usually include vessel elements (open-ended cells that adjoin neighbouring cells to form conduits called vessels), and they may also include tracheids. The size of vessels depends on plant stature, species, and location within the plant, but is generally 20–500 μm in width and from a few millimetres up to many metres in length (Zimmermann, 1983). The vessels experience substantial widening during maturation, pushing other cells out of the rank and file in which they were produced. Hardwoods also often have thick-walled fibres that provide mechanical support. Hardwoods may have a large proportion of axial parenchyma and they lack true resin canals. The radial system has rays that may be much broader (and contain more cells) than those of softwoods. The vessels in many species are highly permeable permitting rapid ingress of fluids for a variety of purposes, while the fibres often resist fluid movement. Ray cells conduct fluids laterally across the living stem. They are responsible for the high rates of drying in the radial direction and they provide pathways to the interior of the stem for invading microorganisms.

The types of cells, the frequency and size of vessels, ray cells and resin canals are all employed to identify wood of individual species. In addition, these differences help explain the variations in treatability (MacLean, 1952; Bailey, 1965), strength (USDA, 1987) and susceptibility to biodeterioration (Scheffer and Cowling, 1966) among the various species.

1.6 Cell Connections

Individual wood cells are interconnected via a series of openings termed pits. Pits consist of thin membranes composed primarily of cellulose and pectin, and regulate flow between cells. The membrane is mostly primary wall and middle lamella, both of which have been somewhat degraded enzymatically during cell maturation (Thomas, 1970; Parham and Baird, 1973; Barnett, 1981). Pits can be simple, semi-bordered or bordered (Figure 1.1). Pits in the sapwood are generally permeable, although older sapwood may have less permeable pits than younger sapwood (Sperry *et al.*, 1991). In heartwood the pits are impermeable because materials manufactured during sap/heart conversion accumulate on the membranes and also because some of the membranes may become pulled permanently to one side (by air pressure), blocking off the permeable part of the membrane (termed pit aspiration).

Understanding the nature of pit permeability as well as the factors that affect this property has long intrigued scientists (Wardrop and Preston, 1950; Côté, 1958; Krahmer, 1961; Côté and Krahmer, 1962). Pit permeability to air or water varies widely among species and within an individual tree. In the living plant, the structure of the pit membrane controls the magnitude of tension that can be tolerated by the water column before it cavitates, that is, before a small air bubble is pulled into the conduit and quickly expands, functionally breaking the water column (Tyree and Sperry, 1988; Sperry and Tyree, 1990; Sperry and Saliendra, 1994; Jarbeau *et al.*, 1995). Thus, the pits actually serve to control which parts of a plant stay alive during a drought (Zimmermann, 1983). The variability of pit membrane structure can affect a variety of properties, most notably wood drying and impregnation (Gregory and Petty, 1973). In lumber, pits represent the major anatomical feature affecting flow of gases and liquids through the wood. In living plants, which operate above fibre saturation point (i.e., the walls are saturated and some water may be present in the lumens), tracheid or vessel diameter is as important as pit characteristics in understanding the movement of liquids through wood.

In addition to the pits, the cell wall of many wood species is characterized by the presence of smaller microcapillaries. The function of these structures in the living tree remains unknown, although they are probably the source of air that seeps into vessels and tracheids during cavitation. They may also play an important role in the

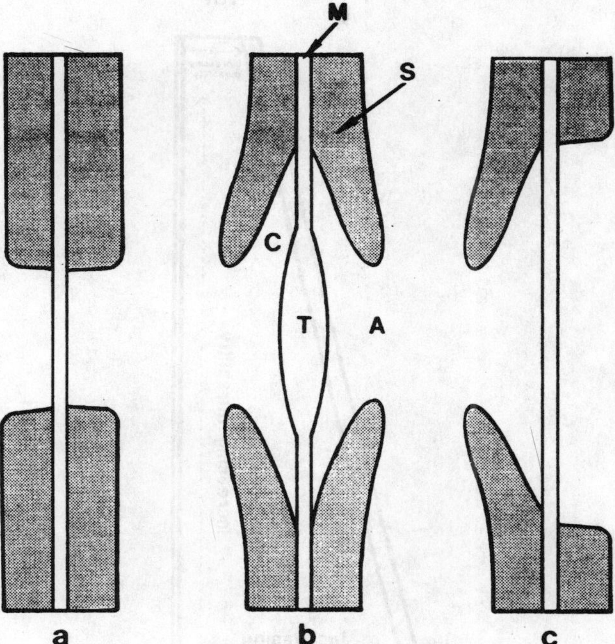

Figure 1.1 Simple (a), bordered (b), or semi-bordered (c) pits. T, torus; M, middle lamella; S, secondary cell wall; A, aperture; C, margo, (reproduced with kind permission from United States Department of Agriculture)

microbial degradation of the wood. A variety of studies have shown that microbial wood degradation enzymes, especially lignolytic enzymes, are present deep within the wood cell wall even at the earlier stages of attack (Ericksson *et al.*, 1990). Micro-capillaries have been implicated as channels permitting the movement of these enzymes into the seemingly impermeable wood cell wall. In green wood of specific gravity 0.45 (expressed per green volume), moisture content 115 per cent (g water/g dry weight of wood), values characteristic of Douglas-fir sapwood, one can calculate that 29 per cent of the volume is occupied by cell wall material, 34 per cent is occupied by water and 37 per cent is occupied by air. In the heartwood, with the same specific gravity and a moisture content of 37 per cent, 11 per cent is occupied by water and 60 per cent is occupied by air. Thus, air is locally abundant within green wood.

1.7 Intra-tree Variability

The structure of wood varies within an individual tree in predictable patterns depending on the species. Radially, the wood is divided into sapwood, which is the outer rings of wood containing live parenchyma cells, and heartwood, which is the interior of older trees that are no longer living. Sapwood and heartwood have differ-ent characteristic moisture contents (sap generally moister than heart in softwoods,

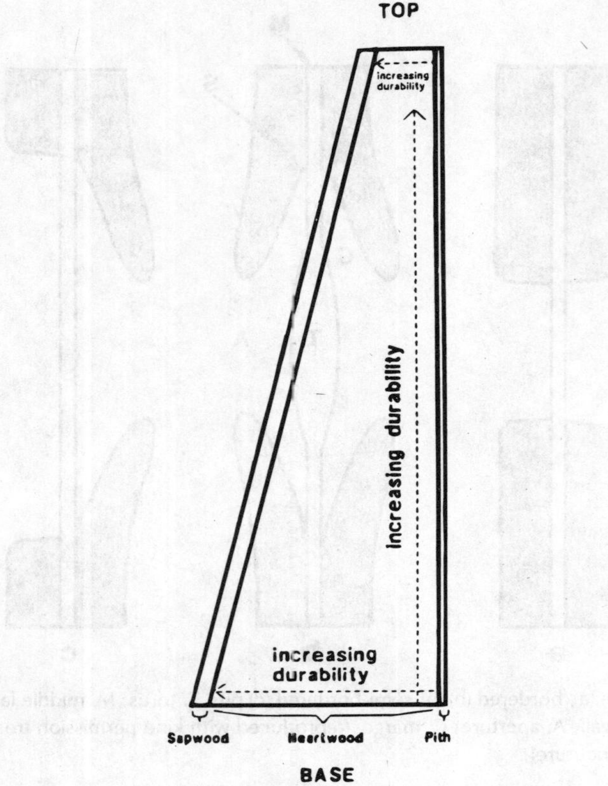

Figure 1.2 Relative decay resistance with stem position (reproduced with kind permission of Theodore Scheffer)

and, though more equal, heart moister than sap in hardwoods) (USDA, 1987). Sapwood extractives consist mostly of proteins, waxes, lipids and other storage materials. Heartwood extractives tend to be more phenolic in nature. The toxicity of these materials tends to be greatest at the heart–sap interface and declines towards the pith. Additionally, where nutrients are not limiting, cell dimensions generally increase radially (going outward from the pith towards the bark), and the microfibrils in the S2 cell walls become more axial (Dadswell, 1958; Megraw, 1985; Clark and Saucier, 1991).

Lignin levels vary with cell wall layer, rendering some layers more susceptible to microbial attack. Lignin types vary between hardwoods and conifers concomitant with other characteristics, making it difficult to determine whether the lignin changes actually alter durability.

Axially, the very base of a tree often has denser wood and more reaction wood (a modification of normal structure that produces stresses that help keep the tree upright) than higher up the stem. For a given growth ring, wood density decreases and cell dimensions increase between the base of the tree and the beginning of the live crown. These trends then reverse with increasing height in the crown. Similarly, decay resistance is often greatest in the base of a tree at the heartwood/sapwood interface and declines slightly radially from that zone (Figure 1.2). While there is relatively little that can be done to limit the effects of these variations on wood properties, a complete understanding of the magnitude of these changes can help to explain variations in biological and physical assessments of wood.

Lastly, the structure of wood varies by the organ of the plant. In general, roots are less dense than trunks, branches are denser than trunks (Fegel, 1941), and knots (branch insertions) are extremely dense (Hakkila, 1969). Roots owe their low density to having wider conducting elements and less latewood than stems. Branches are denser than stems because of their narrower conducting elements (Fegel, 1941). Roots and branches have more parenchyma than do trunks, particularly in hardwoods (Fegel, 1941), probably for storage of organic compounds. Growth rings are much narrower in branches and roots than trunks. Knots are sometimes very decay resistant owing to the deposition of high levels of resin and their high density. Conversely they often serve as invasion points for many heartrot fungi in the living tree.

1.8 Wood Properties

The chemical and cellular composition of wood and the patterns by which the polymers and cells are arranged can have marked effects on wood properties including dimensional stability, hygroscopicity, thermal conductivity, electrical conductivity, permeability and many mechanical properties (Côté, 1986; Schniewind and Berndt, 1991). All of these properties must be considered when contemplating the potential use of wood.

1.8.1 *Hygroscopicity*

The numerous exposed hydroxyl groups present on the various polymers in wood make this material especially susceptible to moisture uptake. Woods higher in extractives tend to be more dimensionally stable because many of the extractives (such as oils and waxes) decrease the wood's hygroscopicity. At lower moisture levels, any moisture absorbed will be bound to the wood, but as the wood moisture content approaches 30 per cent (w/w), free water begins to collect in the lumens.

The presence of free water provides a medium for diffusion of microbial enzymes and provides water for various metabolic processes by organisms that can degrade wood. As a result, architects deliberately attempt to keep the wood moisture content below 20 per cent. Where this is not possible, the use of preservatives to protect the wood or the substitution of alternative materials must be considered. On the opposite end of the scale, increasing the moisture content to fill the cell lumens with water can limit the amount of oxygen available for deterioration by aerobic organisms, also retarding microbial attack. This approach is often used in lumber mills which pond or sprinkle logs with water prior to cutting. Similarly, fungal deterioration is often minimal in ships that have been submerged for long periods either in the mud or in freshwater where marine borers are absent owing to the inhibition of many decay organisms due to low oxygen levels. Attack by anaerobic or microaerophilic bacteria, however, can eventually cause substantial degradation in these environments (Eaton and Hale, 1993).

1.8.2 Dimensional Stability

Moisture relationships also influence the dimensional stability of wood. Unlike steel or concrete, whose dimensions change little in the presence of water, wood tends to swell and shrink with changes in moisture (Table 1.3). Since the sorption occurs between individual longitudinally oriented microfibrils, the swelling tends to be greatest in the radial and tangential directions and least longitudinally. The dimensional changes associated with moisture sorption by wood pose a significant challenge to wood designers, and the development of effective methods for dimensionally stabilizing wood has been the subject of much interest since the early 1900s (Rowell, 1990). Most of the techniques are of limited application because they require the incorporation of extensive quantities of chemicals to block access to the hydroxyl groups by cross-linking or other reactions. Increasing challenges to the use of biocides to protect wood should encourage further exploration of this potentially more environmentally benign method for wood protection.

1.8.3 Permeability

The ability of fluids to move through wood has major implications for uses such as finishing, pulping, gluing and impregnation with preservatives. The pits are the primary influence on permeability in most wood. One would expect fluid movement through sapwood in the living tree to be relatively unobstructed, but that is not always the case. The best example of differences in sapwood permeability occurs between inland and coastal Douglas-fir whose sapwood differ markedly in permeability (Miller and Graham, 1963; Kavanagh *et al.*, 1996). These differences may reflect adaptations on the part of the inland Douglas-fir for the drier site conditions in which it is found. Generally however, sapwood is far more receptive to gas and fluid movement than is heartwood of the same species (Comstock, 1970). There has

Table 1.3 Relative shrinkage during drying from green to oven-dry moisture levels for selected wood species in the radial or tangential directions

Wood species	Radial	Tangential	Volumetric
Populus tremuloides (Aspen)	3.5	6.7	11.5
Ulmus americana (American white elm)	4.2	9.5	15.6
Quercus alba (American white oak)	5.6	10.5	16.3
Pseudotsuga menziesii (Douglas-fir)	4.8	7.6	12.4
Pinus taeda (Loblolly pine)	4.8	7.4	12.3
Thuja plicata (Western red cedar)	2.4	5.0	6.8

Data from USDA (1987)

been a variety of attempts to improve the permeability of heartwood. Researchers have used pectinolytic enzymes to dissolve the pit membranes, have applied mould fungi which grow through the wood cells via the pits (Lindgren, 1952; Graham, 1954), thereby improving permeability, and have explored the effects of drying conditions on subsequent permeability (Morris, 1991; Lebow *et al.*, 1996). In all instances, the results, while interesting, have been too inconsistent to warrant their commercial use.

1.8.4 Thermal and Electrical Conductivity

Because of its high void volume (see Section 1.8.1), wood is an excellent thermal insulator in comparison with other structural materials such as steel or concrete. Changes due to microbial attack can alter this insulating value slightly, but the effects are not substantial unless the wood mass is markedly lost. Similarly, the low levels of metals makes it an excellent insulator for electricity. Wood poles are commonly employed for supporting overhead electric lines because of their combination of high strength/weight and low conductivity. Electrical conductivity can, however, become altered by changes caused by microbial degradation or by changes in wood moisture content. Both of these changes can be assessed by measuring electrical resistance. Commercial resistance type moisture meters have been available for decades, while a resistance type meter for detecting the early stages of decay has been available since the early 1970s (Shigo and Shigo, 1974).

1.8.5 Mechanical Properties

While wood has numerous attractive properties, its ability to support substantial loads/mass places it among our most versatile renewable materials. The highly ordered nature of the cellulose microfibrils coupled with the crystalline nature of cellulose result in a material that has an extremely high tensile strength : mass ratio reflecting the need for a material that can support a leafy canopy without itself adding too much weight to the support column (Beery *et al.*, 1983).

Like any natural material, however, this material can be degraded by a variety of microorganisms. Subtle changes in the microstructure of wood can produce dramatic effects on material properties (Wilcox, 1978). These effects are most evident at the early stages of attack by brown rot fungi, but the nature of disruptions that occur at the early stages of colonization by wood-degrading organisms will require considerable study. Developing an understanding of these early stages of attack has important implications not only for identifying methods for preventing decay, but also for enhancing the activity of organisms in biopulping, feedstock improvement, waste detoxification and a host of other potential biotechnological uses of various wood-inhabiting microorganisms.

Wood is highly ordered on many scales (from polymer to organ) and along two axes (radial, axial). Because of the order but also because of the variety of configurations that can be found at each scale, wood is spatially heterogeneous with respect to chemical composition, decay resistance, mechanical performance, permeability, and thermal and electrical conductivity. These heterogeneities are the features that allow the wood to function for the living plant, and they impart some of the characteristics to wood that we prize, such as grain pattern and ability to absorb shocks. However, they present us with substantial challenges in wood utilization. They require that we have a knowledge of the structure of wood, and how that structure relates to the way in which the plant grew, in order to predict better how a particular specimen of wood will perform structurally, chemically, or as a microbial substrate. Such knowledge will help us use our forests and forest lands more efficiently, whether for fuel, structural lumber, pulp, or recreation.

References

ABRAHAM, L. D., ROTH, A., SADDLER, J. N and BREUIL, C. (1993) Growth, nutrition, and proteolytic activity of the sapstain fungus *Ophiostoma piceae*. *Can. J. Bot.* **71**, 1224–1230.

BAILEY, I. W. and VESTAL, M. R. (1937) The significance of certain wood-destroying fungi in the study of the enzymatic hydrolysis of cellulose. *J. Arnold Arboretum* **18**, 196–205.

BAILEY, P. J. (1965) (2) Some studies on the permeability of wood in relation to timber preservation. *Record of the 15th Annual Convention British Wood Preservers' Association*, Paper 2, 31–66.

BARNETT, J. R. (1981) Secondary xylem cell development. In: Barnett, J. R., ed., *Xylem Cell Development*. Tunbridge Wells, Kent: Castle House Publications, pp. 47–95.

BEERY, W. H., IFJU, G. and MCLAIN, T. E. (1983) Quantitative wood anatomy – relating anatomy to transverse tensile strength. *Wood Fiber Sci.* **15**, 395–407.

BODIG, J. and JAYNE, B. A. (1982) Material organization. In: *Mechanics of Wood and Wood Composites*. New York: Van Nostrand Reinhold, pp. 461–546.

CLARK III, A. and SAUCIER, J. R. (1991) Influence of planting density, intensive culture, geographic location, and species on juvenile wood formation in southern pine. *Georgia Forest Research Paper* **85**, 1–13.

COMSTOCK, G. L. (1970) Directional permeability in softwoods. *Wood Fiber* **1**(4), 283–289.

CÓTÉ, W. A. (1958) Electron microscope studies of pit membrane structure: implications in seasoning and preservation of wood. *Forest Prod. J.* **8**, 296–301.

CÓTÉ, W. A. (1986) *Wood Structure and Behavior – An Intimate Relationship*, Leslie L. Schaffer Forestry Lecture Series, The University of British Columbia.

CÓTÉ, W. A. and KRAHMER, R. L. (1962) The physical significance of pit structure for inter-tracheid liquid movement in coniferous wood. *Fifth International Congress for Electron Microscopy*.

DADSWELL, H. E. (1958) Wood structure variations occurring during tree growth and their influence on properties. *J. Inst. Wood Sci.* **1**, 11–33.

EATON, R. A. and HALE, M. D. C. (1993) *Wood: Decay, Pests, and Protection*. New York: Chapman and Hall.

ERICKSSON, K. E. L., BLANCHETTE, R. A. and ANDER, P. (1990) *Microbial and Enzymatic Degradation of Wood and Wood Components*. New York: Springer-Verlag.

FEGEL, A. C. (1941) Comparative anatomy and varying physical properties of trunk, branch, and root wood in certain northeastern trees. *Bulletin New York State College of Forestry at Syracuse University*, Technical Publication **14**(55), 5–20.

GARTNER, B. L. (1995) Patterns of xylem variation within a tree and their hydraulic and mechanical consequences. In: Gartner, B. L., ed., *Plant Stems: Physiology and Functional Morphology*. San Diego: Academic Press, pp. 125–149.

GRAHAM, R. D. (1954) The preservative treatment of Douglas-fir post sections infected with *Trichoderma* mold. *J. Forest Prod. Res. Soc.* **4**, 164–166.

GREGORY, S. C. and PETTY, J. A. (1973) Valve action of bordered pits in conifers. *J. Exp. Bot.* **24**, 763–767.

HAKKILA, P. (1969) Weight and composition of the branches of large Scots pine and Norway spruce trees. *Commun. Inst. Forestalis Feniae (Helsinki)* **67**(6).

HIGHLEY, T. L. (1970) Decay resistance of four wood species treated to destroy thiamine. *Phytopathology* **60**(11), 1660–1661.

HIGUCHI, T. (ed.) (1985) *Biosynthesis and Biodegradation of Wood Components.* New York: Academic Press.

HIGUCHI, T. (1990) Lignin biochemistry: biosynthesis and degradation. *Wood Sci. Technol.* **24**, 23–63.

HILLIS, W. E. (1987) *Heartwood and Tree Exudates.* New York: Springer-Verlag.

JARBEAU, J. A., EWERS, F. W. and DAVIS, S. D. (1995) The mechanism of water-stress-induced embolism in two species of chaparral shrubs. *Plant Cell Environ.* **18**, 189–196.

KAVANAGH, K. K, YODER, B. J., GARTNER, B. L. and AITKEN, S. N. (1996) Root and shoot vulnerability to cavitation in four populations of Douglas-fir seedlings. *Bull. Ecol. Soc. Am.* **77**(3), 226.

KOLLMAN, F. F. P. and CÔTÉ, W. A. (1968) *Principles of Wood Science and Technology, Volume I. Solid Wood.* Berlin: Springer-Verlag.

KRAHMER, R. L. (1961) Anatomical features of permeable and refractory Douglas-fir. *Forest Prod. J.* **11**(9), 439–441.

LEBOW, S. T., MORRELL, J. J. and MILOTA, M. R. (1996) Western wood species treated with chromated copper arsenate: effect of moisture content. *Forest Prod. J.* **46**(2), 67–70.

LINDGREN, R. M. (1952) Permeability of southern pine as affected by mold growth and other fungus infection. *Proc. Am. Wood Pres. Assoc.* **48**, 158–174.

MACLEAN, L. D. (1952) *Preservative Treatment of Wood by Pressure Methods.* US Department of Agriculture, Agricultural Handbook 40, Washington, DC.

MEGRAW, R. A. (1985) *Wood Quality Factors in Loblolly Pine: The Influence of Tree Age, Position in Tree, and Cultural Practice on Wood Specific Gravity, Fiber Length, and Fibril Angle.* Atlanta, GA: Tappi Press.

MERRILL, W. and COWLING, E. B. (1966) Role of nitrogen in wood deterioration. Amounts and distribution of nitrogen in tree stems. *Can. J. Bot.* **44**, 1555–1580.

MILLER, D. J. and GRAHAM, R. D. (1963) Treatability of Douglas-fir from western United States. *Proc. Am. Wood Pres. Assoc.* **59**, 218–222.

MORRIS, P. I. (1991) Improved preservative treatment of spruce-pine-fir at higher moisture contents. *Forest Prod. J.* **41**(11/12), 29–32.

NIMZ, H. (1973) Chemistry of potential chromophoric groups in beech lignin. *TAPPI* **56**(5), 124–126.

PANSHIN, A. J. and DE ZEEUW, C. (1980) *Textbook of Wood Technology: Structure, Identification, Properties, and Uses of the Commercial Woods of the United States.* New York: McGraw-Hill.

PARHAM, R. A. and BAIRD, W. M. (1973) The bordered pit membrane in differentiating balsam fir. *Wood Fiber* **5**, 80–86.

ROWELL, R. M. (1990) Chemical modification of lignocellulosic fibers to produce high-performance composites. In: Glass, J. E. and Swift, G., eds, *Agricultural and Synthetic Polymers – Biodegradability and Utilization.* ACS Symposium Series 433, Washington, DC: American Chemical Society, pp. 241–258.

SAKA, S. and GORING, D. A. I. (1985) Localization of lignins in wood cell walls. In: Higuchi, T., ed., *Biosynthesis and Biodegradation of Wood Components.* New York: Academic Press, pp. 51–62.

SAUTER, J. J. (1972) Respiratory and phosphatase activities in contact cells of wood rays and their possible role in sugar secretion. *Zeit Pflanzenphysiol.* **67**, 135–145.

SCHEFFER, T. C. and COWLING, E. B. (1966) Natural resistance of wood to microbial deterioration. *Annu. Rev. Phytopathol.* **4**, 147–170.

SCHNIEWIND, A. P. and BERNDT, H. (1991) The composite nature of wood. In: Lewin, M. and Goldstein, I. S., eds, *Wood Structure and Composition*. New York: Dekker, pp. 435–476.

SHIGO, A. L. and SHIGO, A. (1974) Detection of discoloration and decay in living trees and utility poles. USDA Forest Service Research Paper NE-294, Upper Darby, PA.

SPERRY, J. S. and SALIENDRA, N. Z. (1994) Intra- and inter-plant variation in xylem cavitation in *Betula occidentalis*. *Plant Cell Environ.* **17**, 1233–1241.

SPERRY, J. S. and TYREE, M. T. (1990) Water-stress-induced xylem embolism in three species of conifers. *Plant Cell Environ.* **13**, 427–436.

SPERRY, J. S., PERRY, A. and SULLIVAN, J. E. M. (1991) Pit membrane degradation and air-embolism formation in ageing xylem vessels of *Populus tremuloides* Michx. *J. Exp. Bot.* **42**, 1399–1406.

THOMAS, R. J. (1970) Origin of bordered pit margo microfibrils. *Wood Fiber* **2**, 285–288.

TIMELL, T. E. (1986) Wood: chemical composition. In: Bever, M. B., ed., *Encyclopedia of Materials Science and Engineering*. New York: Pergamon Press, pp. 5402–5408.

TSOUMIS, G. (1991) *Science and Technology of Wood: Structure, Properties and Utilization*. New York: Van Nostrand Reinhold.

TYREE, M. T. and SPERRY, J. S. (1988) Do woody plants operate near the point of catastrophic xylem dysfunction caused by dynamic water stress? Answers from a model. *Plant Physiol.* **88**, 574–580.

USDA (1987) *Wood Handbook: Wood as an Engineering Material*. US Department of Agriculture, Forest Service, Agriculture Handbook 72, Washington, DC.

VAN BEL, A. J. E. (1995) The low profile directors of carbon and nitrogen ecology in plants: parenchyma cells associated with translocation channels. In: Gartner, B. L., ed., *Plant Stems: Physiology and Functional Morphology*. San Diego: Academic Press, pp. 205–222.

VANCE, C. P., KIRK, T. K. and SHERWOOD, R. T. (1980) Lignification as a mechanism of disease resistance. *Annu. Rev. Phytopathol.* **18**, 259–288.

WARDROP, A. B. and PRESTON, R. D. (1950) The fine structure of the wall of the conifer tracheid. V. The organization of the secondary wall in relation to the growth rate of the cambium. *Biochim. Biophys. Acta* **6**, 36–47.

WHISTLER, R. L. and CHEN, C. C. (1991) Hemicelluloses. In: Lewin, M. and Goldstein, I. S., eds, *Wood Structure and Composition*. International Fiber Science and Technology Series No. 11, New York: Marcel Dekker, pp. 287–319.

WILCOX, W. W. (1978) Review of the literature on the effects of early stages of decay on wood strength. *Wood Fiber Sci.* **9**(4), 252–257.

WINANDY, J. E. and MORRELL, J. J. (1993) Relationship between incipient decay, strength, and chemical composition of Douglas-fir heartwood. *Wood Fiber Sci.* **25**(3), 278–288.

ZABEL, R. A. and MORRELL, J. J. (1992) *Wood Microbiology: Decay and its Prevention*. San Diego: Academic Press.

ZIMMERMANN, M. H. (1983) *Xylem Structure and the Ascent of Sap*. Berlin: Springer-Verlag.

Biotechnology in the Study of Brown- and White-Rot Decay

TERRY L. HIGHLEY AND WILLIAM V. DASHEK

2.1 Introduction

Many different types of organisms deteriorate wood, but the greatest damage results from fungi. Decay is the most serious form of microbiological deterioration because it can cause structural failure, sometimes very rapidly. Because widespread damage to wood from decay is seldom spectacular, the tremendous economic and resource loss resulting from decay is often overlooked. This oversight is exemplified by a failure to recognize that biodeterioration of wood products is important and that extending the service life of wood is a sure way to prolong the available timber supply. Knowledge about the biochemical systems of decay fungi can serve another purpose as well. Decay fungi can be used for the bioconversion of lignified tissue, such as biopulping or enzymatic treatment of pulps (Kirk and Hammel, 1992). Recent research has advanced our understanding of how wood components are degraded by microorganisms and their enzymes, permitting rapid advances in biotechnology (Wainwright, 1992).

The most important and potent wood-destroying organisms are white- and brown-rot fungi, which attack various components of the wood cell wall. Most white-rot fungi utilize cellulose and hemicelluloses at approximately the same rate relative to the original amounts present, whereas lignin is usually utilized at a somewhat faster relative rate. A few white-rot fungi remove lignin and hemicelluloses preferentially, but ultimately they degrade all wood cell wall components. White-rot fungi cause the wood to become pale, eventually reducing it to a fibrous, whitish mass. Brown-rot fungi utilize cell wall hemicelluloses and cellulose, leaving the lignin essentially undigested. However, brown-rot fungi do modify lignin, as indicated by demethylation and accumulation of oxidized polymeric lignin-degradation products. These fungi cause the wood to darken, shrink and break into brick-shaped pieces that crumble easily into a brown powder.

Brown- and white-rot fungi decay wood by distinctly different mechanisms. However, in spite of considerable research, the biochemical bases for the different morphological and chemical changes are not clear (Figures 2.1 and 2.2). Many factors have hindered progress – in particular, the complexity of the wood substrate

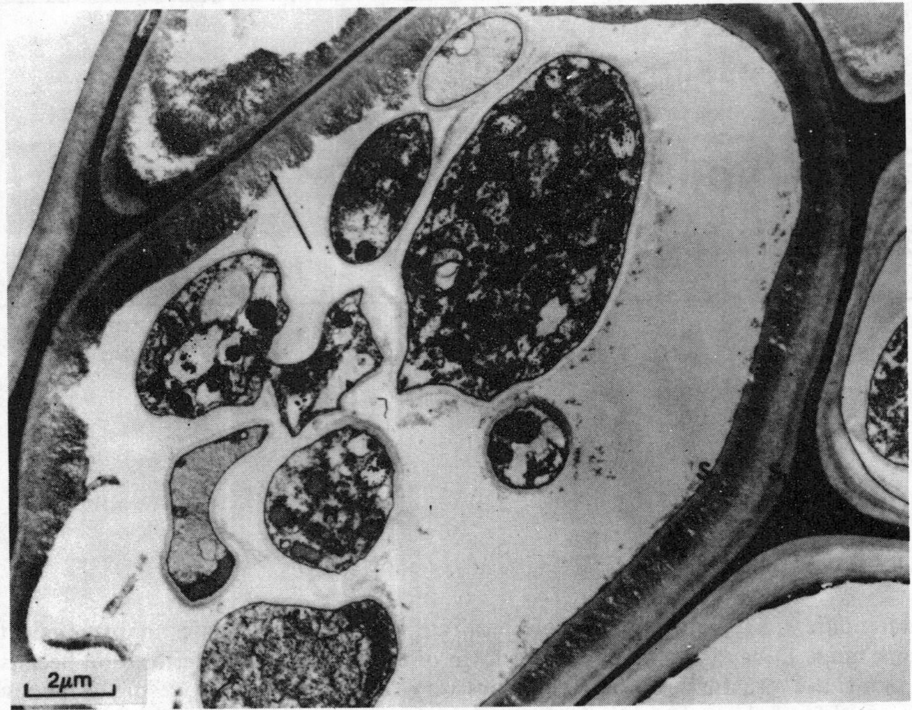

Figure 2.1 Advanced cell wall degradation in sweetgum by hyphae (H) of the brown-rot fungus *Postia placenta*. The S_2 layer was apparently attacked first; the S_1 and S_3 layers appear relatively free of attack. Arrow indicates area where decay affected the entire wall. Aldehyde–OsO_4 fixation, 3000 ×

and the multiplicity of enzymes produced. The applications of classical and molecular genetics to wood-degradative systems of decay fungi have progressed rapidly and can be used to elucidate the mechanisms of wood degradation. This chapter reviews the decomposition of wood by white- and brown-rot fungi and discusses how recent advances in biotechnology and their application to the study of wood decay have augmented knowledge about the fungal mechanisms of wood deterioration. White-rot fungi have received by far the most attention, probably because they produce more enzymes that may have biotechnological application.

2.2 Mechanism of White-Rot Decay

2.2.1 Cellulose Degradation

Phanerochaete chrysosporium has served as a model organism for white-rot degradation studies of wood. The enzyme mechanisms involved in cellulose degradation by this white-rot fungus were extensively investigated by Eriksson (1978). It is well established that the degradation of crystalline cellulose by white-rot fungi, similar to that of other fungal cellulases, is carried out by a multicomponent enzyme complex

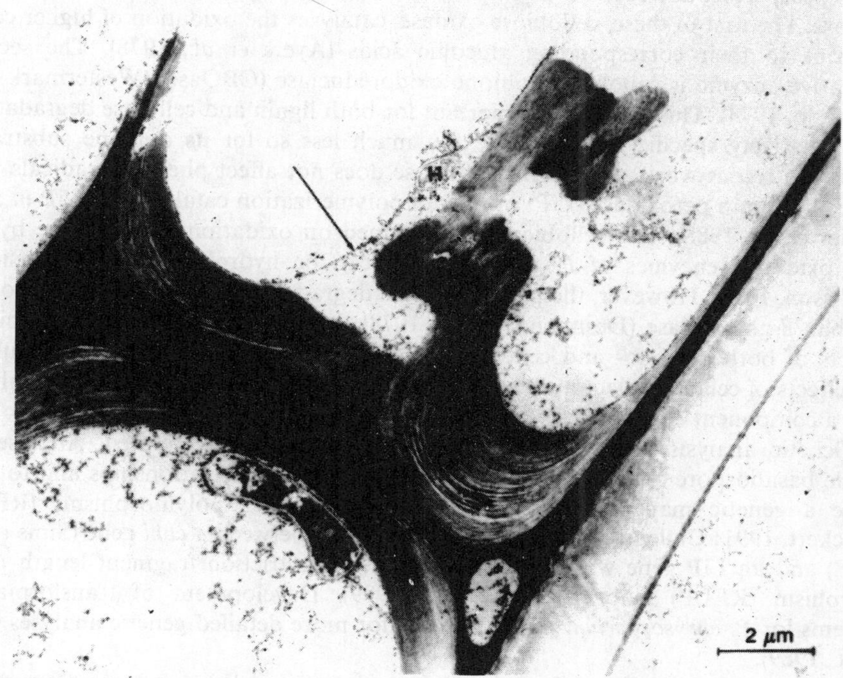

Figure 2.2 Hypha (H) of *Trametes versicolor* attached to sweetgum cell wall by sheath (arrow). Cell wall is degraded beneath the sheath and away from hypha towards the middle lamella. Note localized degradation at cell corners and erosion troughs. $KMnO_4$ fixation, 4100 ×

in which the individual components interact synergistically to degrade cellulose to glucose. Endoglucanases (EGs) act randomly over the exposed surfaces of cellulose microfibrils, exposing non-reducing termini that are hydrolyzed by cellobiohydrolases (CBHs), producing cellobiose. Cellobiose may be cleaved by β-glucosidase, yielding glucose. Eriksson and Pettersson identified five EGs and one CBH (Eriksson and Pettersson, 1975a,b), as well as two glucosidases from *P. chrysosporium*. Multiple CBH-like enzymes were subsequently characterized (Uzcategui *et al.*, 1991).

Relatively little work has been done with regard to cellulase regulation in fungi. The precise mechanism or mechanisms of cellulase induction are unknown. With white-rot fungi, cellulase synthesis is induced by cellulose and repressed by glucose. The most generally accepted view of the induction process is that the organisms produce a basic level or a constitutive amount of cellulase that produces soluble hydrolysis products of cellulose that function as inducers. Cellobiose, a product of cellulase action, both induces and inhibits cellulase of *P. chrysosporium* (Eriksson and Hamp, 1978). This fungus can control cellobiose concentration in at least four ways (Eriksson, 1978). The first way is via the hydrolytic enzyme β-glucosidase, which hydrolyzes cellobiose and oligosaccharides to glucose. Second, cellobiose can also be eliminated by transglucosylation reactions. The other two ways involve at least two oxidative enzymes that have been implicated in cellulose degradation by

P. chrysosporium and seem to be important in preventing enzyme inhibition by cellobiose. The first of these, cellobiose oxidase, catalyzes the oxidation of higher cellodextrins to their corresponding gluconic acids (Ayers *et al.*, 1978). The second oxidative enzyme is cellobiose:quinone oxidoreductase (CBQase) (Westermark and Eriksson, 1974). This enzyme is important for both lignin and cellulose degradation. It is relatively specific for cellobiose but much less so for its quinone substrates. However, recent work showed that CBQase does not affect phenoxyl radicals produced by lignin peroxidase (LiP) or phenol polymerization catalyzed by LiP *in vitro* (Odier *et al.*, 1988). The cellobionic acid formed on oxidation of cellobiose by the two oxidative enzymes of *P. chrysosporium* can be hydrolyzed by β-glucosidase (Eriksson, 1978). However, the product of this degradation, gluconolactone, strongly inhibits β-glucosidase (Deshpande *et al.*, 1978). Lactonase decreases the inhibitory effects of both glucono- and cellobionolactones on β-glucosidase and particularly the effects of cellobionolactone on the complete cellulase system. Lactonase is therefore a component of the synergistic attack of this system.

Genetic analysis of cellulolytic fungi has been extremely limited. Analyses of single basidiospore cultures were used to determine allelic relationships and to generate a genetic map using restriction fragment length polymorphisms (RFLP) (Kockert, 1991; Cullen and Kersten, 1992). Linkage between a *cbhl* gene (Sims *et al.*, 1988) and an LiP gene was established using the restriction fragment length polymorphism (RFLP) map (Raeder *et al.*, 1989). Development of transformation systems for *P. chrysosporium* paved the way for more detailed genetic analyses (Alic *et al.*, 1989).

Studies of bacterial and fungal cellulases have revealed a common structural design composed of discrete functional domains: a catalytic core, a conserved cellulose-binding terminus, and an intervening, highly glycosylated hinge region (Covert *et al.*, 1992a). Cellulases of *P. chrysosporium* also appear to be organized in accordance with this model. For example, as with other cellulases, papain cleavage of *P. chrysosporium* CBH separates the catalytic domain from the hinge and binding domains (Uzcategui *et al.*, 1991). Also, a CBH gene cloned from *P. chrysosporium* ME446 (Sims *et al.*, 1988) is similar in sequence to other fungal CBH genes (Shoemaker *et al.*, 1983; Azevedo *et al.*, 1990).

Restriction mapping and sequence analysis of cosmid clones revealed a cluster of three structurally related CBH genes in *P. chrysosporium*, one of which lacked the cellulose-binding domain common to other microbial cellulases (Covert *et al.*, 1992a,b). *P. chrysosporium cbh1-1* and *cbh1-2* are separated by only 750 bp and are located approximately 14 kb upstream from a cellulase gene previously cloned from *P. chrysosporium* (Sims *et al.*, 1988). Within a well-conserved region, the deduced amino acid sequences of *P. chrysosporium cbh1-1* and *cbh1-2* are, respectively, 80 and 69 per cent homologous to that of the *Trichoderma reesei* CBH I gene. The conserved cellulose-binding domain typical of microbial cellulases is absent from *cbh1-1* gene product. Transcript levels of the three *P. chrysosporium* genes varied substantially, depending on culture conditions. The *cbh1-1* and *cbh1-2* genes were not induced in the presence of cellulose, nor did they appear to be subject to glucose repression. Therefore, Covert *et al.* (1992a,b) concluded that aspects of the chromosomal organization, structure and transcription of these genes are unlike those of previously described cellulase genes.

Eriksson (1981) suggested an additional mechanism for cellulose degradation in white-rot fungi, wherein hydrolysis by cellulases is combined with an oxidative step.

Oxidative enzymes may produce reactants such as superoxide anion (O_2^-), singlet oxygen ($^1O_2$), or hydroxyl radical ($OH^.$) that are involved in the primary attack of crystalline cellulose. $^1O_2$ and $OH^.$ are particularly reactive radicals (Pryor, 1976). These reactants are small enough to penetrate the cellulose microstructure and pre-dispose it to attack by endo- and exoglucanases of white-rot fungi. White-rot fungi have been reported to produce radicals that could react with cellulose. Nakatsubo *et al.* (1982) observed $^1O_2$ formation by *P. chrysosporium* and Eriksson (1981) found that O_2^- is produced extracellularly by this fungus as well as other wood-destroying fungi. Enoki *et al.* (1991) isolated an extracellular substance from *Irpex lacteus* that produced and reduced H_2O_2 to $OH^.$. He concluded that this substance is involved in the degradation of wood cellulose and lignin. The origin of radical-generating oxidases and their involvement in cellulose degradation by white-rot fungi warrant further study.

Immunocytochemical techniques (Polack and Priestly, 1992) have been beneficial in localizing cellulases in white-rotted wood. Ruel *et al.* (1989) localized cellulases in poplar wood decayed by *P. chrysosporium* using a post-embedding immunoelectron microscopic technique. Polyclonal antibodies were formed with a mixture of EG and CBH injected into rabbits. The purified immunoglobulins (IgG) were tagged and used for various localizations of the cellulases in relation to the physiological state of the fungus. These immunoglobulins can be found in intracellular vesicles, concentrated along the plasma membrane, and penetrating a very short distance (0.2 to 0.5 μm) inside the wood cell wall. The presence of gold labelling in wood was only observed in areas of low electron density, which might have been previously degraded by a mechanism that was not necessarily of enzymatic origin. Non-enzymatic degradation of the wood cell wall could effectively be obtained by activated oxygen species generated from ferrous and manganese salts in the presence of hydrogen peroxide.

Transmission electron microscopy (TEM) of immunogold-labelled ultra-thin sections of *Coriolus versicolor* hyphae grown on malt agar and on solidified carboxymethylcellulose (CMC) medium was used by Gallagher and Evans (1990) to localize β-glucosidase of the cellulase complex. β-Glucosidase was mainly detected in the hyphal sheath where dense labelling was observed. Label density was greater in hyphae grown on CMC-containing medium than on malt agar. Labelling, though virtually absent within the cytoplasm, also occurred in the hyphal cell wall and on the plasmalemma. This pattern of labelling revealed that more enzyme was secreted under conditions of growth on cellulose than on malt agar. This substantiates the reported increase in activity of β-glucosidase in the culture medium when cellulose is the carbon source. Synthesis of β-glucosidase may occur at the plasmalemma–cell wall interface since the enzyme is not found intracellularly.

2.2.2 Hemicellulose Degradation

Hemicellulose is structurally more complex than cellulose, which contains only 1,4-β-glycosidic linkages. Hemicelluloses are a group of homo- and heteropolymers consisting largely of anhydro-β-(1 → 4)-D-xylopyranose, mannopyranose, glucopyranose and galactopyranose main-chains with a number of substituents. The enzymes that degrade hemicellulose are similarly complex. Compared with cellulases, little is known about the molecular genetics of hemicellulases. There have been no reports of xylanase clones from fungi (Cullen and Kersten, 1992). However, because hemicellulases have shown promise as bleaching agents in pulp and paper production, research has intensified in recent years.

Degradation of hemicelluloses by white-rot fungi proceeds in a manner roughly analogous to that of cellulose, but the mechanism of attack has been studied in much less detail. The hemicellulose chains are attacked first by endo-enzymes (mannanases, xylanases) that produce progressively shorter chains, which are hydrolyzed to simple sugars by glycosidases (mannosidases, xylosidases, glucosidases). It is not known whether exo-enzymes are involved. The enzymes involved in the removal of side-chain substituents (arabinose, uronic acids, acetyls) have received little attention (Kirk and Cowling, 1984).

As with cellulases, simple sugars repress the production of most hemicellulose-degrading enzymes by white-rot fungi. Cellulose apparently is the only carbon source necessary to induce the formation of hemicellulose-degrading enzymes by these fungi. Eriksson and Goodell (1974) proposed that a single regulatory protein governs the induction of cellulase, mannanase and xylanase in *Bjerkandera adusta* (= *Polyporus adustus*). They further indicated that this single regulatory protein could be used to adjust simply and effectively the rate of wood decay by controlling enzyme induction.

Blanchette *et al.* (1989) used polyclonal and monoclonal antisera to xylanase in conjunction with immunocytochemistry to determine the ultrastructural localization of xylanase within cell walls of wood decayed by white-rot fungi. Labelling with gold-tagged antibodies to xylanase occurred primarily in the inner regions of the S_2 and S_1 layers and middle lamellae. Intercellular regions within the cell corners of the middle lamella were less dense but labelled positive with anti-xylanase gold. The remaining secondary wall exhibited little labelling with the xylanase–gold complex. Erosion troughs that reached the S_1 layer or middle lamella had less labelling in the remaining cell wall.

2.2.3 Lignin Degradation

Unlike cellulose and hemicellulose, lignin is not principally linear but is a complex, heterogeneous, non-stereoregular aromatic polymer composed of phenylpropanoid units. White-rot fungi are the only known organisms that are capable of completely degrading lignin to carbon dioxide and water. Rapid progress has been made in recent years regarding the biochemistry and molecular genetics of lignin biodegradation, primarily through studies of the white-rot fungus *P. chrysosporium*, and this work has been extensively reviewed (Buswell and Odier, 1987; Alic and Gold, 1991; Kuan *et al.*, 1991; Cullen and Kersten, 1992, 1996). This section briefly summarizes some of these findings.

The ligninolytic system of *P. chrysosporium* is not induced by lignin but appears constitutively as cultures enter secondary metabolism; that is, when primary growth ceases because of depletion of some nutrient (Kirk *et al.*, 1978). Secondary metabolism is triggered by nitrogen, carbon or sulphur limitation but not phosphorus limitation (Jeffries *et al.*, 1981). Apparently, the regulation of secondary metabolism, including lignin degradation, is connected to glutamate metabolism (Kirk, 1981). Cultures can obviously cope well with nitrogen limitation, so that sustained lignin degradation occurs. In this connection, Kirk and Fenn (1982) speculated that lignin degradation is a secondary metabolic event because of the very low nitrogen content of wood (Merrill and Cowling, 1966). Soon after a white-rot fungus invades wood, nitrogen becomes limiting and secondary metabolism, including lignin degradation, begins.

Although nitrogen repression of lignin degradation in white-rot fungi is common, it may not always be the rule (Leatham and Kirk, 1983). Thus, addition of nitrogen to certain white-rot fungi in different biotechnical applications utilizing lignin or lignin-related compounds may increase the efficiency of these fungi.

Enzymes thought to play a leading role in lignin depolymerization include two extracellular haem peroxidases, manganese peroxidase (MnP) and LiP, and an H_2O_2-generating system. However, white-rot fungi secrete unique combinations of peroxidases and oxidases (Perie and Gold, 1991). *Trametes* (=*Coriolus*) *versicolor* and *Phlebia radiata* each produce one or more laccases in addition to LiP and MnP (Fahraeus and Reinhammar, 1967; Niku-Paavola *et al.*, 1988). *Pleurotus sajor-caju* secretes an aryl alcohol oxidase (Bourbonnais and Paice, 1988), a laccase and several peroxidases (Fukuzumi, 1987). *Bjerkendera adusta* secretes an aryl alcohol oxidase (Muheim *et al.*, 1990), and *Rigidoporus lignosus* and *Dichomitus squalens* secrete a laccase and a MnP (Galliano *et al.*, 1988). Thus, a variety of oxidative enzymes may be utilized by white-rot fungi for lignin degradation. An essential component of all white-rot ligninolytic systems, however, is a source of H_2O_2. Two enzymes have been proposed to fulfil this task: MnP itself, which can form H_2O_2 from O_2 when NADH, NADPH or glutathione are present; and glyoxal oxidase (GLOX), a novel enzyme produced during secondary metabolism in *P. chrysosporium* and activated by LiP plus veratryl alcohol (Kersten, 1990).

Considerable progress has been made in recent years concerning the molecular genetics of lignin biodegradation by white-rot fungi, primarily with *P. chrysosporium*. Standard methods have been established for auxotroph production, recombination analysis and rapid DNA and RNA purification (Cullen and Kersten, 1992). Since Tien and Tu (1987) first reported cloning of the cDNA-encoding lignin peroxidase H8, much has been learned about the number, structure and organization of the *P. chrysosporium* peroxidase genes. However, there is still considerable uncertainty about the exact number and structure of LiP genes. The number of LiP genes in *P. chrysosporium* has been variously reported – from five to 15 (Gaskell *et al.*, 1994). An RFLP-based genetic map localized LiP genes of *P. chrysosporium* isolate ME446 to two linkage groups (Raeder *et al.*, 1989). Following chromosome separation by clamped homogeneous electrical field (CHEF) electrophoresis, five LiP genes were assigned to a single chromosome (Gaskell *et al.*, 1991). In agreement with the RFLP map, another LiP clone (GLG4) was assigned to the same chromosome as a *cbhl* cluster (Covert *et al.*, 1992a,b; Gaskell *et al.*, 1994). The number of MnP genes and their chromosomal organization have not been reported, although preliminary (unpublished) results show that at least one MnP gene resides on the same chromosome as do five LiP genes (Cullen and Kersten, 1992).

A cDNA-encoding MnP (Pribnow *et al.*, 1989) and its corresponding genomic clone (Godfrey *et al.*, 1990) were isolated from *P. chrysosporium* and sequenced (Pease *et al.*, 1989). The amino sequence homology to LiP H8 is 50 to 60 per cent, and the residues essential for peroxidase activity are conserved.

Research on the transcriptional regulation of peroxidases has been hampered by difficulties in distinguishing closely related genes (Cullen and Kersten, 1992). It is clear, however, that LiP genes are transcriptionally regulated and that expression of

MnP genes is Mn^{2+} dependent (Brown *et al.*, 1991); the specificity of the transcripts observed on Northern blots is questionable (Cullen and Kersten, 1992).

To quantify closely related LiP transcripts, Stewart *et al.* (1992) devised a quantitative polymerase chain reaction (PCR) approach that offers high levels of sensitivity and specificity. The technique is particularly suited to LiP gene families. The authors found that transcriptional regulation of LiP genes GLG4, GLG5 and V4 is dramatically affected by nutrient limitation. Their transcripts are almost mutually exclusive; GLG5 and V4 are present in nitrogen-limited cultures but undetectable or present in very low levels in carbon-limited cultures, whereas the GLG4 level is one-thousand-fold higher in carbon-limited cultures than in nitrogen-limited cultures. Analysis of concentrated culture filtrates with isoelectric focusing (IEF) gels showed a major pI 4.6 band in carbon-limited cultures. The absence of GLG5 from carbon-limited cultures is consistent with the isozyme pattern observed by Glumoff *et al.* (1990) in such cultures. The significance of these patterns of regulation in wood remains to be established, although it is clear that wood is nitrogen limited. In any case, the PCR system of Stewart *et al.* (1992) may be adapted to identify specific transcripts in such complex substrates.

Stewart *et al.* (1992) used Southern blot analysis of CHEF gels to map the GLG4 gene to a dimorphic chromosome separate from the other LiP gene. This chromosome was previously reported by Covert *et al.* (1992a,b) to contain a cellulase gene cluster. Subsequent work by the same researcher indicated that the cellulase genes in this cluster are expressed during carbon limitation (unpublished data). Thus, there may be a link between the organization and regulation of these genes involved in lignocellulose degradation.

Glyoxal oxidase (GLOX) was previously mentioned as a source of the extracellular H_2O_2 that is required by ligninolytic peroxidases. As a first step towards elucidating the molecular genetics of GLOX, Kersten and Cullen (1993) cloned and sequenced a cDNA-clone-encoding GLOX. They demonstrated that GLOX expression is transcriptionally regulated and co-ordinated with LiP and MnP genes. Subsequently, Kersten *et al.* (1995) showed that GLOX is encoded by a single gene with two alleles, that the gene is located on a dimorphic chromosome unlinked to known LiP, MnP and CBH genes, and that GLOX is efficiently expressed in *Aspergillus nidulans*. The results of Kurek and Kersten (1995) suggest that ligninolysis by peroxidase could be regulated by GLOX and influenced by the presence of veratryl alcohol, lignin and lignin degradation products. They further note that such co-ordinated metabolism would influence the kinetics of free radical generation by LiP and, therefore, the overall efficiency of lignin. Ultimately, the catalytic mechanism and interactions of GLOX with peroxidases will be elucidated through crystal structures, site-specific mutagenesis of active sites, and gene disruptions (Kersten and Cullen, 1993). The cloning and sequencing of GLOX is an important step towards this objective.

The molecular genetics of other lignin-degrading fungi has received little attention (Cullen and Kersten, 1992). Saloheimo *et al.* (1989) cloned and sequenced an LiP gene from *Phlebia radiata*. The deduced amino acid sequence of this clone is 62 per cent, identical to that of the *P. chrysosporium* isozyme H8 (ML1). From Southern blot hybridization to the H8 gene, multiple LiPs apparently are present in *Bjerkandera adjusta*, *Coriolus versicolor* and *Fomes lignosus* (Huoponen *et al.*, 1990).

Electron microscopic analyses of *P. chrysosporium* inoculated on wood have suggested various locations for LiP or MnP: in the hyphal wall or slime layer (Daniel *et al.*, 1989, 1990; Ruel and Joseleau, 1991); in the periplasmic space (Forney *et al.*, 1982; Srebotnik *et al.*, 1988; Daniel *et al.*, 1989, 1990); or in association with the plasma membrane and with membranes or luminae of cytoplasmic vesicles (Garcia *et al.*, 1987; Srebotnik *et al.*, 1988; Daniel *et al.*, 1989, 1990). However, these studies, which used wood as a substrate, do not permit a satisfactory spatial and temporal overview of peroxidase secretion with respect to the whole mycelium (Moukha *et al.*, 1993). Also, primary and secondary (idiophasic) growth cannot be distinguished as in artificial media.

Moukha *et al.* (1993) attempted to localize the secretion of LiP and MnP with hyphal growth in *P. chrysosporium* by using cultures sandwiched between perforated polycarbonate membranes. Comparison of sites where newly synthesized proteins and immuno-detected peroxidases were released into the medium suggests that enzyme diffusion from the walls is a limiting step in the release of peroxidases. Microautoradiography of colonies revealed apical growth of thin hyphae and branches in the central secreting area. These secondary hyphae possessed peroxidase activity and reacted with LiP antibodies. The results suggest that LiP and MnP are initially secreted at the apex of secondary growing hyphae and later slowly released into the surrounding medium.

Since the discovery of lignin-degrading enzymes, there has been considerable progress in characterizing their properties and understanding their catalytic mechanisms. However, most of the detailed mechanistic studies have been conducted using simple model compounds. Questions remain as to how lignin-degrading enzymes attack native lignin. In fact, neither the purified lignin-degrading enzymes nor the crude filtrates duplicate the extensive extracellular degradation of lignin observed in intact cultures of ligninolytic fungi. Biotechnological techniques should help identify the apparently missing components of the isolated lignin-degrading machinery.

2.3 Mechanism of Brown-Rot Decay

2.3.1 Cellulose Degradation

Brown-rot fungi are unique among cellulose destroyers because they are the only known microbes that can degrade wood cellulose without first removing the lignin. These fungi leave a brown residue (hence the name) of partially demethylated lignin. Furthermore, brown-rot fungi degrade cellulose in an unusual manner that differs from that of other cellulolytic organisms. Shortly after colonizing wood, the fungi cause a rapid and extensive depolymerization of cellulose to the 'limit' (length of cellulose crystallite) or crystalline degree of polymerization (DP) at low weight loss. Acid hydrolysis has a similar effect on cellulose, as do various strong oxidants. The cellulose that remains has an average DP of 150–200 and is more crystalline than non-depolymerized material because the cleavages occur in the amorphous non-crystalline regions. The biochemical agent – or the system that produces it – that causes this initial depolymerization is clearly a small diffusible agent because enzymes are too large to penetrate wood to reach cellulose (Cowling, 1961; Cowling and Brown, 1969). Furthermore, brown-rot fungi appear to degrade cellulose by a mechanism different from that of the synergistically acting systems involving exo-glucanases present in white-rot fungi and *Trichoderma* (Highley, 1973). Uemura *et al.* (1993) found that the cellulase system from brown-rot fungi gave a negative response towards antibodies to *Trichoderma* CBH, suggesting the absence of homologous sequences and structures with the *Trichoderma* CBH.

Attempts to identify the cellulose depolymerizing agent produced by brown-rot fungi have frustrated researchers for years. More than 25 years ago, Cowling and Brown (1969) recognized that even the smallest cellulases are too large to penetrate the pores of wood. Also, cellulases do not mimic the action of brown-rot fungi in generating cellulose crystallites (Chang *et al.*, 1981; Phillip *et al.*, 1981). Cowling and Brown thus proposed that a non-enzymatic oxidative agent might be involved in depolymerization of cellulose by brown-rot fungi. They noted that Halliwell (1965) had described the degradation of cotton cellulose by Fenton's reagent (H_2O_2/Fe^{2+}), which generates a hydroxyl radical or a similar oxidant (Halliwell and Gutteridge, 1988). Based on these observations, Halliwell was the first to propose the possible existence of a non-enzymatic celluloytic system involving peroxide and iron. Subsequently, Koenigs (1972a,b, 1974a,b, 1975) demonstrated that cellulose in wood can be depolymerized by Fenton's reagent, that brown-rot fungi produce extracellular hydrogen peroxide, and that wood contains enough iron to make Halliwell's hypothesis reasonable.

Enoki *et al.* (1989) reported the ability of brown-rot fungi to oxidize 2-keto-4-thiomethylbutyric acid (KTBA) to ethylene; KTBA is converted to ethylene by one-electron oxidants such as the hydroxyl radical. Ethylene production was correlated with weight loss but not cellulose depolymerization. More recently, Enoki *et al.* (1990) reported the isolation of an extracellular protein from cultures of *G. trabeum* which requires H_2O_2 and is capable of KTBA oxidation. These authors partially purified the protein and reported it to be an iron-containing glycoprotein of molecular weight 1600–2000. Based on their work, Enoki and co-workers suggested the existence of 'a unique wood-component degrading system that participates directly or indirectly in the fragmentation of cellulose as well as of lignin in wood and oxidizes KTBA to give ethylene'. However, it is yet to be established whether the H_2O_2-dependent KTBA-oxidizing ability of this protein is related to cellulose depolymerization. Similar glycoproteins were isolated from both white-rot and soft-rot fungi (Enoki *et al.*, 1991), raising questions about their role in wood decay.

Hydrogen peroxide is an important component of Fenton chemistry. Probably the strongest argument against the occurrence of the Fenton system in brown-rot fungi is that more than two decades of research on H_2O_2 production by brown-rot fungi have not definitely established that these fungi synthesize extracellular H_2O_2 (Highley and Flournoy, 1994). Presumably, the conflicting reports of whether wood decay fungi produce extracellular H_2O_2 can be explained by the transient appearance of H_2O_2 in culture and the lack of a selective assay for the reagent.

The presumptive role of H_2O_2 is the generation of the hydroxyl radical in a reaction with either a metal or a metal chelate. Hydroxyl radical has been detected in liquid media, agar media, or wood by various methods, including *p*-nitrosodium methylaniline (Highley, 1982), desilvering (Veness and Evans, 1989), electron spin resonance (Illman *et al.*, 1989b) and chemiluminescence (Backa *et al.*, 1992). The hydroxyl radical would need to be formed at its site of action because it is very reactive with a very short life time and, therefore, would not diffuse into wood. Lu *et al.* (1994) reported that *G. trabeum* produces low molecular weight phenolate chelators, which the authors propose diffuse into wood and, in the presence of iron and H_2O_2, produce radical species within the wood cell wall. Earlier TEM immuno-labelling studies using an antibody to a low molecular weight fraction from *G.*

trabeum (Jellison *et al.*, 1991) showed that the chelator was present within the decayed and undecayed regions of the wood cell wall. Hyde and Wood (1995) recently proposed a model for attack at a distance from the hyphae of the brown-rot fungus *Coniophora puteana*, based on the formation of H_2O_2 by autoxidation. This fungus produces cellobiose dehydrogenase, which reduces Fe^{3+} to Fe^{2+}. Diffusion of Fe^{2+} from the hyphae in a low pH environment promotes conversion to Fe^{2+}-oxalate and subsequent autoxidation with H_2O_2 as the product. The critical Fe^{2+}/H_2O_2 combination is therefore formed at a distance from the hyphae. Flournoy (1994) noted that because most reports concerning free radicals are phenomenological in nature, it is difficult to assign any meaning to them and little can be concluded regarding their significance in brown-rot decay.

Oxalic acid, an important physiological metabolite of brown-rot fungi, may be produced at concentrations inhibitory to the Fenton reaction. Oxalic acid was proposed by Schmidt *et al.* (1981) to play a role in reduction of Fe^{3+} to Fe^{2+}, which increased cellulose decomposition by the Fenton reaction. However, Schmidt *et al.* (1981) and Tanaka *et al.* (1994) reported that at higher concentrations of oxalic acid, cellulose degradation by the Fenton system is inhibited. *Postia placenta* and *Serpula incrassata* were shown to accumulate oxalic acid in wood (Green *et al.*, 1992b).

Brown-rotted cellulose depolymerization products have been chemically characterized and compared with other depolymerized cellulose samples (Kirk *et al.*, 1989, 1991). The following depolymerized cellulose samples were prepared from pure cotton cellulose: acid-hydrolyzed (HCl) to the limit DP; $H_2O_2/FeSO_4$-oxidized (Fenton-oxidized); HIO_4/Br_2-oxidized; and brown-rotted (*P. placenta*). These samples were characterized as to molecular size distribution, yield of glucose on complete acid hydrolysis and carboxyl, uronic acid and carbonyl contents, as well as sugar acids released on acid hydrolysis. Consistent with earlier results, the Fenton system, but not the other oxidation system, mimicked the brown-rot system in nearly all measured characteristics. The acid-hydrolyzed sample also possessed similar characteristics. Glyceric, erythronic, arabonic and gluconic acids were identified by GC/MS in the hydrolysates of the brown-rotted and Fenton-oxidized samples. These results are consistent with the depolymerizing agent being related to the Fenton system, but Flournoy (1994) noted that the authors did not establish that the fungi employed such a system. Shimada (1993) stated that:

> the identification of these aliphatic acids does not always provide proof of the involvement of Fenton's system in physiological wood decay processes, since enzymatic evidence for the production of these acids has not yet been offered. Furthermore, Fenton's system yields such a destructive OH radical to living matter as to degrade their own cell polymers.

The Fenton system is powerful enough to decompose cellulose to carbon dioxide under appropriate conditions as long as hydrogen peroxide is continuously supplied (Schmidt *et al.*, 1981).

Flournoy (1994) found several pitfalls in the studies of Highley (1977) and Kirk *et al.* (1991). The authors did not establish a correlation between the oxidation of cellulose and depolymerization. It is unknown, for instance, whether oxidation of the cellulose precedes depolymerization, whether oxidation is a result of post-depolymerization modification, or whether oxidation and depolymerization are coupled. The cellulose used in these studies was highly degraded. Samples from early decay were not examined, which would be the samples of most interest in understanding the depolymerization mechanism. In summary, although the literature indicates that chemical alterations in brown-rotted lignin are oxidative in nature, oxidative changes in cellulose as a result of brown-rot decay have not been unequivocally demonstrated.

Assays for metals in brown-rot wood decay have been performed in an attempt to clarify their role in cellulose breakdown by metal-catalyzed oxidation. Electron spin resonance (ESR) spectrometry was used to detect and follow changes in the oxidative states of paramagnetic metals during brown-rot decay (Illman *et al.*, 1989a). Alterations in low-spin iron could not be detected at room temperature, and changes in high-spin iron were not tested at low temperature (Illman *et al.*, 1989a). A comprehensive study of iron oxidation states using ESR has yet to be made.

Changes in manganese were observed with ESR after inoculation of susceptible species of wood with the brown-rot fungus *P. placenta* (Illman *et al.*, 1989a). These alterations were manifested as increases in the size of the sextet spectra specific for manganese (Mn^{2+}). The ESR spectra for Mn^{2+} were taken over a 4-week period from fungal-inoculated white fir, Douglas-fir, sweetgum and redwood. The increases in Mn^{2+} signals correlated with wood susceptibility to brown-rot decay. Little or no increase was found in wood species resistant to brown-rot decay (Illman *et al.*, 1989a). The chemical basis of the Mn^{2+} change was not determined.

In treating white fir slivers with oxalic acid, Illman and Englebert (unpublished data) found that the ESR signal for Mn^{2+} increased with increasing oxalic acid concentration, although the Mn^{2+} signal did not increase as much as that in *P. placenta* decayed wood. The enhanced Mn^{2+} signal may be a result of lowered pH with acid solubilization of the metal. Oxalic acid from several species of wood-decay fungi can mobilize calcium from glass and concrete. Alternatively, the increase in the Mn^{2+} signal may be due to chelation of the metal. Oxalate is a known chelator of several elements, including manganese, calcium and potassium. The chemical basis of possible oxalic acid effects on wood needs further investigation.

2.3.2 *Hemicellulose Degradation*

Brown-rot fungi secrete a number of hemicellulose-degrading enzymes, and the mechanism of hemicellulose breakdown appears similar to that for white-rot fungi (King, 1966; Keilich *et al.*, 1970; Highley, 1976). However, little definitive work has been done on the hemicellulases of brown-rot fungi with respect to specificity, molecular size and other properties such as active sites. The few hemicellulose-degrading enzymes that have been isolated and 'purified' from brown-rot fungi are not substrate specific. A purified β-glucosidase from the brown-rot fungus *G. trabeum* also hydrolyzed β-xyloside (Herr *et al.*, 1978); the purified xylanase from the brown-rot fungus *Tyromyces palustris* also had endoglucanase but no glycosidase activities (Ishihara *et al.*, 1978). The purified endoglucanase from the brown-rot fungus *Polyporus schwenitzii* was accompanied by mannanase and xylanase activities (Keilich *et al.*, 1970). A multiple glycan and glycoside hydrolase were isolated from the brown-rot fungus *Poria placenta* (Highley *et al.*, 1981). However, glycanase activities were later separated from glycosidase activities (Green *et al.*, 1989).

Brown- and white-rot fungi differ in their regulation of hemicellulase synthesis. Many brown-rot fungi exhibit good hemicellulase as well as cellulase activities during growth on simple sugars, whereas white-rot fungi do not (Eriksson and

Goodell, 1974; Highley, 1976). Regulation in wood-decay fungi, however, has not been examined in great detail and warrants additional study.

Green *et al.* (1991b, 1992a) utilized gold-labelled monoclonal antibodies to xylanase to localize xylanase of *Postia placenta* grown on agar medium. Xylanase was associated with the hyphal surface, within the hyphal sheath matrix, and on fibrillar elements of the sheath structure.

Oxalic acid may also be involved in the solubilization and hydrolysis of hemicellulose, thus making the cellulose fibres more accessible to cellulases (Bech-Anderson, 1987). Green *et al.* (1991a, 1992b) expanded on this hypothesis, suggesting that the acidic conditions in wood, caused by oxalic acid production, are responsible for early acid hydrolysis and depolymerization of hemicellulose and amorphous cellulose, thereby increasing wood porosity. Enzymes and other degrading agents in the hyphal sheath would then have access to the remaining cellulose and cause its final removal.

2.3.3 Lignin Degradation

Chemical analysis of brown-rotted wood indicates that brown-rot fungi do not utilize lignin to an appreciable extent (Cowling, 1961; Kirk and Highley, 1973). The main effect of the fungi on lignin is demethylation of aryl methoxyl groups (Kirk and Adler, 1970), although oxidative changes occur, including some cleavage of aromatic rings (Kirk, 1975). Cowling (1961) also showed that lignin decayed by *P. placenta* had appreciably greater solubility in water and 1 per cent NaOH than did lignin in sound wood. Haider and Trojanowski (1980) demonstrated that brown-rot fungi can be induced to metabolize lignin to some extent. They found that isolated lignins were degraded to carbon dioxide to a limited extent by brown-rot fungi in liquid culture. Microscopic studies (Highley *et al.*, 1985) suggest that brown-rot fungi cause degradation of the cell wall, including the lignin-rich middle lamella and cell corners. Thus, brown-rot fungi may have a greater ligninolytic capacity than previously thought. The ligninolytic agent produced by brown-rot fungi is possibly the same or similar to the degrading agent that initiates the rapid depolymerization of cellulose.

Harvey *et al.* (1986) discussed the possibility that brown-rot fungi degrade lignin by a single-electron oxidation similar to that of white-rot fungi. Both types of fungi demethylate methoxyl groups in phenolic- and non-phenolic-containing lignin structures (Kirk and Adler, 1970; Kirk and Chang, 1974) as well as hydroxylate aromatic rings (Kirk and Adler, 1970). Both types of decay fungi promote $C\alpha$ side-chain oxidations and aromatic ring cleavage reactions (Kirk and Adler, 1970; Kirk, 1975; Kirk and Chang, 1975). Lignin peroxidase production has been reported by the brown-rot fungus *Polyporous ostruformis*, but at a low level compared with that by white-rot fungi (Dey *et al.*, 1991). The oxidative processes caused by brown-rot fungi, in contrast to white-rot fungi, result in the formation of polymeric lignin fractions (Kirk, 1975). Harvey *et al.* (1986) proposed that brown-rot fungi repolymerize lignin because highly unstable radical cation intermediates formed from phenolic-containing aromatic compounds are oxidized by a single electron. These phenoxy radicals undergo further reactions, including oxidative carbon-to-carbon and carbon-to-oxygen coupling reactions, to produce higher molecular weight products that the brown-rot fungi evidently cannot metabolize.

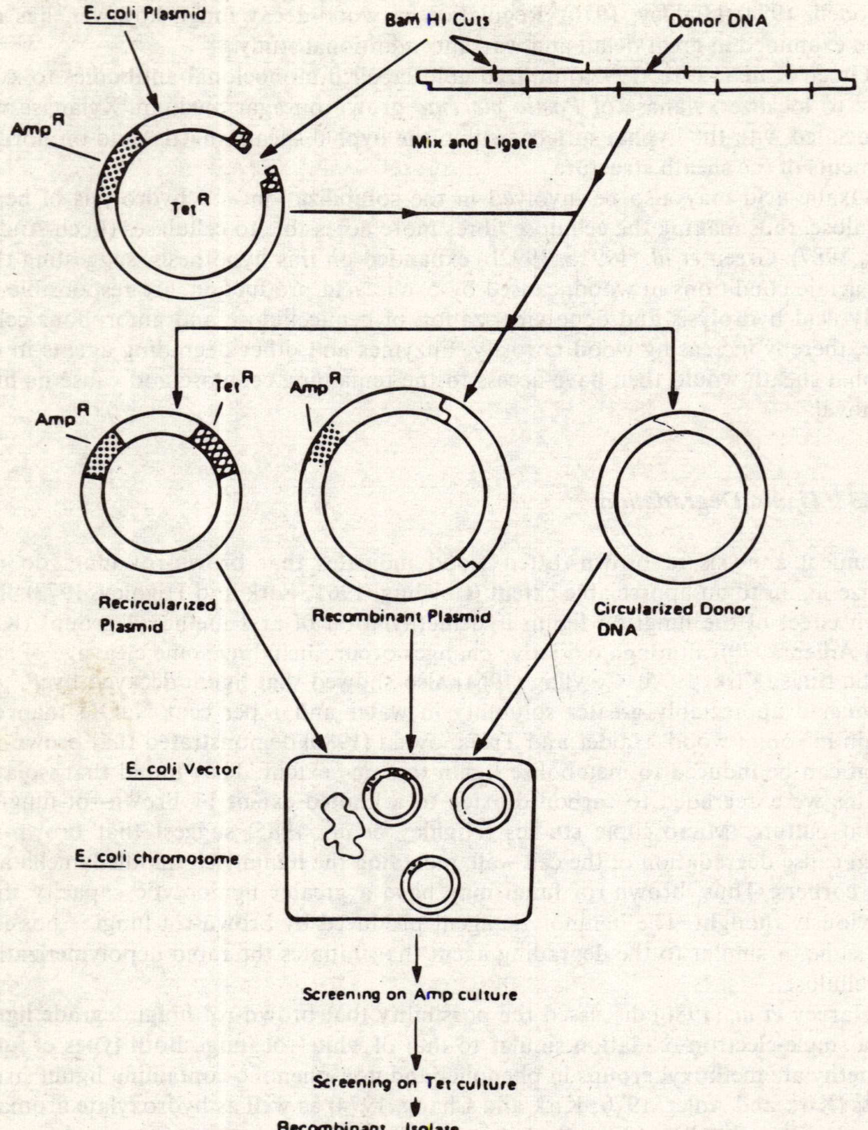

Figure 2.3 Generalized foreign gene insertion. Individual steps for making plasmid receptive to gene insertion and antibiotic screening for selection of properly transformed cells (reprinted from Kammermeyer and Clark (1989), by courtesy of Marcel Dekker, Inc)

2.4 Summary

Because lignocelluloytic enzymes are of commercial value to the pulp and paper industry (Zadrazil and Reninger, 1988) and the agricultural community (Vander Meer *et al.*, 1987), an inexpensive and 'readily available' supply of these enzymes is highly desirable. Mechanistic studies also require large quantities of these enzymes.

Limiting amounts of lignocelluloytic enzymes are produced by wood decay fungi, which impede their commercial use and also prevent true assessment studies of various potential applications. Biotechnology techniques such as cloning, mutation and overproduction of lignocellulolytic enzymes have helped to overcome these limitations.

An example of this approach is that of Williams *et al.* (1991), who applied biotechnological techniques to enhance production of polyphenol oxidase by the white-rot fungus *Trametes versicolor*. Polyphenol oxidase converts *o*-diphenols to *o*-diquinones and oligomerizes syringic acid. Because *T. versicolor* can be 'batch cultured' (Fahraeus and Reinhammar, 1967), overproduction and enhanced secretion of these enzymes by molecular genetics, for example, recombinant DNA technology, are feasible. Briefly, the technology involves isolation of fungal genomic DNA, restriction endonuclease treatment of the genomic DNA, ligation of the restriction fragments into a plasmid, and transformation of *E. coli* with the recombinant plasmid (Williams *et al.*, 1991) (Figure 2.3). This technology should be applicable to the overproduction of more commercially important cellulases and ligninases (Desoretz, 1993). In this connection, Kammermeyer and Clark (1989) published an extensive monograph regarding foreign gene insertion and cell transformation. Recently, MnP was successfully expressed in *Aspergillus oryzae* (Stewart *et al.*, 1996). The recombinant MnP is secreted into culture medium in active form, which will provide large quantities of pure enzyme suitable for mechanistic studies.

References

ALIC, M. and GOLD, M. (1991) Genetics and molecular biology of the lignin-degrading basidiomycete *Phanerochaete chrysosporium*. In: Bennett J. and Lasue, L., eds, *More Gene Manipulations in Fungi*, New York: Academic Press, pp. 319–341.

ALIC, M., KORNEGAY, J. R., PRIBNOW, D. and GOLD, M. H. (1989) Transformation by complementation of an adenine auxotroph of the lignin-degrading basidiomycete *Phanerochaete chrysosporium*. *Appl. Environ. Microbiol.* **55**, 406–411.

AYERS, A. R., AYERS, S. B. and ERIKSSON, K. E. (1978) Cellobiose oxidase, purification and partial characterization of a hemoprotein from *Sporotrichum pulverulentum*. *Eur. J. Biochem.* **90**, 171–181.

AZEVEDO, M., FELIPE, M., ASTOLFI-FILHO, S. and RADFORD, A. (1990) Cloning, sequence and homologies of the *cbh-1* (exoglucanase) gene of *Humicola grisea* var. *thermoidea*. *J. Gen. Microbiol.* **136**, 2569–2576.

BACKA, S., GRIRER, J., REITBERGER, T. and NILSSON, T. (1992) Hydroxyl radical activity in brown-rot fungi by a new chemiluminescence method. *Holzforschung* **46**, 61–67.

BECH-ANDERSON, J. (1987) *Production, Function and Neutralization of Oxalic Acid Produced by the Dry Rot Fungus and Other Brown-Rot Fungi*, Document IRG/WP/1330, International Research Group on Wood Preservation.

BLANCHETTE, R. A., ABAD, A. R., FARRELL, R. L. and LEATHERS, T. D. (1989) Detection of lignin peroxidase and xylanases by immunocytochemical labelling in wood decayed by basidiomycetes. *Appl. Environ. Microbiol.* **55**, 1457–1465.

BOURBONNAIS, R. and PAICE, M. G. (1988) Veratryl alcohol oxidases from the lignin-degrading basidiomycete *Pleurotus sajor-caju*. *Biochem. J.* **255**, 445–450.

BROWN, J. A., ALIC, M. and GOLD, M. H. (1991) Manganese peroxidase gene transcription in *Phanerochaete chrysosporium*: activation by manganese. *J. Bacteriol.* **173**, 4101–4106.

BUSWELL, J. A. and ODIER, E. (1987) Lignin biodegradation. *CRC Crit. Rev. Biotechnol.* **6**, 1–60.

CHANG, M. M., CHOU, T. C. and TSAO, G. T. (1981) Structure pretreatment and hydrolysis of cellulose. In: Ziechter, A., ed. *Advances in Biochemical Engineering*, Berlin: Springer-Verlag, pp. 15–42.

COVERT, S., BOLDUC, J. and CULLEN, D. (1992a) Genomic organization of a cellulase gene family in *Phanerochaete chrysosporium. Curr. Genet.* **22**, 407–413.

COVERT, S. F., WYMELENGERG, A. V. and CULLEN, D. (1992b) Structure, organization and transcription of a cellobiohydrolase gene cluster from *Phanerochaete chrysosporium. Appl. Environ. Microbiol.* **58**, 2168–2175.

COWLING, E. B. (1961) *Comparative Biochemistry of the Decay of Sweetgum Sapwood by White-Rot and Brown-Rot Fungi*, Tech. Bull. No. 1258, Washington, DC.: US Department of Agriculture.

COWLING, E. B. and BROWN W. (1969) Structural features of cellulosic materials in relation to enzymatic hydrolysis. In: Hajny. G. J. and Reese, E. T., eds, *Cellulases and their Applications, Advanced Chemical Series* **95**, 152–187.

CULLEN, D. and KERSTEN, P. (1992) Fungal enzymes in lignocellulose degradation. In: Kinghorn, R. and Turner, G., eds. *Applied Molecular Genetics of Filamentous Fungi*, London: Chapman and Hall, pp. 100–131.

CULLEN, D. and KERSTEN, P. J. (1996) Enzymology and molecular biology of lignin degradation. In: Bramble, R. and Marzluf, G., eds, *The Mycota III*, Berlin: Springer-Verlag, pp. 297–314.

DANIEL, G., NILSSON T. and PETTERSON, B. (1989) Intra- and extracellular localization of lignin peroxidase during the degradation of solid wood and wood fragments by *Phanerochaete chrysosporium* by using transmission electron microscopy and immuno-gold labelling, *Appl. Environ. Microbiol.* **55**, 871–881.

DANIEL, G., PETTERSON, B., NILSSON. T. and VOLC, J. (1990) Use of immunogold cytochemistry to detect Mn(II)-dependent and lignin peroxidases in wood degraded by the white rot fungi *Phanerochaete chrysosporium* and *Lentinula edodes. Can. J. Bot.* **68**, 920–933.

DESHPANDE, V., ERIKSSON, K. E. and PETTERSSON, B. (1978) Production, purification and partial characterization of 1,4-β-glucosidase enzymes from *Sporotrichum pulverulentum. Eur. J. Biochem.* **90**, 191–198.

DESORETZ, C. G. (1993) Overproduction of lignin peroxidase by *Phanerochaete chrysosporium* (BKM-F-1767). *Appl. Environ. Microbiol.* **59**, 1919–1926.

DEY, S., MAITI, T. K. and BHATTACHACHARYYA, B. C. (1991) Lignin peroxidase production by a brown-rot fungus *Polyporous ostriformes. J. Ferment. Bioeng.* **72**, 402–404.

ENOKI, A., TANAKA, H. and FUSE, G. (1989) Relationship between degradation of wood and production of H_2O_2-producing one-electron oxidases by brown-rot fungi. *Wood Sci. Technol.* **23**, 1–12.

ENOKI, A., YOSHIOKA, S., TANAKA, H. and FUSE, G. (1990) *Extracellular H_2O_2-Producing and One-Electron Oxidation System of Brown-Rot Fungi*, Document IRG/WP/1445, International Research Group on Wood Preservation.

ENOKI, A., FUSE, G. and TANAKA, H. (1991) *Extracellular H_2O_2-Producing and H_2O_2-Reducing Compounds of Wood Decay Fungi*, Document IRG/WP/1516, International Research Group on Wood Preservation.

ERIKSSON, K. E. (1978) Enzyme mechanisms involved in cellulose hydrolysis by the rot fungus *Sporotrichum pulverulentum. Biotechnol. Bioeng.* **20**, 317–332.

ERIKSSON, K. E. (1981) Microbial degradation of cellulose and lignin. *Proceedings of International Symposium on Wood and Pulping Chemistry*, Stockholm, June, Vol. 3, pp. 60–65.

ERIKSSON, K. E. and GOODELL, B. (1974) Pleiotropic mutants of the wood-rotting fungus *Polyporus adustus* lacking cellulase, mannanase and xylanase. *Can. J. Microbiol.* **20**, 371–378.

ERIKSSON, K. E. and HAMP, S. G. (1978) Regulation of endo-1,4-β-glucanase production in *Sporotrichum pulverulentum. Eur. J. Biochem.* **90**, 183–190.

ERIKSSON, K. E. and PETTERSSON, B. (1975a) Extracellular enzyme system utilized by the fungus *Sporotrichum pulverulentum* (*Chrysosporium lignorum*) for the breakdown of cellulose. 1. Separation, purification and physico-chemical characterization of five endo-1,4-β-glucanases. *Eur. J. Biochem.* **51**, 193–206.

ERIKSSON, K. E. and PETTERSSON, B. (1975b) Extracellular enzyme system utilized by the fungus *Sporotrichum pulverulentum* (*Chrysosporium lignorum*) for the breakdown of cellulose. 3. Purification and physico-chemical characterization of an exo-1,4-β-glucanase. *Eur. J. Biochem.* **51**, 213–218.

FAHRAEUS, G. and REINHAMMAR, B. (1967) Large-scale production and purification of laccase from cultures of *Polyporus versicolor* and some properties of laccase A. *Acta Chem. Scand.* **21**, 2367–2378.

FLOURNOY, D. S. (1994) Chemical changes in wood components and cotton cellulose as a result of brown-rot: is Fenton chemistry involved? In: *Biodeterioration Research* Vol. 4, New York: Plenum Press, pp. 257–294.

FORNEY, L. J., REDDY, C. A. and PANKRATZ, H. S. (1982) Ultrastructural localization of hydrogen peroxide production in ligninolytic *Phanerochaete chrysosporium* cells. *Appl. Environ. Microbiol.* **44**, 732–736.

FUKUZUMI, T. (1987) Ligninolytic enzymes of *Pleuortis sajorcoya.* In: Odier, E., ed., *Lignin: Enzymic and Microbial Degradation*, Paris: Institute National de la Recherche Agronomique, pp. 137–142.

GALLAGHER, I. M. and EVANS, C. S. (1990) Immunogold-cytochemical labelling of β-glucosidase in the white-rot fungus *Coriolus versicolor. Appl. Microb. Biotechnol.* **32**, 588–593.

GALLIANO H., GAS, G. and BOUDET, A. (1988) Biodegradation of *Hevea brasiliensis* lignocellulose by *Rigidoporus lignosus*: influence of culture conditions and involvement of oxidizing enzymes. *Plant Physiol. Biochem.* **26**, 619–627.

GARCIA, S., LATGE, J. P., PREVOST, M. C. and LEISOLA, M. (1987) Wood degradation by white rot fungi: cytochemical studies using lignin peroxidase–immunoglobulin–gold complexes. *Appl. Environ. Microbiol.* **53**, 2384–2387.

GASKELL, J., DIEPERINK, E. and CULLEN, D. (1991) Genomic organization of lignin peroxidase genes of *Phanerochaete chrysosporium. Nucl. Acids Res.* **19**, 599–603.

GASKELL, J., STEWART, P., KERSTEN, P. J., COVERT, S. F., REISER, J. and CULLEN, D. (1994) Establishment of genetic linkage by allele-specific polymerase chain reaction application to the lignin peroxidase gene family of *Phanerochaete chrysosporium. Biotechnology* **12**, 1372–1375.

GLUMOFF, T., HARVEY, P., MOLINARI, S., GOBLE, M., FRANK, G., PALMER, J., SMIT, J. and LEISOLA, M. (1990) Lignin peroxidase from *Phanerochaete chrysosporium*, molecular and kinetic characterization of isozymes. *Eur. J. Biochem.* **187**, 515–520.

GODFREY, B. J., MAYFIELD, M. B., BROWN, J. A. and GOLD, M. H. (1990) Characterization of a gene encoding a manganese peroxidase from *Phanerochaete chrysosporium. Gene* **93**, 119–124.

GREEN, F., CLAUSEN, C. A., MICALES, J. A., HIGHLEY, T. L. and WOLTER, K. E. (1989) Carbohydrate-degrading complex of the brown-rot fungus *Poria placenta*: purification of β-1,4-xylanase. *Holzforschung* **43**, 25–31.

GREEN, F. III, LARSEN, M. J., WINANDY, J. E. and HIGHLEY, T. L. (1991a) Role of oxalic acid in incipient brown-rot decay. *Mat. u. Org.* **26** (3), 191–213.

GREEN, F. III, CLAUSEN, C. A., LARSEN, M. J. and HIGHLEY, T. L. (1991b) Ultrastructural characterization of the hyphal sheath of *Postia placenta* by selective removal and immunogold labelling. In: *Proceedings, 8th International Biodeterioration and Biodegradation symposium*, pp. 530–532.

GREEN, F. III, CLAUSEN, C., LARSEN, M. and HIGHLEY, T. L. (1992a) Immuno-scanning electron microscopic localization of extracellular polysaccharidases within the fibrillar sheath of the brown-rot fungus *Postia placenta. Can. J. Bot.* **38**, 898–904.

GREEN, F., LARSEN, M. J., HACKNEY, J. M., CLAUSEN, C. A. and HIGHLEY, T. L. (1992b) Acid-mediated depolymerization of cellulose during incipient brown-rot decay by *Postia placenta*. In: Kuwahar, M. and Shimada, M., eds, *Biotechnology in Pulp and Paper Industry*, 5th International Conference, Tokyo, Japan: Uni Publishers.

HAIDER, K. and TROJANOWSKI, J. (1980) A comparison of the degradation of ^{14}C-labelled DHP and corn stalk lignins by micro- and macrofungi and by bacteria. In: Kirk, T. K., Higuchi, T. and Chang, H. M., eds, *Lignin Biodegradation: Microbiology, Chemistry and Potential Applications*, Vol. I, Boca Raton, FL: CRC Press, p. 111.

HALLIWELL, G. (1965) Catalytic decomposition of cellulose under biological conditions. *Biochem. J.* **95**, 35–40.

HALLIWELL, B. and GUTTERIDGE, J. M. C. (1988) Iron as a biological pro-oxidant. *ISI Atlas of Science, Biochemistry* **1**, 48–52.

HARVEY, P. J., SCHOEMAKER, H. E. and PALMER, J. M. (1986) Lignin degradation by wood degrading fungi, Document IRG/WP/1310, International Research Group on Wood Preservation.

HERR, D., BAUMER, F. and DELLWEG, H. (1978) Purification and properties of an extra-cellular endo-1,4-β-glucanase from *Lenzites trabea. Arch. Microbiol.* **117**, 287–292.

HIGHLEY, T. L. (1973) Influence of carbon source on cellulase activity of white-rot and brown-rot fungi. *Wood Fiber* **5**, 50–58.

HIGHLEY, T. L. (1976) Hemicellulases of white- and brown-rot fungi in relation to host preferences. *Mat. u. Org.* **11**, 33–46.

HIGHLEY, T. L. (1977) Requirements for cellulose degradation by a brown-rot fungus. *Mat. u. Org.* **12**, 25–36.

HIGHLEY, T. L. (1982) Is extracellular hydrogen peroxide involved in cellulose degradation by brown-rot fungi? *Mat. u. Org.* **17**, 205–214.

HIGHLEY, T. L. and FLOURNOY, D. S. (1994) Decomposition of cellulose by brown-rot fungi. In: Garg, K. L., Gerg, N. and Mukerji, K. G., eds, *Recent Advances in Biodeterioration and Biodegradation*, Vol. II, Naya Prokash, 206, Bidhan Serani, Calcutta-6, India, pp. 191–221.

HIGHLEY, T. L., WOLTER, K. E. and EVANS, F. (1981) Polysaccharide-degrading complex produced in wood and in liquid media by the brown-rot fungus, *Poria placenta. Wood Fiber* **13**, 265–274.

HIGHLEY, T. L., MURMANIS, L. L. and PALMER, J. G. (1985) Micromorphology of degradation in western hemlock and sweetgum by the brown-rot fungus *Poria placenta. Holzforschung* **39**, 73–78.

HUOPONEN, K., OLLIKKA, P., KALIN, M., WALTHER, I., MANTSALA, P. and REISER, J. (1990) Characterization of lignin peroxidase-encoding genes from lignin-degrading basidiomycetes. *Gene* **89**, 145–150.

HYDE, S. M. and WOOD, P. M. (1995) A model for attack at a distance from the hyphae based on studies with the brown-rot *Coniophora puteana*, Document IRG/95/10104, International Research Group on Wood Preservation.

ILLMAN, B. L., MEINHOLTZ, D. C. and HIGHLEY, T. L. (1989a) Manganese as a probe of fungal degradation of wood. In: *Biodeterioration Research II*, New York: Plenum Press, pp. 485–496.

ILLMAN, B. L., MEINHOLTZ, D. C. and HIGHLEY, T. L. (1989b) Oxygen free radical detection in wood colonized by the brown-rot fungus, *Postia placenta*. In: *Biodeterioration Research II*, New York: Plenum Press, pp. 497–509.

ISHIHARA, M., SHIMIZU, K. and ISHIHARA, T. (1978) Hemicellulases of brown rotting fungus, *Tyromyces palustris*, III. Partial purification and mode of action of an extracellular xylanase. *Mokuzai Gakkaishi* **24**, 108–115.

JEFFRIES, T. W., CHOI, S. and KIRK, T. K. (1981) Nutritional regulation of lignin degradation by *Phanerochaete chrysosporium. Appl. Environ. Microbiol.* **42**, 290–296.

JELLISON, J., CHANDHOKE, V., GOODELL, B. and FEKETE, F. (1991) The isolation and immunology of iron-binding compounds produced by *Gloeophyllum trabeum. Appl. Microbiol. Biotechnol.* **35**, 805–809.

KAMMERMEYER, K. and CLARK, V. L. (1989) *Genetic Engineering Fundamentals, An Introduction to Principles and Applications,* New York: Marcel Dekker.

KEILICH, G., BAILEY, P. and LIESE, M. (1970) Enzymatic degradation of cellulose, cellulose derivatives and hemicelluloses in relation to the fungal decay of wood. *Wood Sci. Technol.* **4**, 273–283.

KERSTEN, P. (1990) Glyoxal oxidase of *Phanerochaete chrysosporium*: its characterization and activation by lignin peroxidase. *Proc. Natl. Acad. Sci. USA* **87**, 2936–2940.

KERSTEN, P. J. and CULLEN, D. (1993) Cloning and characterization of cDNA encoding glyoxal oxidase, a H_2O_2-producing enzyme from the lignin-degrading basidiomycete, *Phanerochaete chrysosporium. Proc. Natl. Acad. Sci. USA* **90**, 7411–7413.

KERSTEN, P. J., WITEK, C., WYMELENBERG, A. V. and CULLEN, D. (1995) *Phanerochaete chrysosporium* glyoxal oxidase is encoded by two allelic variants: structure, genomic organization and heterologous expression of glxl and glx2, *J. Bacteriol.* **177**, 6106–6110.

KING, N. J. (1966) The extracellular enzymes of *Coniophora cerebella. Biochem. J.* **100**, 784–792.

KIRK, T. K. (1975) Effects of the brown-rot fungus, *Lenzites trabea*, on lignin in ʹspruce wood. *Holzforschung* **29**, 99–107.

KIRK, T. K. (1981).Toward elucidating the mechanism of action of the ligninolytic system in basidiomycetes. In: Hollaender, A., Rabson, R., Rogers, A., San Pietro, A., Valentine, R. and Wolfe, R., eds, *Trends in the Biology of Fermentations*, New York: Plenum Press, pp. 131–149.

KIRK, T. K. and ADLER, E. (1970) Methoxyl-deficient structural elements in lignin of sweetgum decayed by a brown-rot fungus. *Acta Chem. Scand.* **24**, 3379–3390.

KIRK, T. K. and CHANG, H-M. (1974) Decomposition of lignin by white-rot fungi, I. Isolation of heavily degraded lignins from decayed spruce. *Holzforschung* **28**, 217–222.

KIRK, T. K. and CHANG, H-M. (1975) Decomposition of lignin by white-rot fungi, II. Characterization of heavily degraded lignins from decayed spruce. *Holzforschung* **29**, 56–64.

KIRK, T. K. and COWLING, E. B. (1984) Biological decomposition of solid wood. In: Rowell, R. M., ed, *Chemistry of Solid Wood*, Advances in Chemistry Series, 207, Washington, DC: American Chemical Society Press, pp. 455–487.

KIRK, T. K. and FENN, P. (1982) Formation and action of the ligninolytic system in basidiomycetes. In: Frankland, J. C., Hedger, J. N. and Swift, M. J., eds, *Decomposer Basidiomycetes*, British Mycological Society Symposium 4, Cambridge University Press, pp. 67–90.

KIRK, T. K. and HAMMEL, K. C. (1992) What is the primary agent of lignin degradation in white rot fungi. In: Kurwaloa, M. and Shida, M., eds, *Biotechnology in Pulp and Paper Industry*, Proceedings of 5th International Conference in Biotechnology in Pulp and Paper Industry, Japan: Uni Publishers, pp. 535–540.

KIRK, T. K. and HIGHLEY, T. L. (1973) Quantitative changes in structural components of conifer woods during decay by white- and brown-rot fungi. *Phytopathology* **63**, 1338–1342.

KIRK, T. K., SCHULTZ, E., CONNORS, W. J., LORENZ, L. F. and ZEIKUS, J. G. (1978) Influence of culture parameters on lignin metabolism by *Phanerochaete chrysosporium. Arch. Microbiol.* **117**, 277–285.

KIRK, T. K., HIGHLEY, T. L., IBACH, R. and MOZUCH, M. D. (1989) *Identification of Terminal Structures in Cellulose Degraded by the Brown-Rot Fungus* Postia placenta,

Document IRG/WP/1389, International Research Group on Wood Preservation.

KIRK, T. K., IBACH, R. E., MOZUCH, M. D., CONNER, A. H. and HIGHLEY, T. L. (1991) Characterization of cotton cellulose depolymerized by a brown-rot fungus, by acid or by chemical oxidants. *Holzforschung* **45**, 239–244.

KOCKERT, G. (1991) Restriction fragment length polymorphism in plants and its implications. In: *Subcellular Biochemistry: Plant Genetic Engineering*, Vol. 17, New York: Plenum Press, pp. 167–190.

KOENIGS, J. W. (1972a) Effects of hydrogen peroxide on cellulose and on its susceptibility to cellulase. *Mat. u. Org.* **7**, 133–147.

KOENIGS, J. W. (1972b) Production of extracellular hydrogen peroxide by wood-rotting fungi. *Phytopathology* **62**, 100–110.

KOENIGS, J. W. (1974a) Hydrogen peroxide and iron: a proposed system for decomposition of wood by brown-rot basidiomycetes. *Wood Fiber Sci.* **6**, 66–80.

KOENIGS, J. W. (1974b) Production of hydrogen peroxide by wood-rotting fungi in wood and its correlation with weight loss, depolymerization and pH changes. *Arch. Microbiol.* **99**, 129–145.

KOENIGS, J. W. (1975) Hydrogen peroxide and iron: a microbial cellulolytic system? *Biotechnology and Bioengineering Symposium*, No. 5, pp. 151–159.

KUAN, I-C., CAI, D. and TIEN, M. (1991) The lignin-degrading system of the wood-destroying fungus *Phanerochaete chrysosporium*. In: Pell, E. and Steffen, K., eds, *Active Oxygen/Oxidative Stress and Plant Metabolism*, Rockville, MD: American Society of Plant Physiologists, pp. 180–192.

KUREK, B. and KERSTEN, P. J. (1995) Physiological regulation of glyoxal oxidase from *Phanerochaete chrysosporium* by peroxidase systems. *Enzyme Microb. Technol.* **17**, 751–756.

LEATHAM, G. and KIRK, T. K. (1983) Regulation of ligninolytic activity by nutrient nitrogen in white-rot basidiomycetes. *FEMS Microbiol. Lett.* **16**, 65–67.

LU, J., GOODELL, B., LIU, J., ENOKI, A., JELLISON, J. and FEKETE, F. (1994) *The Role of Oxygen and Oxygen Radicals in One-Electron Oxidation Reactions Mediated by Low-Molecular Weight Compounds Isolated from* Gloeophyllum trabeum, Document IRG/WP95-1457, International Research Group on Wood Preservation.

MERRILL, W. and COWLING, E. B. (1966) Rate of nitrogen in wood deterioration: amounts and distribution of nitrogen in tree stems. *Can. J. Bot.* **44**, 1555–1580.

MOUKHA, S. M., WOSTEN, H. A. B., ASTHER, M. and WESSELS, J. G. H. (1993) *In situ* localization of the secretion of lignin peroxidases in colonies of *Phanerochaete chrysosporium* using a sandwiched mode of culture. *J. Gen. Microbiol.* **139**, 969–978.

MUHEIM, A., WALDNER, R., LEISOLA, M. S. A. and FIECHTER, A. (1990) An extracellular aryl alcohol oxidase from the white rot fungus *Bjerkendera adusta*. *Enzyme Microb. Technol.* **12**, 204–209.

NAKATSUBO, F., REID, I. D. and KIRK, T. K. (1982) Incorporation of $^{18}O_2$ and absence of stereospecificity in primary product formation during fungal metabolism of a lignin model compound. *Biochem. Biophys. Acta* **719**, 284–291.

NIKU-PAAVOLA, M. L., KARHUNEN, E., SALOLA, P. and RAUNIO, V. (1988) Lignolytic enzymes of the white-rot fungus *Phlebia radicata*. *Biochem. J.* **254**, 877–884.

ODIER, E., MOZUCH, M. D., KALYANARAMAN, B. and KIRK, T. K. (1988) Ligninase-mediated phenoxy radical formation and polymerization unaffected by cellobiose: quinone oxidoreductase. *Biochemica* **70**, 847–852.

PEASE, E. A., ANDRAWIS, A. and TIEN, M. (1989) Manganese-dependent peroxidase from *Phanerochaete chrysosporium*: primary structure deduced from cDNA sequence. *J. Biol. Chem.* **264**, 13531–13535.

PERIE, F. H. and GOLD, M. H. (1991) Manganese regulation of manganese peroxidase expression and lignin degradation by the white-rot fungus *Dichomitus squalens*. *Appl. Environ. Microbiol.* **57**, 2240–2245.

PHILLIP, B., DAN, D. C. and FINK, H. P. (1981) Acid and enzymatic hydrolysis of cellulose in relation to its physical structure. *Proceedings of International Symposium on Wood and Pulping Chemistry*, Stockholm, pp. 79–83.

POLACK, J. M. and PRIESTLY, J. V. (1992) *Electron Microscopic Immunocytochemistry Principles and Technologies*, Oxford: Oxford University Press.

PRIBNOW, D., MAYVIELD, M. B., NIPPER, V. J., BROWN, J. A. and GOLD, M. H. (1989) Characterization of a cDNA encoding a manganese peroxidase, from the lignin-degrading basidiomycete *Phanerochaete chrysosporium. J. Biol. Chem.* **264**, 5036–5040.

PRYOR, W. A. (1976) *Free Radicals in Biology*, Vol. I, San Francisco, London: Academic Press.

RAEDER, U., THOMPSON, W. and BRODA, P. (1989) RFLP-based genetic map of *Phanerochaete chrysosporium* ME446: lignin peroxidase genes occur in clusters. *Mol. Microbiol.* **3**, 911–918.

RUEL, K. and JOSELEAU, J. P. (1991) Involvement of an extracellular glucan sheath during degradation of *Populus* wood by *Phanerochaete chrysosporium. Appl. Environ. Microbiol.* **57**, 374–384.

RUEL, K., ODIER, E. and JOSELEAU, J.-P. (1989) Observation of wood cell wall degradation by immunocytochemical techniques. In: Kirk, T. K. and Chang, H.-M., eds, *Biotechnology in Pulp and Paper Manufacture*, Boston: Butterworth-Heinemann, pp. 83–97.

SALOHEIMO, M., BARAJAS, V., NIKIO-PAAVOLA, M. and KNOWLES, J. (1989) A lignin peroxide-encoding complementary DNA from the white-rot fungus *Phlebia radiata*: characterization and expression in *Trichoderma reesei. Gene* **85**, 343–351.

SCHMIDT, C. J., WHITTEN, B. K. and NICHOLAS, D. D. (1981) A proposed role for oxalic acid in nonenzymatic wood decay by brown-rot fungi. *Proc. Am. Wood Pres. Assoc.* **77**, 157–164.

SHIMADA, M. (1993) Biochemical mechanisms for the biodegradation of wood. *Curr. Jap. Mat. Res.* **II**, 207–222.

SHOEMAKER, S., SLCHWEICKART, V., LADNER, M., GELFAND, D., KWOK, S., MYAMBO, K. and INNIS, M. (1983) Molecular cloning of exo-cellobiohydrolase I derived from *Trichoderma reesei* strain L27. *Bio/Technology* **1**, 691–696.

SIMS, P., JAMES, C. and BRODA, P. (1988) The identification, molecular cloning and characterisation of a gene from *Phanerochaete chrysosporium* that shows strong homology to the exocellobiohydrolase I gene from *Trichoderma reesei. Gene* **74**, 411–422.

SREBOTNIK, E., MESSNER, K., FOISNER, R. and PETTERSON, B. (1988) Ultrasturctural localization of ligninase of *Phanerochaete chrysosporium* by immunogold labelling. *Curr. Microbiol.* **16**, 221–227.

STEWART, P., KERSTEN, P., VANDEN WYMELENBERG, A., GASKELL, J. and CULLEN, D. (1992) The lignin peroxidase gene family of *Phanerochaete chrysosporium*: complex regulation by carbon and nitrogen limitation, and the identification of a second dimorphic chromosome. *J. Bacteriol.* **174**, 5036–5042.

STEWART, P., WHITWAM, R. E., KERSTEN, P. J., CULLEN, D. and TIEN, M. (1996) Efficient expression of *Phanerochaete chrysosporium* manganese peroxidase gene in *Aspergillus oxyzae. Appl. Environ. Microbiol.* **62**, 860–864.

TANAKA, N., AKAMATSU, Y., HATTORI, T. and SHIMADA, M. (1994) Effect of oxalic acid on the oxidative breakdown of cellulose by the Fenton reaction. *Wood Res.* **81**, 8–10.

TIEN, M. and TU, C. P. D. (1987) Cloning and sequencing of a cDNA for a ligninase from *Phanerochaete chrysosporium. Nature* **326**, 520–523.

UEMURA, S., MITSURO, M. and JELLISON, J. (1993) Differential responses of wood-rot fungi cellulases towards antibodies against *Trichoderma viride* cellobiohydrolase I. *Appl. Microbiol. Biotechnol.* **39**, 788–794.

UZCATEGUI, E., RUIZ, A., MONTESINO, R., JOHANSSON, G. and PETTERSSON, G. (1991) The 1,4-β-D-glucan cellobiohydrolases from *Phanerochaete chrysosporium* I. A system of synergistically acting enzymes homologous to *Trichoderma reesei*. *J. Biotechnol.* **19**, 271–286.

VANDER MEER, J. M., RIJKES, B. A. and FERIARNTI, M. P. (1987) Degradation of lignocelluloses. In: *Ruminants and or Industrial Processes*, Amsterdam: Elsevier Science Publishers.

VENESS, R. G. and EVANS, C. S. (1989) The role of hydrogen peroxide in the degradation of crystalline cellulose by basidiomycete fungi. *J. Gen. Microbiol.* **135**, 2799–2806.

WAINWRIGHT, M. (1992) *An Introduction to Fungal Biotechnology*, New York: Wiley and Sons.

WESTERMARK, U. and ERIKSSON, K. E. (1974) Cellobiose: quinone oxidoreductase, a new wood-degrading enzyme from white-rot fungi. *Acta Chem. Scand.* **B28**, 209–214.

WILLIAMS, A. L., WILLIAMS, A. C., MOORE, N. L. and DASHEK, W. V. (1991) Biotechnology of wood deteriorating enzymes synthesized and secreted by *Coriolus versicolor*, a white-rot basidiomycete. In: Rossmore, H. W., ed, *Biodeterioration and Biodegradation*, London: Elsevier Applied Science, pp. 547–549.

ZADRAZIL, F. and RENINGER, P. (1988) *Treatment of Lignocelluloses With White-Rot Fungi*, Amsterdam: Elsevier Science Publishers.

3

Developments in the Study of Soft Rot and Bacterial Decay

GEOFFREY DANIEL AND THOMAS NILSSON

3.1 Soft Rot – A Definition

Soft rot is a form of microbiological wood degradation caused by fungi. The term 'soft rot' was originally proposed by Savory (1954) to be used for 'decay caused by cellulose-destroying microfungi to distinguish it from the brown and white rots caused by wood-destroying basidiomycetes'. The term was based on the observation that wood surfaces become very soft when attacked by microfungi. Findlay and Savory (1954) and Savory (1954) described the typical cavity chains within wood cell walls associated with this form of decay but did not however report any observations of erosion of cell walls. Such attack was later described by Courtois (1963a,b) and Corbett (1965) for a number of microfungi that also formed typical cavities. Corbett observed that one species, *Camarosporium ambiens*, exclusively eroded the cell walls in birch. Nilsson (1973) found during an extensive study on wood degradation by microfungi that most cavity-forming species also eroded the cell walls in birch wood and that several species exclusively caused an erosion form of attack.

The softening of the wood surface observed by Savory (1954) was related to very wet wood. Later studies, especially on soft rot in salt treated transmission poles, have shown that the wood is actually hard in comparison to that decayed by brown or white rot. Soft rot is thus not a very suitable term. Soft rot is generally associated with cavity formation, but the reports that some typical white rot fungi also form cavities complicates this concept (Duncan, 1960a; Daniel *et al.*, 1992; Schwarze *et al.*, 1995). If the proposal by Savory (1954) is followed, all forms of wood decay caused by microfungi should be referred to as soft rot. We realize that the term soft rot is so widely accepted that it would be difficult to introduce a new concept. Nilsson (1988) proposed the term soft rot to be used for all forms of decay caused by ascomycetes and deuteromycetes. He also included *Daldinia*, *Hypoxylon* and *Xylaria* species among the soft rotters. This was justified by observations that several *Xylaria* and one *Hypoxylon* species formed typical soft rot cavities in birch and pine wood (Nilsson *et al.*, 1989). We suggest that this proposal is followed until better and more precise classification schemes are defined.

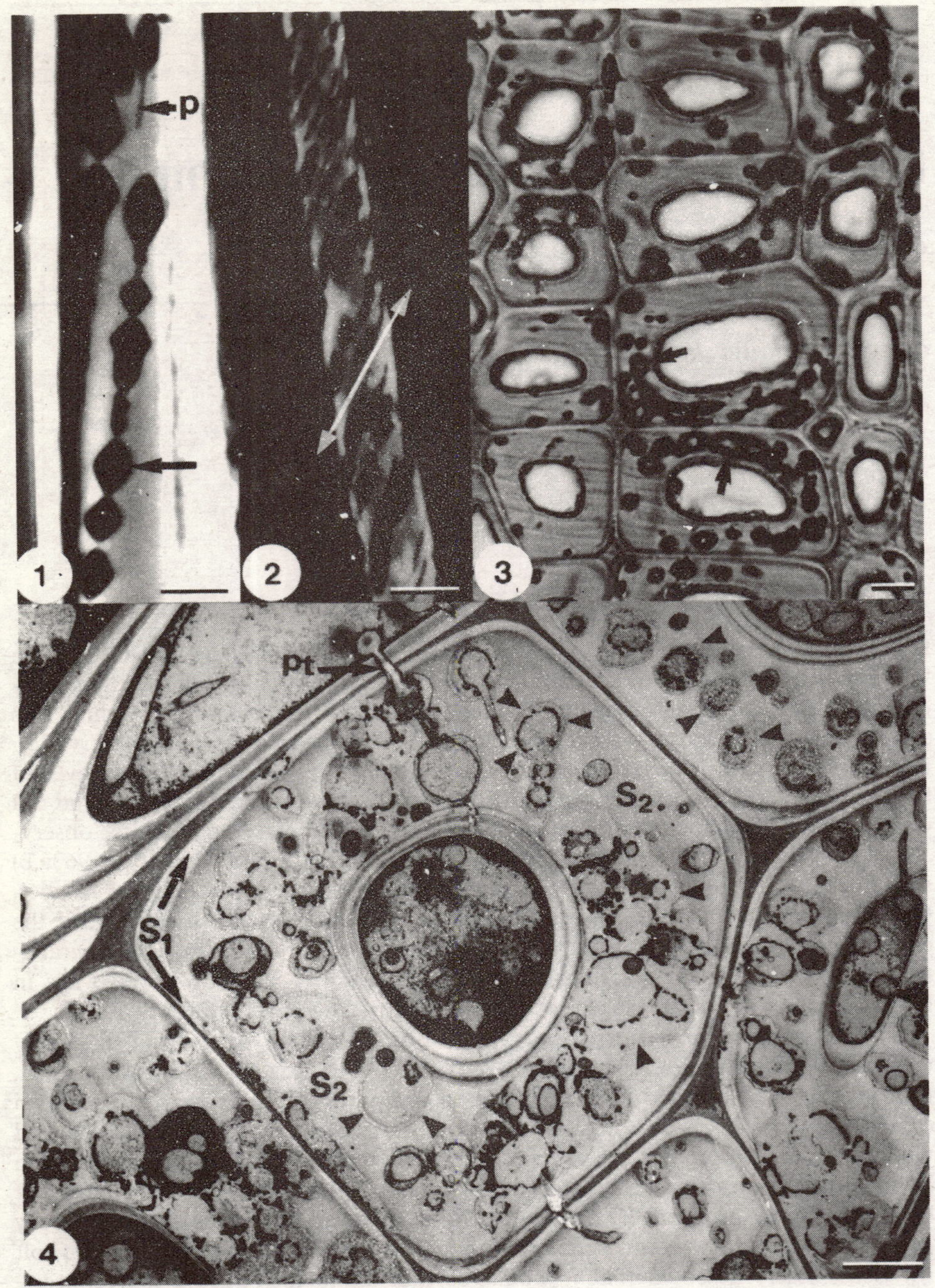

3:1.1 *Micromorphology of Soft Rot Attack*

In the following text the micromorphology of soft rot attack will be described for wood fibres in sapwood of typical temperate hardwoods and softwoods. The main difference in importance for soft rot attack between these wood types seems to be the type and content of lignin. Hardwoods have a lignin content around 18–20 per cent, while that of softwoods is slightly higher at 25–28 per cent. Softwood lignin is however composed almost exclusively of guaiacyl lignin, whereas hardwood lignin is characterized by both guaiacyl and syringyl units. It has been generally observed that the rate of attack by white and soft rot is higher in hardwoods (Savory, 1954; Nilsson, 1973; Eslyn and Highley, 1976; Nilsson *et al.*, 1989). This is due to the fact that soft rot susceptibility is inversely correlated with wood lignin content. Lignin type may also influence the micromorphology of attack.

Cavity Formation

Formation of chains of discrete cavities with conical ends within the wood fibre walls is characteristic for a large number of soft rot fungi (Figures 3.1, 3.2). In transverse sections the cavities within the S_2 are generally observed as rounded holes (Figures 3.3, 3.4). Most of the soft rot species also cause an erosion form of attack in hardwoods (Nilsson, 1973). There appears however to be no apparent difference between hardwoods and softwoods in the cavity formation process. The number of cavities formed after a fixed time for a certain volume of wood is, however, substantially higher in hardwoods indicating that cavities are more easily formed in hardwoods. Cavity formation in vessel element walls is also delayed compared with fibres, possibly due to the more guaiacyl rich lignin present.

Corbett (1965) referred to cavity formation as Type 1 attack. She also described how cavities were initiated by hyphae penetrating into wood fibre walls where they formed a T-shape branch that aligned the hypha in the longitudinal direction of the fibre (Corbett and Levy, 1963). Dissolution of the wood cell wall material around the aligned hypha then results in cavity formation (Figure 3.1–3.4). The most detailed studies on cavity formation were carried out by Hale and Eaton (1985a,b,c). The process can be summarized as follows. After colonization of the wood, a fungal

---

Figure 3.1–3.4 Typical soft rot cavities formed within wood fibres

Figures 3.1, 3.2 Polarized light microscopic photos showing typical cavity chains and a hyphal proboscis (p) formed in *Pinus sylvestris* L. by soft rot fungi. The cavities are aligned with the cellulose microfibril orientation of the secondary S_2 wall layer and show typical variations in the shape with both diamond and rhomboid-like cavities apparent. The fibre in Figure 3.2 has been delignified to highlight the helical arrangement (arrow) of cavity development

Figure 3.3 Light microscopic photo of pine wood cells in cross-section showing typical appearance of cavities at various stages of development

Figure 3.4 TEM micrograph showing cavities in *Homalium foetium* fibres. Note the presence of cavities (arrowheads) in both S_2 and S_1 layers, cell wall penetration (pt), and the presence of electron dense materials (presumably lignin remains and melanin) associated with fungal hyphae within cavities

Bars: 3.1, 25 μm; 3.2, 50 μm; 3.3, 5.0 μm; 3.4, 1.0 μm

hypha in the cell lumen penetrates into the wood fibre wall. This hypha may pass directly through the wall and the adjacent wall to emerge in the adjacent fibre lumen. In such a case no cavity is formed but rather a fine borehole. Cavity formation seems to require that the penetrating hypha is aligned along the cellulose microfibrils in the fibre wall. This is achieved either through the formation of a T-branch or a simple bending of the hypha (L-bend). Multiple branching is also sometimes observed. When the hypha is aligned, growth stops and a cavity begins to form. After some time when a cavity has developed, new rapid growth is intitiated, either from one or both ends of the cavity. The newly formed hypha, called a proboscis (Figure 3.1), is extremely thin and appears to lack a true cell wall. Growth then stops after a seemingly preset distance and a new cavity is formed at one or both ends of the first cavity. This process is then repeated and leads to an interconnected chain of cavities (Figure 3.1, 3.2). New cavities and cavity chains may also be initiated through branching of hyphae within the first formed cavities. The rate of widening of cavities is thought to be related to wood cell wall characteristics (Hale and Eaton, 1984).

It has been discussed whether T-branching or L-bending occurs randomly or as a response to certain chemical or morphological structures in the wood fibre wall. Nilsson and Daniel (1983a) and Daniel and Nilsson (1996) observed that in certain timbers cavities were preferentially formed in, or next to, concentric lamellae in the secondary cell walls. It is believed that these lamellae have a chemical structure different from that of the surrounding cell wall as shown with the tropical hardwood *Homalium foetium* (Daniel and Nilsson, 1996). A similar observation was reported by Schmitt and Peek (1996).

Using transmission electron microscopy (TEM) it has been observed that an electron dense material is left within the cavities (Figure 3.4) (Daniel and Nilsson, 1989a) with the amount remaining apparently dependent on the fungal species. This material probably represents lignin remnants and melanin (Daniel and Nilsson, 1989a). Continued cavity formation eventually leads to almost complete destruction of the S_2 layer, but the middle lamella, and particularly in softwoods the S_3 layer, remains seemingly intact (Meier, 1955). Middle lamella degradation has, however, been reported (Courtois, 1963b). The S_3 layer often becomes coloured very dark brown or almost black. We believe that this may be due to melanin which penetrates the S_3 layer. In birch wood which has not been demonstrated to possess a true S_3 layer, there is still a resistant layer which often becomes very dark (Daniel and Nilsson, 1989a). The dark colouration depends on fungal species and does not occur with all species of soft rot fungi. The degradation of cell wall substance also involves the innermost layers of the secondary cell wall in certain cases. This means that cavities will open to the cell lumen and can be seen as open cavity chains when viewed by scanning electron microscopy (SEM) (Greaves, 1977). This pattern should not be confused however with true erosion of the cell wall.

Cavity formation has not only been observed in the S_2 secondary cell wall layers but also in S_1 layers (for example, Figure 3.4), particularly in softwoods and especially in compression wood tracheids where the S_1 layer is thicker compared with that in normal tracheids. Although it has not been studied in any detail, cavities appear to be formed preferentially in the outermost parts in fibres of certain durable heartwoods (Nilsson and Daniel, unpublished observation). This possibly indicates that the concentration of decay-inhibiting extractives is lower in this area. Courtois (1963b) attributed many forms of attack to soft rot, but as most of his studies concerned wood exposed in a cooling tower, it is not clear what the causative microbes were.

The enlargement of cavities is obviously related to enzymes being secreted by the cavity-forming hyphae. It has been suggested that these enzymes are tightly bound to hyphae, although evidence is lacking. On the contrary it can be seen that where the surrounding cell wall has been damaged, enzymes are able to diffuse into the damaged area to cause degradation of the cell wall substance. Nilsson (1974b) also demonstrated that enzymes are able to diffuse out of small wood blocks attacked by soft rot when placed on cellulose agar to produce clearing zones. The fact that the degradation occurs in the form of cavities with well defined borders is most probably related to the fact that the enzymes are unable to diffuse through the intact cell wall. Migration of enzymes may also occur with the simultaneous degradation of the cell wall substance.

The fact that bore holes perpendicular to the cell wall remain small and that cavities are formed as discrete entities rather than as a continuous cylinder of cell wall dissolution remains unexplained. Possible explanations are: absence of enzyme secretion from hyphal segments involved, or incorrect stereochemistry of cell wall polymer groups hindering enzyme activity. The fact that cavity-forming hyphae appear to align themselves along the cellulose microfibrils in fibre walls (Bailey and Vestal, 1937; Nilsson, 1974c; Sulaiman and Murphy, 1995) suggests that this arrangement induces the production of cell wall-degrading enzymes.

It has been observed that soft rot cavities are not only formed in wood fibres but also in a wide variety of other plant fibres such as cotton, seed hairs, bast fibres (Nilsson, 1974a,c) and bamboo (Sulaiman and Murphy, 1995). Soft rot cavities have also been reported frcm roofing material such as reed (Bosman, 1985; Kirby and Rayner, 1989), pampas grass (Fukuda, 1991) and also from decomposing coniferous needles (Gourbière *et al.*, 1989). An ordered (parallel) microfibrillar arrangement and a cell wall layer of sufficient thickness appears to be a prerequisite. Soft rot cavities are formed in most types of wood elements such as tracheids, fibres, axial and radial parenchyma cells, ray tracheids and vessel elements.

Erosion Forms of Attack

Erosion of cell walls is accomplished through enzyme secretion from hyphae growing in the wood cell lumina. Corbett (1965) referred to erosion attack as Type 2 attack. The erosion morphology is distinctly different in hard- and softwoods. In hardwoods the erosion is a form of removal of the cell wall layers from the lumen progressing towards the middle lamella (Figures 3.5–3.7). Erosion may have a striped appearance or may be in the form of erosion troughs of varying sizes. In softwoods, where the S_3 layer is usually highly lignified, the enzymes appear to diffuse through the S_3 layer to cause attack on the adjacent S_2. It is not known whether the S_3 layer becomes partially degraded in this process. In later stages hyphae may be observed within the degraded S_2 layer, and may have entered the wall when already softened by degradation (Figure 3.7). Erosion will eventually cause decay of the S_2 layer while the middle lamella and particularly the S_1 layer in softwoods will remain largely unaffected (Daniel and Nilsson, 1989a; Nilsson *et al.*, 1989). Soft rot erosion in hardwoods such as birch is difficult to distinguish from erosion caused by white rot fungi. Extensive middle lamella degradation will, however, indicate white rot.

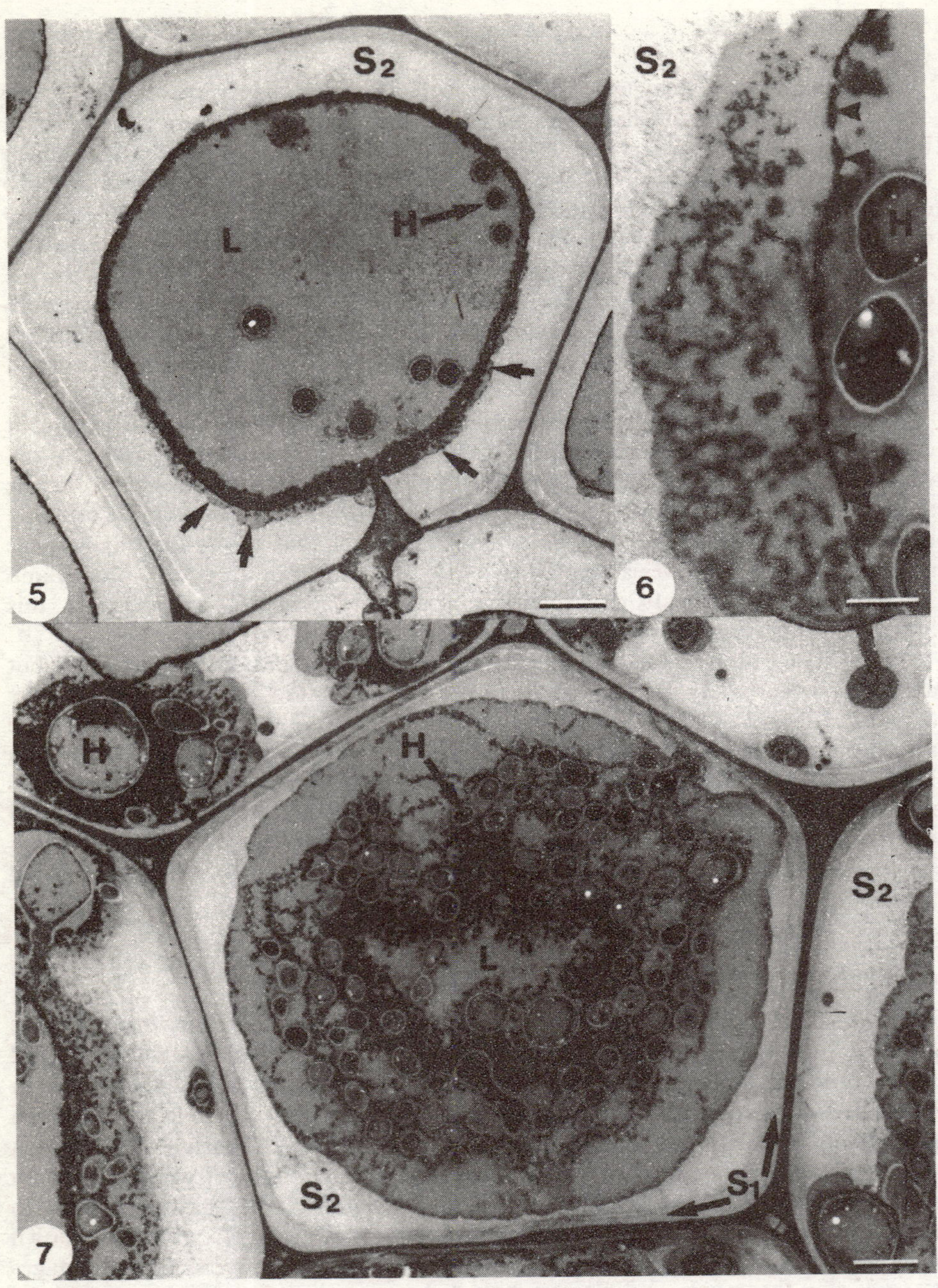

It appears that most cellulolytic microfungi are able to cause at least some erosion of fibre walls in hardwoods (Nilsson, 1973). Even fungi that are regarded as typical moulds of the genera *Aspergillus*, *Gliocladium*, *Penicillium* and *Trichoderma* cause slight erosion attack in birch wood. Weight losses are typically low however, though Nilsson (1973) reported weight losses of 10 to nearly 30 per cent for *Aspergillus fumigatus* and *Gliocladium catenulatum*. Cavity formation by *Aspergillus*, *Penicillium* and *Trichoderma* species has been reported (Liese and von Pechman, 1959; Courtois, 1963b; Fukuda and Haraguchi, 1975) but this has not been substantiated by later studies. Most species of soft rot that cause extensive erosion in hardwoods cause much less erosion in softwoods. This is probably due to differences in lignin type and content rather than differences in anatomical structure. This is supported by the fact that hardwoods with a high lignin content but a low syringyl:guaiacyl (S/G) ratio are also quite resistant to erosion attack. *Daldinia concentrica* was exceptional in producing very high weight losses (60 per cent) in birch wood but failed to cause any erosion in pine wood even after very extended incubation times (Nilsson *et al.*, 1989). The more lignified S_3 layer in softwoods also appears to act as a barrier to erosion.

Chemical delignification, especially of softwoods, leads to a dramatic increase in susceptibility to soft rot (Courtois, 1963a; Zainal, 1975; Takahashi and Nishimoto, 1976; Morrell and Zabel, 1987). The degradation of delignified wood appears to occur primarily through erosion and not through cavity formation. The latter seems to be greatly impeded by the chemical treatment (Nilsson, 1974c; Zainal, 1976; Morrell and Zabel, 1987). Erosion of non-lignified cellulose in wood such as the gelatinous layers in the tension wood fibres of hardwood has also been described (Encinas and Daniel, 1997).

Diffuse Cavity Pattern

When transverse sections of wood with soft rot cavities are observed, the edges of cavities are usually very sharp and no degradation can be observed beyond the edges (Figures 3.3, 3.4). We have, however, noted that diffuse degradation sometimes occurs which extends beyond the cavity edge. This appears to happen more frequently when wood has decayed under quite wet conditions. This diffuse attack may be quite extensive and extend across the cell wall to the middle lamella. No chemical analyses have been carried out, but observations using polarized light suggest that the cellulose is degraded in a way reminiscent of brown rot attack. This pattern has

Figure 3.5–3.7   TEM micrographs showing examples of soft rot erosion caused by *Phialophora mutabilis* in copper–chromium–arsenic (CCA) treated *Betula verrucosa* fibres
Figures 3.5, 3.6 Hyphae (H) are causing characteristic erosion (arrows) of the secondary cell wall beneath the luminal (L) wall and layer of precipitated CCA (arrowheads) preservative. The remaining electron dense materials within the degraded areas probably represent partially degraded and modified lignin
Figure 3.7 Micrograph showing a combination of advanced cavity and erosion decay with only the S_1 and outer part of the S_2 remaining
Bars: 3.5–3.7, 10 μm

been studied in more detail by Anagnost *et al.* (1994) who suggested the term 'diffuse Type 1' to describe this form of soft rot.

Nilsson (1973, 1974a,b,c) discovered that a small number of cavity-forming fungi did not produce any cellulose degradation through hyphae not aligned with the cellulose microfibrils within a fibre cell wall. This observation supports the idea that enzyme production is induced by a stereospecific arrangement. The results also raise the question of whether enzymes produced by luminal hyphae are different from those produced by cavity hyphae.

3.1.2 Fungi Causing Soft Rot

Soft rot, by definition, is caused by ascomycetes and deuteromycetes. Phycomycetes are not known to attack lignified wood cells. A large number of species have been reported to cause soft rot, either in the form of cavities or erosion or both. Studies concerning cellulase production and wood attack suggest that most cellulolytic species are able to cause at least some erosion of hardwoods such as birch and aspen. Cavity formation is limited however to a smaller number of species.

A large number of typical soil-inhabiting fungi are known to cause soft rot. Typical examples are species from genera such as *Acremonium, Chaetomium, Doratomyces, Coniothyrium, Humicola, Phialocephala, Phialophora, Phoma* and *Trichocladium*. Cavity-forming *Phialophora* species have been reported to have a worldwide distribution occurring in preservative treated timber in ground contact (Nilsson and Henningsson, 1978; Zabel *et al.*, 1991; Wong *et al.*, 1992) and several *Phialophora* species were reported colonizing heartwood of *Sequioa sempervirens* in a cooling tower (Morrell and Smith, 1988). A large number of marine species have been reported to cause soft rot of Type 1 and Type 2 (Curran, 1979; Leightley, 1980; Mouzouras, 1989). Typical examples are *Ceriporiopsis halima, Humicola alopallonella, Lulworthia* spp. and *Monodictys pelagica*. Except for the occurrence in the marine environment, they have also been observed in timber of water cooling towers cooled with either estuarine or fresh water (Eaton and Jones, 1971; Eaton, 1972). Several typical freshwater fungi have also been reported to cause soft rot (Zare-Maivan and Shearer, 1988). Truly thermophilic soft rot fungi, such as *Allescheria terrestris, Chaetomium thermophilum, Sporotrichum thermophilum* and *Thermoascus aurantiacus* have been reported from wood chip piles.

3.1.3 Environmental Factors and the Occurrence of Soft Rot

Oxygen

Oxygen is required for wood degradation by soft rot fungi. Oxygen is, however, rarely a limiting factor in nature except for in waterlogged wood. Soft rot fungi are, in contrast to most white and brown rot fungi, able substantially to degrade wood completely submerged in water. When oxygen becomes limiting, as for example in the interior of large pieces of waterlogged timber, activity will stop. This is illustrated in archaeological waterlogged wood which often has an outer shell of heavily soft rot degraded fibres, but the interior is mainly degraded by erosion bacteria which seem to survive on very limited amounts of oxygen.

Moisture

It is a common misconception that soft rot only occurs in very wet wood. We have observed soft rot attack in relatively dry wood, one example being around *Camponotus* ants' tunnels in dry building timber. The moisture emanating from the ants creeping through the tunnels was obviously enough for soft rot to occur. Soft rot like most other forms of microbial wood decay occurs in certain ecological niches where the fungi can be active mainly due to lack of competition from other micro-organisms. This is why soft rot is commonly observed in wet timber where the more aggressive white and brown rot fungi are less competitive. Other factors such as toxic heartwood extractives, wood preservatives and temperature may also prevent competition from basidiomycetes and this will allow soft rot fungi to develop.

Temperature

No specific studies appear to have been done on the effect of low temperature, but with the knowledge of temperature relations of microfungi in general it may be assumed that soft rot is active in wood just above 0°C. Thermophilic soft rot fungi from wood chip piles may however be actively degrading wood up to 60°C (Nilsson, 1973; Ofosu-Asiedu and Smith, 1973).

pH

Few studies have been done on the influence of pH on soft rot activity. The reports by Sharp and Eggins (1970) and Duncan (1960a) indicate that soft rot activity occurs within a broad pH range (3.7–8.6). This is supported by the fact that soft rot is also observed in acid coniferous forest soils as well as neutral soils (Nilsson and Daniel, 1990).

Nitrogen

It has been observed that addition of nitrogen to wood substrates greatly increases the rate of attack by soft rot fungi (Worrall and Wang, 1991). This is also reflected by the fact that soft rot attack is quicker and more extensive in fertile soils compared with soils poor in nitrogen. It has also been reported that nitrogen which has been redistributed during drying to the surface of wood will significantly increase soft rot attack when the wood is incubated in unfertile soil (King *et al.*, 1989). The reasons for the demand for nitrogen are probably that soft rot fungi, in contrast to brown and white rot fungi, have a less efficient system for re-utilizing the nitrogen of their enzyme proteins and that some nitrogen remains firmly bound for a very long time to hyphae within cavities. Furthermore, complete degradation of the fibre walls requires a large number of cavities each containing a hypha. Hyphae have been observed to persist for an extremely long time but it is unknown whether their nitrogen can be reallocated to other parts of the mycelium in the wood.

3.1.4 Tolerance of Toxic Compounds

It has been observed that soft rot fungi tend to attack timber which, due to the content of toxic extractives or preservative treatment, has become resistant to basidiomycete attack. Transmission poles treated with copper–chromium–zinc or copper–chromium–arsenic (CCA) compounds are preferentially attacked by soft rot fungi. Soft rot has also been reported from the outer layers of poles treated with pentachlorophenol and creosote (Lew and Wilcox, 1981; Zabel *et al.*, 1985; Wylde and Dickinson, 1988). Detailed studies have found that many soft rot fungi, particularly from the genus *Phialophora* are relatively resistant to copper compounds (Daniel and Nilsson, 1988). It has also been observed that softwood timbers are more easily protected from soft rot than hardwoods (Purslow and Williams, 1979). This is one explanation for the many reports of soft rot attack in CCA treated eucalypt and other hardwoods (Greaves, 1977; Levy, 1978; Leightley, 1978, 1981). A puzzling fact is that the amount of toxic compounds affording protection to soft rot in softwoods is much lower than that required to prevent growth in agar tests (Daniel and Nilsson, 1988). This led Nilsson (1982a) to suggest that the protection of softwoods by CCA was a combination of toxic effect and some form of chemical modification of the wood structure. It was also suggested that this phenomenon was related to the high lignin content of softwoods and the high amount of guaiacyl lignin. It was later shown that certain hardwoods could be protected if they had a high lignin content and a low S:G ratio. This is supported by laboratory experiments and from field data (Nilsson *et al.*, 1988). A comparatively high tolerance to pentachlorophenol (Savory, 1954), sodium arsenate, sodium chromate, zinc chloride and creosote has also been reported (Duncan, 1960b). Heartwoods with natural durability to basidomycete attack will often be attacked by soft rot when exposed in ground contact (Liese, 1961; Thornton and Johnson, 1988).

3.1.5 *Chemistry of Soft Rot Attack*

There are only a few studies on the chemical changes occurring in wood following soft rot attack. It seems clear, however, that attack on lignin is limited compared with white rot fungi. This is also illustrated by the fact that soft rot fungi appear unable to degrade the middle lamella. It seems that lignin degradation may occur and may even be quite extensive, but the lignin degrading capacity seems to be dependent on fungal and timber species.

Several reports describe changes in chemical composition following attack by *Chaetomium globosum*. Savory and Pinion (1958) found that the fungus preferentially removed the polysaccharide fraction in beech wood. The loss of lignin was only 6.3% per cent at a weight loss of 69.6 per cent. Levi and Preston (1965) and Seifert (1966) reported considerably higher losses of lignin in beech wood. The remaining lignin had a lower methoxyl content and was more acid-soluble compared with lignin in sound wood. Eslyn *et al.* (1975) and Nilsson *et al.* (1989) reported significant losses of lignin from hardwoods and softwoods attacked by soft rot fungi, but the carbohydrates were preferentially degraded by most fungi. Nilsson *et al.* showed that the syringylpropane units of the lignin in birch wood were removed selectively.

Soft rot fungi that exclusively degrade wood through erosion vary in their capacity to degrade from slight erosion to complete removal of the secondary cell wall. This probably reflects variations in their lignin degrading ability. Fungi with a more extensive attack of fibre cell walls probably have a more efficient system for degrading at least hardwood lignin. We are, however, not aware of any studies on lignin degrading enzymes of the soft rot fungi except for one report by Durán *et al.* (1987) describing ligninase production by the ascomycete *Chrysonilia sitophila*. Phenolic and lignin related compounds are degraded by a number of soft rot fungi (Haider and Trojanowski, 1975; Buswell *et al.*, 1982; Eriksson *et al.*, 1984; Bugos *et al.*, 1988) but there appears to be no correlation between the degradation of these compounds and the degradation of lignin in wood.

3.2 Wood Degrading Bacteria

3.2.1 *Characteristics of Wood Degrading Bacteria*

The occurrence of bacteria in wood has been known for a long time (Liese, 1950). However it is only in the last 20 years or so that evidence has been obtained to show unequivocally that bacteria can degrade lignified wood cell walls. This evidence has been provided primarily by correlated light, SEM and TEM electron microscopic observations. Previous to the application of electron microscopy, decay patterns detected in wood were often ascribed to bacteria or actinomycete decay without true evidence. By use of microscopy and by comparing patterns produced in wood samples removed from natural environments with patterns produced by mixed cultures of bacteria grown under laboratory conditions, major features on the micromorphology of bacterial degradation of wood have been established. This descriptive approach forms the basis for current detection of bacterial decay in wood. Previously it was often difficult for researchers to distinguish bacterial decay from fungal decay patterns and thus bacteria went 'unnoticed'. The ability of bacteria to help increase the permeability of refractory wood species such as spruce during 'ponding' or storage of poles under water spraying has been utilized and known for a long time. During the latter processes, aerobic and anaerobic bacteria cause selective degradation of non-lignified cells (for example parenchyma cells in rays) and pit membranes (for example bordered and window pits (Daniel *et al.*, 1993)) allowing subsequent increased penetration of preservative fluids into the wood structure (Bauch *et al.*, 1970). This method for example has been proven for prior treatment of poles before creosote preservation (Bergman *et al.*, 1975). However the bacteria causing such degradation have been shown to possess primarily cellulolytic and pectinolytic activity and thus are unable to penetrate into wood cell walls in which the cellulose and hemicelluloses are protected by the lignin matrix. A variety of other bacteria may also exist as secondary feeders but these have not been proven to be able to degrade lignified wood cell walls and probably act as scavengers causing only slight cell wall modification (Eriksson *et al.*, 1990; Schmidt and Liese, 1994).

True wood degrading bacteria in the present context are those bacteria which have been conclusively shown by light and electron microscopic methods to cause significant attack and decay of lignified fibres and tracheids in hard- and softwoods. Currently only two major forms have been recognized which have been ascribed as 'tunnelling' (TB) and 'erosion' (EB) bacterial decay. These two forms are recognized from the micromorphological decay patterns which they cause during the degradation of wood cells and they do not represent any form of taxonomic classification. Varying decay patterns are produced within each group as will be outlined below. In recent years there has been an upsurge in the interest in bacterial degradation of

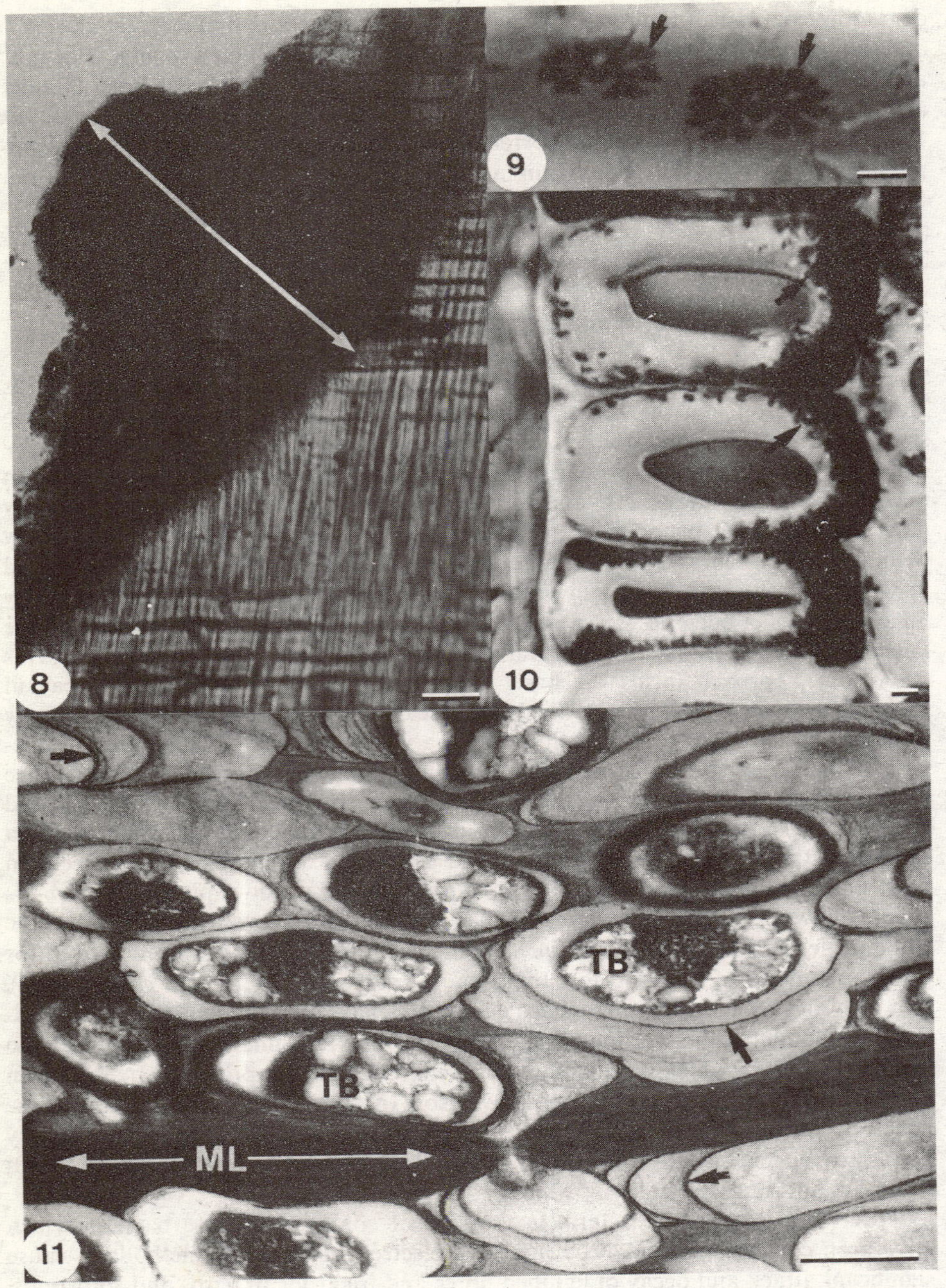

9
8
10
TB
TB
ML
11

wood and it is being reported in the literature more frequently. This reflects primarily the greater ability of researchers to identify bacterial decay patterns.

3.2.2　*Ecology of Wood Degrading Bacteria*

The examination of lignocellulose (i.e. hard- and softwoods) material removed for a wide range of environments (Figures 3.8–3.15), for example, seawater (Eaton and Dickinson, 1976; Kohlmeyer, 1980; Mouzouras *et al.*, 1986), terrestrial (Greaves, 1968; Liese and Karnop, 1968; Holt, 1983; Willoughby and Leightley, 1984), and freshwater (Holt *et al.*, 1979; Holt, 1981; Nagashima *et al.*, 1990), strongly support the cosmopolitan distribution of wood degrading bacteria around the world. Wood degrading bacteria may be found occurring together with fungal attack produced by the three major groups including soft, brown and white rot decay. For example, under seawater and freshwater conditions bacterial decay may be recognized together with soft rot, while under terrestrial conditions it may be found together with any of the major groups (for example, sandy soils with brown rot; forest soils with white rot; and clay wet soils and fertile soils with soft rot). The absence of bacterial decay with white and brown rot under aquatic situations is however more a reflection of the absence or reduced occurrence of these major fungal rot types rather than the greater importance of wood degrading bacteria.

Concerning the ecology of wood degrading bacteria two features are apparent. First, wood degrading bacteria apparently cannot compete with fungal decay unless environmental conditions are such that fungal development is in some way suppressed so as to produce a comparatively competitive-free environment for the bacteria. For example, under conditions of oxygen limitation for fungi (for example, Singh *et al.*, 1990) such as buried shipwrecks (Blanchette *et al.*, 1990; Kim and Singh, 1994), archaelogical artefacts (Kim, 1989; Blanchette *et al.*, 1990, Blanchette, 1995) and building foundations (Bouteljie and Bravery, 1968), erosion bacteria are able to proliferate in the absence of fungal competition. Furthermore, woody materials which are otherwise more resistant to fungi through wood protective treatments by high loadings of wood preservatives such as CCA, copper–chrome–boron (CCB) (Nilsson, 1984; Daniel and Nilsson, 1986) creosote (Pitman *et al.*,

---

Figure 3.8–3.11　Light and TEM micrographs showing aspects of tunnelling bacterial attack of wood

Figure 3.8 TB attack of durable greenheart (*Ocotea rodiaei*) timber from marine gate seals on the south coast of the UK. A distinct decay zone (arrow) formed entirely from TB attack is apparent

Figure 3.9 Polarized light micrograph of an LS section of pine showing a typical decay pattern (arrows) from one type of TB attack. The pattern is very similar to the habit form reported for *Nevskia ramosa* by Famintzin (1892) for 'bacterial neusten'

Figure 3.10 Cross-section of spruce tracheids chemically modified with 8% butylene oxide showing presence of TB attack (arrows) of the S_1 and S_2 cell wall layers

Figure 3.11 TEM micrograph showing TB within a heavily degraded pine cell wall. Note the characteristic nature of TB tunnels with concentric slime secretions (arrows) behind the bacteria. ML, middle lamella)

Bars: 3.8, 10.0 μm; 3.9–3.11, 1.0 μm

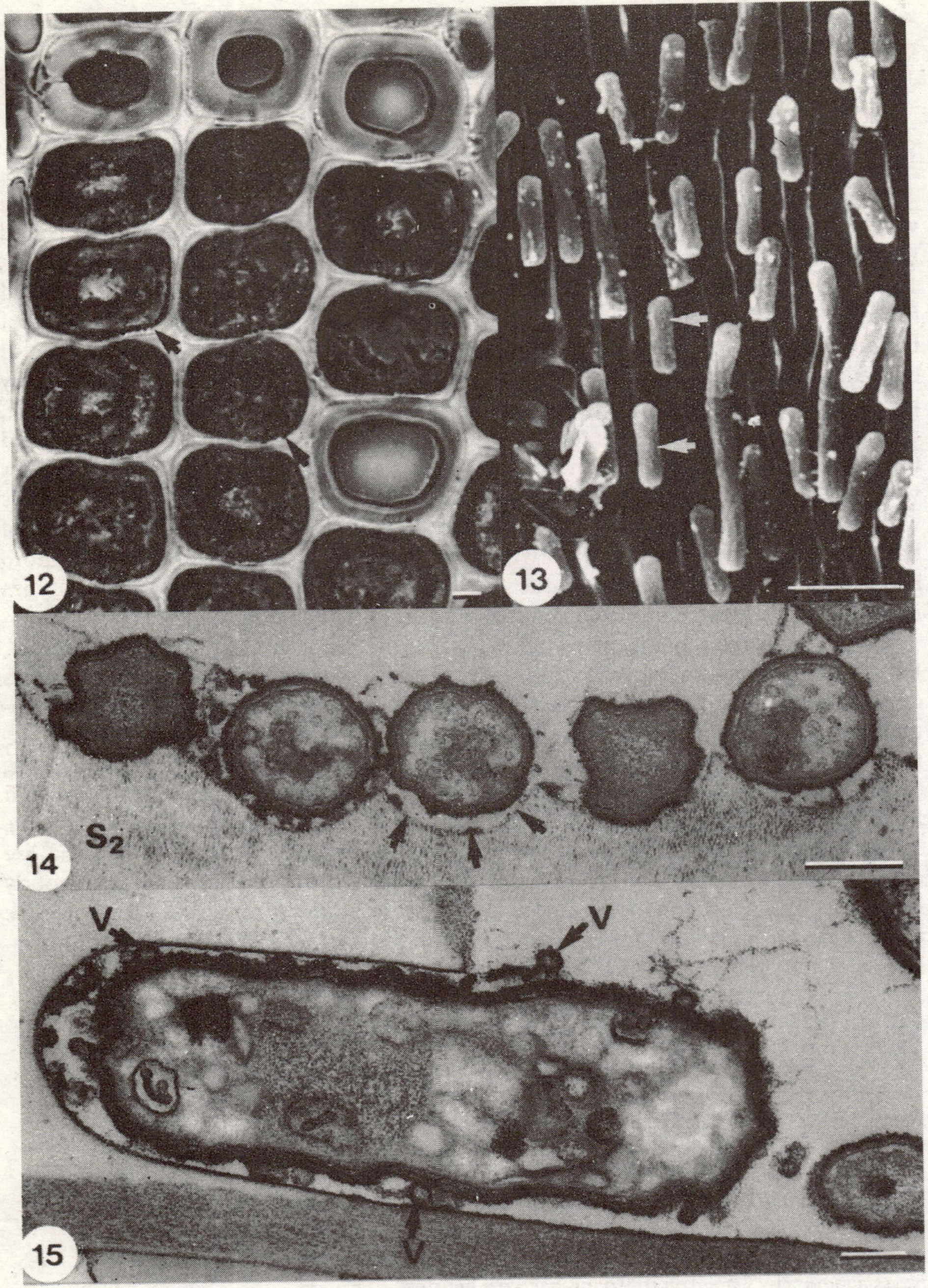
12
13
S$_2$
14
V
V
15
V

1995), or by chemical modification (Nilsson and Daniel, unpublished observation) or are naturally durable via high lignin (Singh *et al.*, 1987) or extractive content (Nilsson *et al.*, 1992; Pitman *et al.*, 1995) provide a niche in which fungi are suppressed by one or other features allowing for successful development of bacterial decay, particularly due to tunnelling bacteria which appear to be more widespread than erosion bacteria. This is readily shown by the use of hydrofluoric treated stakes placed in terrestrial environments. The acid treatment partly dissolves out the wood polysaccharides and sugars and modifies the lignin (Daniel and Nilsson, 1989b) so as to raise the lignin:carbohydrate ratio in the wood. This suppresses soft rot decay which is severely affected by lignin type and content producing a competition-free environment for the bacteria. The technique works well and has been used both as a baiting method for obtaining a source of wood degrading bacteria for isolation, and for ecological studies on the distribution of bacteria.

Under certain conditions where fungal decay is suppressed, bacterial decay may represent the only form of true wood degradation and in itself represent a major form of biodegradation to man-made structures. For example, in studies on the durable timber greenheart, used as moorings in lock gates, tunnelling bacteria were a major biodeteriogen (Pitman *et al.*, 1995). Similarly, under terrestrial conditions in kiwi orchards in New Zealand, tunnelling bacteria were found to be the major cause of serious degradation of support pilings (Nilsson, 1984). These few examples help to show the major problems which wood degrading bacteria may pose to man and preservative systems. Currently there are no preservative standards in use against bacterial decay, which serves to highlight both the continued lack of understanding of these microbes and their potential threat.

The lack of pure cultures has hampered physiological studies on the effects of various physical conditions on bacterial decay. However from both the observations of woody materials exhibiting bacterial decay removed from widely differing environments and studies performed on mixed cultures under laboratory conditions, it is apparent that bacteria have an ability to survive a wide range of environmental conditions (Blanchette *et al.*, 1990). For example, bacterial decay of wood has been reported in samples removed from 5000 m deep in the sea (Kohlmeyer, 1980), and has been produced by mixed cultures in the laboratory at 45°C, suggesting a wide tolerance to temperature. Similarly, bacterial decay has been reported from various soils suggesting tolerance to a wide range of pH, although TB appear to prefer more

---

Figure 3.12–3.15 Light and electron micrographs showing aspects of bacterial erosion attack of wood

Figure 3.12 Erosion decay of pine wood excavated from the Swedish battleship *Kronan* which sunk in the Baltic sea during 1676. The bacteria degrade the tracheids towards the middle lamella regions and a serrated edge (arrows) is noted in the outermost regions of attack. Note the irregularity of attack with both highly degraded and undegraded tracheids apparent

Figure 3.13 SEM micrograph showing erosion decay of *Betula verrucosa* with the bacteria (arrows) longitudinally aligned within troughs formed in the S_2 cell layer

Figure 3.14 TEM micrograph showing bacterial alignment within troughs formed in S_2

Figure 3.15 TEM micrograph showing a longitudinal section through an erosion bacterium and the presence of extracellular vesicles (V)

Bars: 3.12, 5.0 μm; 3.13, 3.14, 1.0 μm; 3.15, 0.5 μm

alkaline soils and have not been found below pH 4.5. The occurrence of bacterial decay under near anaerobic conditions further suggests a tolerance of low O_2 conditions although evidence for true anaerobic TB or EB forms has not so far been obtained. More recently Rogers and Baecker (1988) reported the degradation of *Pinus patula* wood chips by bacteria under strictly anaerobic conditions which is contradictory to current opinion that lignin degradation is an oxidative process. More research is therefore required on the physiology of wood degrading bacteria.

3.2.3 *Micromorphology of Bacterial Wood Degradation*

These observations have been carried out using light and electron microscopy. Observations on wood samples degraded with mixed bacterial cultures and of samples removed from natural situations show that bacteria utilize rays just as fungi do for deeper penetration into the wood from where they radiate out into adjacent wood cells. This colonization process applies to all the wood degrading bacteria currently known.

Tunnelling bacteria have so far been recognized as globose unicellular bacteria which are Gram negative, motile, and exhibit considerable cell wall plasticity (Figure 3.11). They produce characteristic polysaccharide secretions by a peculiar type of stop–start degradative action behind the bacteria during decay through wood cell walls and other solid substrates such as cellulose (Daniel and Nilsson, 1985) (Figure 3.11). These features were first demonstrated via TEM by Nilsson and Daniel (1983b) and further characterized by Daniel and Nilsson (1985) and have since been shown in subsequent studies on TB (Singh *et al.*, 1987; Singh and Butcher, 1991). The secretions, which appear fibrous after dehydration, are also produced around the bacteria when they adhere to solid surfaces such as the wood cell lumina (Daniel *et al.*, 1987). Recent observations with a tentatively pure culture of TB suggest that the bacteria were nearly always contained within a capsule or slime tube and that this slime probably acts as a buffer against the immediate environment allowing for the passage or support of wood degrading enzymes against the wood cell wall, binding and detoxification of heavy metals during decay of preservative treated wood (Daniel and Nilsson, 1985) and resistance to desiccation. The slime tubes are presumed to act and function rather like extracellular glycocalyx of other bacteria and provide the medium for motility. That the extracellular slime tubes exist during wood decay was recently confirmed using cryo-SEM. During colonization a typical TB cell 'sinks' into the wood cell wall by enzymatic activity (Daniel *et al.*, 1987) from where, by periodic cell division and decay, distinct colonies are produced (Holt, 1983). This is most easily recognized using polarized light microscopy after staining the bacteria with lactophenol blue (Figures 3.9, 3.10). The tunnels radiate out in all directions within the wood cell wall and no particular cell wall layer appears to be resistant to decay. Thus in contrast to erosion bacteria the lignin-rich middle lamella regions are also penetrated and degraded (Figure 3.11) as well as durable wood cell walls containing high extractive and lignin levels (Singh *et al.*, 1987; Nilsson *et al.*, 1992). TB do not align themselves with the wood cell wall cellulose microfibrils as seen with EB. That a variety of decay patterns may be recognized in the same wood species suggests that an unknown number of TB species probably exist. In some wood samples the bacteria remain in rather tightly bound colonies whereas in others they spread over several wood cell wall layers (Figure 3.8). Within colonies the bacteria normally attack all the available wood substance (Daniel and Nilsson, 1985), although it is unusual for more than one bacterium to remain within the same slime tube after cell division. With progressive decay the bacterial colonies coalesce and the entire fibres become degraded.

Detailed electron microscopic observations have shown TB and EB to produce abundant extracellular vesicles by apparent budding of the cell membrane (Daniel and Nilsson, 1986; Daniel *et al.*, 1987) (Figure 3.15). These vesicles are presumed to contain extracellular enzymes (for example, cellulases, hemicellulases and pectinases) as described for rumen bacteria (Forsberg *et al.*, 1981) and to be involved in the decay process although definite evidence has so far not been obtained. Nevertheless the vesicles appear morphologically very similar to those produced by cellulolytic bacteria such as *Clostridium* and may represent similar types of 'cellulosome' (Lamed *et al.*, 1983). The vesicles may fuse with the surrounding extracellular slime materials which provides the medium for direct attack on the wood cell wall. The use of general lignin stains at the TEM level has also indicated mineralization of lignin since very little lignin seems to remain in the tunnels after decay. Studies with chromogens incorporated into agar has further indicated the ability of TB to produce extracellular H_2O_2 and peroxidases, both components consistent with an ability to be able to degrade lignin (Daniel and Nilsson, unpublished observation).

Erosion bacteria are typically rod-shaped (about 1–2 μm long) with pointed ends (Figures 3.13, 3.15) and frequently produce a characteristic 'stripy' decay pattern in lignified cell walls. This stripy pattern is best seen in longitudinal sections cut from degraded wood material using polarized light microscopy in which sites of decay are recognized by loss of birefringence indicating attack of the cellulose microfibrils and hemicellulose components. The micromorphology of erosion bacterial decay has several characteristics in common with soft rot fungi. First, during decay the bacteria frequently show characteristic alignment with the underlying wood cell wall cellulose microfibrils (Figure 3.13). Second, the bacteria apparently cannot degrade the high lignin-containing middle lamella region which remains in highly degraded wood samples (Figure 3.12) and have difficulties in degrading the S_3 layer in softwoods which is also highly lignified. Third, very often the bacteria produce a discrete decay pattern with angular cavities as seen with soft rot. Correlated scanning and transmission electron microscopy (Daniel and Nilsson, 1986) indicates that decay results from bacteria which produce and actively move along discrete channels aligned with the underlying wood cell wall cellulose microfibrils. The channels are aligned with each other so that all the wood substance becomes attacked (Figures 3.13, 3.14). The bacteria readily divide within the channels so that the colony progressively enlarges end-wise. Variations in the speed by which the bacteria progressively 'sink' into the wood cell walls create variability in the angular type patterns recognized. The nature of the decay process as with TB indicates that erosion bacteria possess cellulolytic, hemicellulolytic and ligninolytic ability. Electron microscopy has further indicated that EB have an ability to 'bleach' middle lamella regions suggesting the presence of a low molecular weight redox system (Daniel and Nilsson, 1986). As with TB, electron microscopy observations and staining with lignin stains (for example, $KMnO_4$) has suggested that these bacteria decay the lignin within the channels. Physiological studies with radiolabelled lignin model compounds (Daniel *et al.*, 1987) have similarly shown EB to mineralize lignin although not to the same degree as TB (Nilsson and Daniel, 1986).

There is a further form of bacterial attack which has been referred to as cavitation decay (CB). This type of decay was first described in preservative treated vineyard posts in kiwi orchards in New Zealand (Nilsson and Singh, 1984). Decay results in characteristic angular cavities produced within the S_2 cell wall, the cavities frequently developing perpendicularly outwards from sites of bordered pits in softwoods; the latter suggests colonization and decay via the bordered pit chambers. The nature of cavitation bacteria is still unknown, as the bacteria have not been frequently observed in cavities during the decay process, but rather after the event. The micromorphology of cavitation bacterial decay has been described from a few other sites, but much less is known of their occurrence. Erosion bacteria have been described to cause a similar pattern of decay (Daniel and Nilsson, 1986) so it remains to be determined whether CB really constitute an additional group.

The importance of actinomycetes (i.e. filamentous bacteria) and evidence for their decay of wood still remains a controversal issue. Actinomycetes are abundant in soils and are associated with decaying plant and lignocellulose materials, and thus it has been widely accepted that they may play a major role in the recycling of woody materials. However despite the large numbers of actinomycetes which may be isolated from decaying wood (for example, Calvalcante, 1981; Safo-Sampah, 1985), very few, if any, of these isolates have been confirmed as true wood degraders. In the early 1980s, a *Streptomyces* sp. was reported to cause soft rot decay in lime wood (Baecker and King, 1981) and *Streptomyces xanthochromogenus* specifically caused 12 per cent dry weight loss in lime (Baecker *et al.*, 1983), although subsequent studies have not confirmed or been able to repeat this work. Actinomycetes however can readily cause the degradation of non-lignified tissues in wood such as parenchyma cells and have been shown to degrade the phloem in hard- and softwoods and the cellulosic G-layer in tension wood (Sutherland *et al.*, 1979; Eriksson *et al.*, 1990). Certain microbial patterns found in wood samples placed in acid soils have been attributed to actinomycete decay (Nilsson *et al.*, 1990) although the organisms responsible have not so far been isolated in pure culture. Actinomycetes have been more easily isolated than true wood degrading unicellular bacteria and it has therefore been possible to carry out a wide range of physiological and biochemical studies which have not so far been possible with true wood degrading bacteria. They have thus been shown to produce a variety of cellulosic, hemicellulosic, pectinolytic and even lignin peroxidases (Ramachandra *et al.*, 1988; Adhi *et al.*, 1989) considered to be involved in lignin decay. Their biotechnological potential is thus currently greater than that known for wood degrading bacteria.

3.2.4 Isolation and Culture of Wood Degrading Bacteria

The importance of wood decay fungi was recognized very early and consequently methods for their isolation were rapidly developed and Koch's postulates fulfilled. The situation was helped by the fact that many important wood rotting fungi of economic importance were easily isolated using traditional mycological methods. The isolation of wood decay bacteria has however been much more difficult and methodology for their isolation is still in its infancy. This reflects both the greater interest in and considered importance of wood decay fungi and also the fact that wood decay bacteria appear to be much more difficult to isolate using traditional bacteriological methods.

Methodology for the isolation of wood decay bacteria has primarily involved the removal of thin slivers of wood from samples supporting decay and the plating out or spreading of these slivers onto agar growth medium containing a fungicide (for example, natamycin) to suppress fungal development which otherwise quickly out-competes bacterial growth. Acid-treated wood stakes as described above have often been used since the bacterial inoculum is usually high and the fungal level low in comparison. A wide variety of rich and weak growth media have been tried over the years but of particular success has been the use of Bak-38, a comparatively weak growth medium with added vitamins and trace elements in either liquid or solid agar cultures (Nilsson, 1982b; Nilsson and Daniel, 1992). Using this methodology it has been easy to establish mixed cultures of wood degrading bacteria which have then been used to reproduce the decay patterns found in samples removed from the natural environment in added wood shavings, pulp fibres or pure cellulosic substrates. This possibly was the greatest step forward in our understanding of bacterial wood decay as, together with convincing evidence from electron microscopy, it demonstrated beyond doubt that bacteria were able to produce true decay of wood cell walls. These mixed cultures are quite resilient and can be maintained for many years; they will even survive periodic drying out of the wood samples, a feature which certainly reflects natural situations. Subsequent establishment of pure cultures from such mixed inocula has been problematic in that traditional methods of bacterial isolation (for example, streaking, serial dilution, etc.) have not been successful. The problems would appear to be related to the following. First, true wood degrading bacteria appear to be slow growing and during wood decay are usually found together in a consortium containing bacteria which may be dependent on wood degrading bacteria, but which themselves cannot degrade wood cell walls. Use of rich media or incorrect media tends to select for these secondary bacteria or other more rapidly growing bacteria present in wood but which are not true wood degraders. Second, wood degrading bacteria tend to have copious extracellular slime secretions associated with them. This makes separation via serial dilution and streaking difficult without prior dissolution or disruption of the slime. Third, a possibility remains that wood degrading bacteria support a life cycle similar to that reported for some of the cytophaga groups. If such a situation exists, it is highly possible that pure cultures were developed many years ago but that the stage of the life cycle isolated is not that involved in true wood decay and that conditions for completion of the cycle have not been established. It should also be noted that many bacteria which are available from culture collections have an ability to degrade lignin model compounds but are unable to degrade wood and therefore cannot be classified in the same category.

3.2.5 Taxonomy and Classification of True Wood Degrading Bacteria

Taxonomy and classification has been seriously hampered by the lack of pure cultures and it is therefore difficult to extrapolate an apparently pure culture to represent true wood degrading bacteria even after taxonomic classification if Koch's postulates have not been fulfilled. Certain pure cultures have been taxonomically described over the last 15 years (for example, Drysdale *et al.*, 1986; Schmidt *et al.*, 1995) using traditional keys and chemical methods. For example, Drysdale *et al.*

(1986) considered the bacteria they isolated from CCA treated horticultural posts to be *Bacillus* (sp. *cereus* or *cereus* var. *mycoides*) or *Pseudomonas* (*maltophilia* or *pickettia*) while Schmidt *et al.* (1995) described the strain they isolated from lake water in which trees were ponded to be *Aureobacterium luteolum* (DSM 20143). While it is highly likely that a wide range of wood degrading bacteria probably exists from the large number of environments in which evidence for bacterial degradation has been found, none of these isolates appear consistent with those currently known from features for TB and EB described from microscopic studies. The taxonomic classification of wood degrading bacteria may lie within the cytophagas or sporocytophagas or more likely within the little known genus *Nevskia* (Famintzin, 1892). This genus of motile Gram-negative bacteria which have been isolated from aquatic neuston communities produces characteristic hyaline, dichotomously branched stalks. In particular, the habit form of *N. ramosa* shows a great many morphological similarities with tunnelling bacteria patterns (see Figure 3.9) including the characteristic slime secretions produced behind the bacteria. It is likely that each morphological group of wood degrading bacteria so far described is composed of many different species and it is possible that a large number of species and decay patterns are yet to be discovered.

References

ADHI, T. P., KORUS, R. A. and CRAWFORD, D. L. (1989) Production of major extra-cellular enzymes during lignocellulose degradation by two *Streptomyces* in agitated submerged culture. *Appl. Environ. Microbiol.* **55**, 1165–1168.

ANAGNOST, S. E., WORRALL, J. J. and WANG, C. J. K. (1994) Diffuse cavity formation in soft rot of pine. *Wood Sci. Tech.* **28**, 199–208.

BAECKER, A. A. W. and KING, B. (1981) Soft rot in wood caused by *Streptomyces*. *J. Inst. Wood Sci.* **9**, 65–71.

BAECKER, A. A. W., DYKER, R. M. P. and KING, B. (1983) The role of actinomycetes in the biodeterioration of wood. In: Oxley, T. A. and Barry, S., eds, *Biodeterioration 5*, New York: J. Wiley & Sons, pp. 64–74.

BAILEY, I. W. and VESTAL, M. R. (1937) The significance of certain wood-destroying fungi in the study of the enzymatic hydrolysis of cellulose. *J. Arnold Arboretum* **18**, 196–205.

BAUCH, J., LIESE, W. and BERNDT, H. (1970) Biological investigations for the improvement of the permeability of softwoods. *Holzforschung* **24**, 199–205.

BERGMAN, Ö., HENNINGSSON B. and PERSSON, E. (1975) Water storage – a method to reduce bleeding of creosote treated poles. *Svenska Träskyddsinstitutet Meddelande*, Report No. 132.

BLANCHETTE, R. A. (1995) Biodeterioration of archaeological wood. *Biodeterioration Abstracts* **9**, 113–126.

BLANCHETTE, R. A., NILSSON, T., DANIEL, G. F. and ABAD, A. (1990) Biological degradation of wood. In: Rowell, R. M. and Barbour, R. J., eds, *Advances in Chemistry Series No. 225, Archaeological Wood: Properties, Chemistry and Preservation*, Washington, DC: American Chemical Society, pp. 141–174.

BOSMAN, M. T. M. (1985) Some effects of decay and weathering on the anatomical structure of *Phragmites australis* Trin. ex Steud. *IAWA Bull.* **6**, 165–170.

BOUTELJE, J. B. and BRAVERY, A. F. (1968) Observations on the bacteriological attack of piles supporting a Stockholm building. *J. Inst. Wood Sci.* **20**, 47–57.

BUGOS, R. C., SUTHERLAND, J. B. and ADLER, J. H. (1988) Phenolic compound utilization by the soft rot fungus *Lecytophora hoffmannii*. *Appl. Environ. Microbiol.* **54**, 1882–1885.

BUSWELL, J. A., ERIKSSON, K.-E., GUPTA, J. K., HAMP, S. G. and NORDH, I. (1982) Vanillic acid metabolism by selected soft-rot, brown-rot and white-rot fungi. *Arch. Microbiol.* **131**, 366–374.

CALVALCANTE, M. S. (1981) The role of actinomycetes in timber decay. PhD Thesis, Department of Biological Sciences, Portsmouth University, England.

CORBETT, N. H. (1965) Micro-morphological studies on the degradation of lignified cell walls by ascomycetes and fungi imperfecti. *J. Inst. Wood Sci.* **14**, 18–29.

CORBETT, N. H. and LEVY, J. F. (1963) Penetration of tracheid walls of *Pinus sylvestris* L. (Scots pine) by *Chaetomium globosum* Kunz. *Nature* **198**, 1322–1323.

COURTOIS, H. (1963a) Beitrag zur Frage holzabbauender Ascomyceten und Funi imperfecti. *Holzforschung* 17, 176–183.

COURTOIS, H. (1963b) Mikromorphologische Befallsymptome beim Holzabbau durch Moderfäulepilze. *Holzforschung u. Holzverwertung* **15**, 88–101.

CURRAN, P. M. T. (1979) Degradation of wood by marine and non-marine fungi from Irish coastal waters. *J. Inst. Wood Sci.* **8**, 114–120.

DANIEL, G. F. and NILSSON, T. (1985) *Ultrastructural and TEM-EDAX Studies on the Degradation of CCA-treated Radiata Pine by Tunnelling Bacteria*, Document No. IRG/WP/1283, The International Research Group on Wood Preservation.

DANIEL, G. F. and NILSSON, T. (1986) *Ultrastructural Observations on Wood Degrading Erosion Bacteria*, Document No. IRG/WP/1283, The International Research Group on Wood Preservation.

DANIEL, G. F. and NILSSON, T. (1988) Studies on preservative tolerant *Phialophora* species. *Int. Biodet.* **24**, 327–335.

DANIEL, G. and NILSSON, T. (1989a) Interactions between soft rot fungi and CCA preservatives in *Betula verrucosa*. *J. Inst. Wood Sci.* **11**, 162–171.

DANIEL, G. and NILSSON, T. (1989b) Effect of hydrofluoric acid pretreatment on the degradation of wood by soft rot fungi. *Mat. u. Org.* **24**, 121–138.

DANIEL, G. and NILSSON, T. (1996) Polylaminate concentric cell wall layering in fibres of *Homalium foetium* and its effect on degradation by soft rot fungi. In: Donaldson, L. A., Singh, A. P., Butterfeld, B. G. and Whitehouse, L. J., eds, *Recent Advances in Wood Anatomy*, IRG, pp. 369–373.

DANIEL, G. F., NILSSON, T. and SINGH, A. P. (1987) Degradation of lignocellulosics by unique tunnel-forming bacteria. *Can. J. Microbiol.* **33**, 943–948.

DANIEL, G., VOLC, J. and NILSSON, T. (1992) Soft rot and multiple T-branching by the basidiomycete *Oudemansiella mucida*. *Mycol. Res.* **96**, 49–54.

DANIEL, G., ELOWSON, T., NILSSON, T., SINGH, A. and LIUKKO, K. (1993) *Water Sprinkled Pine Wood: A Microscope Study on Boards Showing Streaking*, Document No. IRG/WP/93 10033, The International Research Group on Wood Preservation.

DRYSDALE, J. A., RULAND, P. J. and BUTCHER, J. A. (1986) *Isolation and Identification of Bacteria from Degraded Wood – A Progress Report*. Document No. IRG/WP/192, The International Research Group on Wood Preservation.

DUNCAN, C. G. (1960a) *Wood-attacking Capabilities and Physiology of Soft Rot Fungi*, Report No. 2173, Forest Products Laboratory, US Department of Agriculture, Madison, WI.

DUNCAN, C. G. (1960b) Soft-rot in wood, and toxicity studies on causal fungi. *Proc. Am. Wood Preservers Assoc.* **56**, 27–35.

DURÁN, N., FERRER, I. and RODRIGUEZ, J. (1987) Ligninases from *Chrysonilia sitophila* (TFB 27441 strain). *Appl. Biochem. Biotech.* **16**, 157–163.

EATON, R. A. (1972) Fungi growing on wood in water cooling towers. *Int. Biodet. Bull.* **8**, 39–48.

EATON, R. A. and DICKINSON, D. J. (1976) The performance of copper–chrome–arsenic treated wood in the marine environment. *Mat. u. Org.* **3**, 521–529.

EATON, R. A. and JONES, E. B. G. (1971) The biodeterioration of timber in water cooling towers. I. Fungal ecology and the decay of wood at Connah's Quay and Ince. *Mat. u. Org.* **6**, 51–80.

ENCINAS, O. and DANIEL, G. (1997) Degradation of the gelatinous-layer in aspen and rubber wood by the blue stain fungus *Lasiodiploidia theobromae*. *IAWA J.* **18**, 107–115.

ERIKSSON, K.-E., GUPTA, J. K., NISHIDA, A. and RAO, M. (1984) Syringic acid metabolism by some white-rot, soft-rot and brown-rot fungi. *J. Gen Microbiol.* **130**, 2457–2464.

ERIKSSON, K.-E., BLANCHETTE, R. A. and ANDER, P. (1990) *Microbial and Enzymatic Degradation of Wood and Wood Components*, Heidelberg: Springer-Verlag.

ESLYN, W. E. and HIGHLEY, T. L. (1976) Decay resistance and susceptibility of sapwood of fifteen tree species. *Phytopathology* **66**, 1010–1017.

ESLYN, W. E., KIRK, T. K. and EFFLAND, M. J. (1975) Changes in the chemical composition of wood caused by six soft-rot fungi. *Phytopathology* **65**, 473–476.

FAMINTZIN, A. (1892) Eine neue Bacterienform: *Nevskia ramosa. Bull Acad. Sci. St. Peterb.* New ser. **2**, 48–486.

FINDLAY, W. P. K. and SAVORY, J. G. (1954) Moderfäule. Die Zersetzung von Holz durch niedere Pilze. *Holz Roh u. Werkstoff* **12**, 293–296.

FORSBERG, C. W., BEVERIDGE, T. J. and HELLSTRÖM, A. (1981) Cellulase and xylanase release from *Bacteroides succinogenes* and its importance in the rumen environment. *Appl. Environ. Microbiol.* **42**, 886–896.

FUKUDA, K. (1991) *Deterioration and Preservation of Japanese Pampas Grass as a Roofing Material*, Document No. IRG/WP/1490, The International Research Group on Wood Preservation.

FUKUDA, K. and HARAGUCHI, T. (1975) Wood decay by soil fungi. *J. Jap. Wood Res. Soc.* **21**, 635–638.

GOURBIÈRE, F., PÉPIN, R. and BERNILLON, D. (1989) Microscopie de la mycoflore des aigilles de Sapin blanc (Abies alba). IV. Décomposition de la cuticule, de l'hypoderme et de l'épiderme. *Can. J. Bot.* **67**, 933–939.

GREAVES, H. (1968) Occurrence of bacterial decay in copper–chrome–arsenic-treated wood. *Appl. Microbiol.* **16**, 1599–1601.

GREAVES, H. (1977) An illustrated comment on the soft rot problem in Australia and Papua New Guinea. *Holzforschung* **31**, 71–79.

HAIDER, K. and TROJANOWSKI, J. (1975) Decomposition of specifically ^{14}C-labelled phenols and dehydropolymers of coniferyl alcohol as models for lignin degradation by soft and white rot fungi. *Arch. Microbiol.* **105**, 33–41.

HALE, M. D. and EATON, R. A. (1984) *Soft Rot Cavity Widening – A Consideration of the Kinetics*, Document No. IRG/WP/227, The International Research Group on Wood Preservation.

HALE, M. D. and EATON, R. A. (1985a) The ultrastructure of soft rot fungi. I. Fine hyphae in wood cell walls. *Mycologia* **77**, 447–463.

HALE, M. D. and EATON, R. A. (1985b) Oscillatory growth of fungal hyphae in wood cell walls. *Trans. Br. Mycol. Soc.* **84**, 277–288.

HALE, M. D. and EATON, R. A. (1985c) The ultrastructure of soft rot fungi. II. Cavity-forming hyphae in wood cell walls. *Mycologia* **77**, 594–605.

HOLT, D. M. (1981) Bacterial breakdown of timber in aquatic habitats and their relationship with wood degrading fungi. PhD Thesis CNAA, Portsmouth Polytechnic.

HOLT, D. M. (1983) Bacterial degradation of lignified wood cell walls in aerobic aquatic habitats: decay patterns and mechanisms proposed to account for their formation. *J. Inst. Wood Sci.* **9**, 212–223.

HOLT, D. M., JONES, E. B. G. and FURTADO, S. E. I. (1979) Bacterial breakdown of wood in aquatic habitats. *Rec. Ann. Br. Wood Pres. Assoc.*, pp. 13–24.

KIM, Y. S. (1989) Micromorphology of degraded archaeological pine wood in waterlogged situations. *Mat. u. Org.* **24**, 271–286.

KIM, Y. S. and SINGH, A. (1994) Ultrastructural aspects of bacterial attacks on a submerged ancient wood. *Mokuzai Gakkaishi* **40**, 554–562.

KING, B., SMITH, G. M., BRISCOE, P. A. and BAECKER, A. A. W. (1989) Influence of surface nutrients in wood on effectiveness and permanence of CCA. *Mat. u. Org.* **24**, 179–192.

KIRBY, J. J. H. and RAYNER, A. D. M. (1989) The deterioration of thatched roofs. *Int. Biodet.* **25**, 21–26.

KOHLMEYER, J. (1980) Bacterial attack on wood and cellophane in the deep sea. In: Oxley, T. A., Allsopp, D. and Becker, G., eds, *Proceedings of the 4th Biodeterioration Symposium*, London: Pitman, pp. 187–192.

LAMED, R., SETTER, E. and BAYER, E. A. (1983) Characterization of a cellulase-containing complex in *Clostridium thermocellum. J. Bacteriol.* **156**, 828–836.

LEIGHTLEY, L. E. (1978) *Soft Rot Fungi Found in Copper/Chrome/Arsenic Treated Hardwood Power Transmission Poles in Queensland*, Document No. IRG/WP/85, The International Research Group on Wood Preservation.

LEIGHTLEY, L. E. (1980) Wood decay activities of marine fungi. *Botanica Marina* **23**, 387–395.

LEIGHTLEY, L. E. (1981) *Soft-rot Studies on CCA Treated Eucalypt Power Transmission Poles*, Document No. IRG/WP/1132, The International Research Group on Wood Preservation.

LEVI, M. P. and PRESTON, R. D. (1965) A chemical and microscopic examination of the action of the soft-rot fungus *Chaetomium globosum* on beechwood (*Fagus sylvatica.*). *Holzforschung* **19**, 183–190.

LEVY, C. R. (1978) Soft rot. *Proc. Am. Wood Preserver Assoc.* **74**, 145–164.

LEW, J. D. and WILCOX, W. W. (1981) The role of selected deuteromycetes in the soft-rot of wood treated with pentachlorophenol. *Wood and Fiber* **13**, 252–264.

LIESE, J. (1950) Zerstörung des Holzes durch Pilze und Bakterien. In: Mahike, F., Troschel, R. and Liese, J., eds, *Handbuch der Holzkanservierun*, 3rd edition, Berlin, pp. 44–111.

LIESE, W. (1961) Uber die naturliche Dauerhaftigkeit einheimischer und tropischer Holzarten gegenuber Moderfäulepilzen. *Mitt. Dtsch. Ges. Holzforschung* **48**, 18–28.

LIESE, W. and KARNOP, G. (1968) Uber den Befall von Nadelholz durch Bakterien. *Holz als Roh u. Werkstoff* **26**, 202–208.

LIESE, W. and VON PECHMAN, H. (1959) Untersuchungen uber den Einfluss von Moderfäulepilzen auf die Holzfestigkeit. *Forstwiss. Cbl.* **78**, 271–278.

MEIER, H. (1955) Uber den Zellwandabbau durch Holzvermorschungspilze und die submikroskopische Struktur von Fichtentracheiden und Birkenholzfasern. *Holz als Roh u. Werkstoff* **13**, 323–338.

MORRELL, J. J. and SMITH, S. M. (1988) Fungi colonizing redwood in cooling towers: identities and effects on wood properties. *Wood Fiber Sci.* **20**, 243–249.

MORRELL, J. J. and ZABEL, R. A. (1987) Partial delignification of wood: its effect on the action of soft rot fungi isolated from preservative-treated southern pines. *Mat. u. Org.* **22**, 215–224.

MOUZOURAS, R. (1989) Soft rot decay of wood by marine microfungi. *J. Inst. Wood Sci.* **11**, 193–201.

MOUZOURAS, R., JONES, E. B. G., VENKATASAMY, R. and HOLT, D. (1986) Microbial decay of lignocellulose in the marine environment. In: Thompson, M. F., Sarojini, R. and Nagabhushanam, R., eds, *Marine Biodeterioration: Advanced Techniques Applicable to the Indian Ocean*, Oxford: IBH Publishing Co.

NAGASHIMA, Y., FUKADA, K., SATO, S. and HARAGUCHI, T. (1990) Morphological changes of wood-degrading erosion bacteria in aquatic habitats. *Mokuzai Gakkaishi* **36**, 1076–1083.

NILSSON, T. (1973) Studies on wood degradation and cellulolytic activity of microfungi. *Studia Forestalia Suecica*, No. 104.

NILSSON, T. (1974a) Formation of soft rot cavities in various cellulose fibers by *Humicola alopallonella* Meyers & Moore. *Studia Forestalia Suecica*, No. 112.

NILSSON, T. (1974b) The degradation of cellulose and the production of cellulase, xylanase, mannanase and amylase by wood-attacking microfungi. *Studia Forestalia Suecica*, No. 114.

NILSSON, T. (1974c) Microscopic studies on the degradation of cellophane and various cellulosic fibres by wood-attacking micro-fungi. *Studia Forestalia Suecica*, No. 117.

NILSSON, T. (1982a) *Comments on Soft Rot Attack in Timbers Treated with CCA Preservatives: A Document for Discussion*, Document No. IRG/WP/1167, The International Research Group on Wood Preservation.

NILSSON, T. (1982b) Bacterial degradation of untreated and preservative treated wood. In: *Proceedings 16th Convention of Deutsche Gesellschaft für Holzforschung Munster*, Westfalen.

NILSSON, T. (1984) *Occurrence and Importance of Various Types of Fungal and Bacterial Decay in CCA-treated Horticultural Posts in New Zealand*, Document No. IRG/WP/1234, The International Research Group on Wood Preservation.

NILSSON, T. (1988) *Defining Fungal Decay Types – Final Proposal*, Document No. IRG/WP/1355, The International Research Group on Wood Preservation.

NILSSON, T. and DANIEL, G. (1983a) *Formation of Soft Rot Cavities in Relation to Concentric Layers in Wood Fibre Walls*, Document No. IRG/WP/1185, The International Research Group on Wood Preservation.

NILSSON, T. and DANIEL, G. F. (1983b) *Tunnelling Bacteria*, Document No. IRG/WP/1186, The International Research Group on Wood Preservation.

NILSSON, T. and DANIEL, G. (1986) Lignolytic activity of wood-degrading bacteria. In: *Proceedings of Biotechnology in Pulp and Paper Industry*, Stockholm, 16–19 June 1986, pp. 54–57.

NILSSON, T. and DANIEL, G. (1990) *Decay Types Observed in Small Stakes of Pine and* Alstonia scholaris *Inserted in Different Types of Unsterile Soil*, Document No. IRG/WP/1443, The International Research Group on Wood Preservation.

NILSSON, T. and DANIEL, G. (1992) *Attempts to Isolate Tunnelling Bacteria Through Physical Separation from other Bacteria by use of Cellophane*, Document No. IRG/WP/2394-92, The International Research Group on Wood Preservation.

NILSSON, T. and HENNINGSSON, B. (1978) *Phialophora* species occurring in preservative treated wood in ground contact, *Mat. u. Org.* **13**, 297–313.

NILSSON, T. and SINGH, A. (1984) *Cavitation Bacteria*, Document No. IRG/WP/1235, The International Research Group on Wood Preservation.

NILSSON, T., OBST, J. R. and DANIEL, G. (1988) *The Possible Significance of the Lignin Content and Lignin Type on the Performance of CCA-treated Timber in Ground Contact*, Document No. IRG/WP/1357, The International Research Group on Wood Preservation.

NILSSON, T., DANIEL, G., KIRK, T. K. and OBST, J. R. (1989) Chemistry and microscopy of wood decay by some higher ascomycetes. *Holzforschung* **43**, 11–18.

NILSSON, T., DANIEL, G. and BARDAGE, S. (1990) *Evidence for Actinomycete Degradation of Wood Cell Walls*, Document No. IRG/WP/1444, The International Research Group on Wood Preservation.

NILSSON, T., SINGH, A. and DANIEL, G. (1992) Ultrastructure of the attack of *Eusideroxylon zwageri* wood by tunnelling bacteria. *Holzforschung* **46**, 361–367.

OFOSU-ASIEDU, A. and SMITH, R. S. (1973) Degradation of three softwoods by thermophilic and thermotolerant fungi. *Mycologia* **65**, 240–244.

PITMAN, A. J., CRAGG, S. and DANIEL, G. (1995) *The Attack of Naturally Durable and Creosoted Treated Timbers by* Limnoria tripunctata *Menzies*, Document NO. IRG/WP/

95-10132, The International Research Group on Wood Preservation.

PURSLOW, D. F. and WILLIAMS, N. A. (1979) A laboratory examination of the comparative resistance to decay of certain timbers treated with a copper/chrome/arsenic preservative. *Mat. u. Org.* **14**, 117–130.

RAMACHANDRA, M., CRAWFORD, D. L. and HERTEL, G. (1988) Characterization of an extracellular lignin peroxidase of the lignocellulolytic actinomycete *Streptomyces viridosporus. Appl. Environ. Microbiol.* **54**, 3057–3063.

ROGERS, G. M. and BAECKER, A. A. W. (1988) *A New Method for the Study of Micromorphlogical Decay of Wood in a Strictly Anaerobic Environment*, Document No. IRG/WP/2319, The International Research Group on Wood Preservation.

SAFO-SAMPAH, S (1985) The role of actinomycetes in the terrestrial degradation of wood. PhD Thesis, University of California, Berkeley.

SAVORY, J. G. (1954) Breakdown of timber by ascomycetes and fungi imperfecti. *Ann. Appl. Biol.* **44**, 336–347.

SAVORY, J. G. and PINION, L. C. (1958) Chemical aspects of decay of beech wood by *Chaetomium globosum. Holzforschung* **12**, 99–103.

SCHMIDT, O. and LIESE, W. (1994) Occurrence and significance of bacteria in wood. *Holzforschung* **48**, 271–277.

SCHMIDT, O., MORETH, U. and SCHMITT, U. (1995) Wood degradation by a bacterial pure culture. *Mat. u. Org.* **20**, 289–293.

SCHMITT, U. and PEEK, R.-D. (1996) A note on the fine structure of soft rot decay in the polylamellate fibre walls of kempas (*Koompassia malaccensis* Maing. ex Benth.). *Holz als Roh u. Werkstoff* **54**, 42.

SCHWARZE, F. W. M. R., LONSDALE, D. and FINK, S. (1995) Soft rot and multiple T-branching by the basidiomycete *Inonotus hispidus* in ash and London plane. *Mycol. Res.* **99**, 813–820.

SEIFERT, K. (1966) Die chemische Veränderung der Buchenholz-Zellwand durch Moderfäule (*Chaetomium globosum* Kunze). *Holz als Roh u. Werkstoff* **24**, 185–189.

SHARP, R. F. and EGGINS, H. O. W. (1970) The ecology of soft rot fungi. I. Influence of pH. *Int. Biodet. Bull.* **6**, 53–64.

SINGH, A. P. and BUTCHER, J. A. (1991) Bacterial degradation of wood cell walls. A review of degradation patterns. *J. Inst. Wood Sci.* **12** 143–157.

SINGH, A. P., NILSSON, T. and DANIEL, G. F. (1987) Ultrastructure of the attack of wood of two high lignin tropical hardwood species, *Alstonia scholaris* and *Homalium foetidum*, by tunnelling bacteria. *J. Inst. Wood Sci.* **11**, 26–42.

SINGH, A. P., NILSSON, T. and DANIEL, G. F. (1990) Bacterial degradation of *Pinus sylvestris* wood under near-anaerobic conditions. *J. Inst. Wood Sci.* **11**, 237–249.

SULAIMAN, O. and MURPHY, R. J. (1995) Ultrastructure of soft rot decay in bamboo cell walls. *Mat. u. Org.* **29**, 241–253.

SUTHERLAND, J. B., BLANCHETTE, R. A., CRAWFORD, D. L. and POMETTO III, A-L. (1979) Breakdown of Douglas-fir phloem by a lignocellulose-degrading *Streptomyces. Curr. Microbiol.* **2**, 123–126.

TAKAHASHI, M. and NISHIMOTO, K. (1976) Action of soft rot- and white rot fungi on partially delignified softwoods. *Wood Res.* No. 59/60, 19–32.

THORNTON, J. D. and JOHNSON, G. C. (1988) An in-ground natural durability field test of Australian timbers and exotic reference species. IV. Incidence of white, brown and soft rot in hardwood stakes after approximately 15 years' exposure. *Mat. u. Org.* **23** 115–123.

WILLOUGHBY, G. A. and LEIGHTLEY, L. E. (1984) *Patterns of Bacterial Decay in Preservative Treated Eucalypt Power Transmission Poles*, Document No. IRG/WP/1223, The International Research Group on Wood Preservation.

WONG, A. H. H., PEARCE, R. B. and WATKINSON, S. C. (1992) *Fungi Associated with Groundline Soft Rot Decay in Copper–Chrome–Arsenic Treated Heartwood Utility Poles of Malaysian and Hardwoods*, Document No. IRG/WP/1567-92, The International

Research Group on Wood Preservation.

WORRALL, J. J. and WANG, C. J. K. (1991) Importance and mobilization of nutrients in soft rot of wood. *Can. J. Microbiol.*, **37**, 864–868.

WYLDE, A. and DICKINSON, D. J. (1988) *The Evaluation of the Occurrence of Soft Rot in Creosoted Wooden Poles*, Document No. IRG/WP/1368, The International Research Group on Wood Preservation.

ZABEL, R. A., LOMBARD, F. F., WANG, C. J. K. and TERRACINA, F. (1985) Fungi associated with decay in treated southern pine utility poles in the eastern United States. *Wood Fiber Sci.* **17**, 75–91.

ZABEL, R. A., WANG, C. J. K. and ANAGNOST, S. E. (1991) Soft-rot capabilities of the major microfungi, isolated from Douglas-fir poles in the northeast. *Wood Fiber Sci.* **23**, 220–237.

ZAINAL, A. S. (1975) The soft rot fungi: effect of lignin. Internat. Symp. Berlin-Dahlem 1975. *Mat. u. Org.* Beiheft 3, 121–127.

ZAINAL, A. S. (1976) The effect of a change in the cell wall constituents on decay of Scots pine by a soft rot fungus. *Mat. u. Org.* **11**, 295–301.

ZARE-MAIVAN, H. and SHEARER, C. A. (1988) Extracellular enzyme production and cell wall degradation by freshwater lignicolous fungi. *Mycologia* **80**, 365–375.

Biopulping

KURT MESSNER

4.1 Introduction

The main steps in paper making are to separate the fibres from the plant tissue, to bleach them and to rearrange them to form a paper sheet. The fibres are separated either mechanically by stone-grinding whole logs or by disc-refining wood chips. Both processes are extremely energy consuming. In mechanical pulping processes the wood tissues break apart mainly at the middle lamella or at the interface of the middle lamella and the wood cell wall. In chemi-thermo-mechanical pulping (CTMP) the refining process is assisted by chemicals and high temperature. As mechanical pulps still contain most of the lignin, their yield is very high (88–96 per cent). They lead to an opaque paper, mainly used for newsprint, books and magazines. Besides the high energy consumption, mechanical pulps suffer from low strength and low brightness (Biermann, 1993).

In chemical pulping processes such as kraft or sulphite pulping, the fibres are separated by dissolution of the middle lamella. Modern sulphur-free processes are being developed. Due to its high strength, unbleached kraft pulp is used for making bags, wrappings and linerboard; after bleaching the pulp could be used to make white papers. Kraft pulp is more resistant to modern totally chlorine-free (TCF) bleaching processes than sulphite pulps, which are weaker. Kraft pulp is used for newsprint, fine paper and tissues. With sulphite pulp, wood resins may pose problems in paper making by leading to sticky deposits. The yield of bleached chemical pulps is about 50 per cent.

Many parameters are used to evaluate the paper making process. The amount of delignification after pulping and bleaching is monitored by the kappa number, an indirect method measuring the amount of permanganate consumed by lignin. The pulp and paper properties are determined by optical parameters such as brightness and colour; and physical parameters such as tensile index measured on paper strips using a constant force, and tear index, measured by the energy required to propagate an initial tear through several sheets of paper (Biermann, 1993).

The developments in pulp and paper production during the 1990s have been mainly aimed at reducing environmental impact. While no major improvement has been made to reduce the energy input in mechanical pulping, elemental chlorine-free (ECF) and TCF bleaching processes drastically added to the decrease of toxicity and colour of bleach plant effluents from chemical pulping plants. Unfortunately, the new oxygen-, hydrogen peroxide- or ozone-based bleaching technologies reduce the fibre strength and result in lower brightness levels and a higher lignin content of the bleached pulp. One of the strategies to overcome these problems is extended 'cooking' aimed at extracting more lignin during the cooking process.

In its strict sense, biopulping is defined as the pretreatment of wood chips with selectively delignifying white-rot fungi prior to mechanical or chemical pulping. It takes advantage of the diffusible lignin- and hemicellulose-degrading agents excreted by these fungi during incubation of wood chips to cleave the chemical bonds in the wood tissue which must be broken up in mechanical or chemical pulping. This leads to energy reduction and a lower lignin content after cooking – this is the same objective as the extended cooking process. In a broader sense, the term biopulping is also used for any biochemical assistance to the pulping process such as the use of non-wood decay fungi for resin degradation or the use of enzyme preparations in pulp and paper production, for example for bleaching, resin degradation, dewatering etc. For a clearer definition, the prefix 'bio' should be used only for whole fungal processes and 'enzymatic' when isolated enzymes are applied.

This chapter describes the biochemical and ultrastructural background of fungal activities in biopulping, the technology, and achievements to date; it then discusses how biopulping could help to solve the problems mentioned above.

4.2 Fungi Used for Biopulping

When wood is decayed by white-rot fungi in nature, it is mostly bleached and may even fall apart into cellulose fibres, strongly resembling pulp. This effect is exerted by some species of fungi and is caused by the selective degradation of lignin and hemicellulose in the middle lamella and also in the wood cell wall. It seems logical to use these fungi for biotechnological processes, either to improve the forage digestibility or to pretreat wood for pulping or other purposes. In fact, wood, delignified by *Ganoderma australe* and other microorganisms, is traditionally used as cattle feed in Southern Chile and is called *palo podrido* (Philippi, 1893). There are no records of the use of naturally delignified wood to produce paper, but early attempts were inspired by naturally white-rotted wood (Lawson and Still, 1957; Reis and Libby, 1960; Kawase, 1962). Further biopulping studies were undertaken with *Rigidoporus ulmarius* at the Forest Products Laboratory, Madison, WI (Kirk *et al.*, 1990). More detailed studies started with the discovery made by Henningsson *et al.* (1972) describing *Phanerochaete chrysosporium* as a thermophilic basidiomycete which caused defibration of wood. Mechanical pulp was produced from wood chips pretreated with *P. chrysosporium* by Ander and Eriksson (1975) at STFI in Stockholm, Sweden. Mutants of this fungus, reduced in cellulase activity, were produced and the first biopulping patent was filed (Eriksson *et al.*, 1976). Adamski *et al.* (1987) showed that pretreatment of wood chips with the white-rot fungi *Phellinus pini* and *Stereum hirsutum* resulted in a decrease of the refining energy in kraft pulping.

A systematic screening for selective lignin degrading fungi was made by Otjen *et al.* (1987) and Blanchette *et al.* (1988) who analyzed the relative amount of the structural components of wood blocks after 3 months cultivation. A new screening method, based on Simons stain, proved to be successful to predict energy savings in mechanical pulping (Blanchette *et al.*, 1992a). As a result of about 200 tested fungal strains, *P. chrysosporium* was found to be the best biopulping fungus for hard wood and *Ceriporiopsis subvermispora* for hard and soft wood when mechanical pulp was produced (Akhtar *et al.*, 1992a; Blanchette *et al.*, 1992b). A screening programme to evaluate the best selective white-rot fungi for pretreating wood chips prior to sulphite pulping also identified *C. subvermispora* as the best choice with *Phlebia tremellosa*, *Phlebia brevispora* and *Dichomitus squalens* ranking next (Messner *et al.*, 1992).

Early strategies of biopulping were aimed at a high degree of delignification of wood chips. Such type of process would lead to a considerable yield loss and additionally to a decrease in paper strength due to the polysaccharide depolymerizing enzymes also excreted by selectively delignifying fungi. A new concept of biopulping was created when it was found that with *C. subvermispora* and other selectively delignifying fungi, after a relatively short incubation period of 2 weeks and a weight loss of less than 2 per cent with no visible attack on the wood cell walls, a considerably lower kappa number, corresponding to a lower lignin content, can be gained after sulphite cooking (Messner *et al.*, 1993; Messner and Srebotnik, 1994).The same also applied to refiner mechanical pulping, where high energy savings are reached after incubation times of 2 weeks (Kirk *et al.*, 1993).

Based on morphological studies of wood, and also on the basic understanding of the penetration of enzymes into wood cell walls, the high effectiveness of biopulping is hard to understand. As postulated earlier (Cowling and Brown, 1969; Stone *et al.*, 1969), molecules of the molecular mass of ligninolytic enzymes (between approx. 45 and 80 kDa) cannot penetrate into the cell walls until rather late stages of decay due to their high molecular weight, resulting in a high hydrodynamic diameter in relation to a low average pore size of the wood cell wall. This was proved by infiltrating wood cell walls with marker proteins and subsequent immunolabelling by Srebotnik *et al.* (1988) and Daniel *et al.* (1989). Recently, Blanchette *et al.* (1997) indicated that after incubation of pine wood with *C. subvermispora* for 2 weeks – the time when strong biopulping effects are already evident – only the very small marker protein insulin (molecular mass 5730 kDa) penetrated the secondary wall in a narrow band around the circumference of the lumen. None of the larger proteins were able to penetrate.

It was shown by differential staining and light microscopy (Srebotnik and Messner, 1994) that after 2 weeks, only the cell walls of parenchyma cells, colonized first by hyphae, are delignified. Total delignification of the cell walls and fibre separation appeared only after 6 weeks cultivation, corresponding to 15–20 per cent weight loss. The delignification process mainly started at the lumen surface and progressed towards the middle lamella.

When birch wood chips were incubated with *C. subvermispora*, until dissolution of the middle lamella and separation of the fibres, they still appeared more or less undegraded in transmission electron microscopy. Infiltrating these fibres with lignin peroxidase, no enzymes were detected within the wood cell wall by immunoelectron microscopy (Messner and Srebotnik, 1994). Taking into account that hyphal growth and excretion of enzymes takes place in the lumen, one would assume that they would have penetrated into the middle lamella, leading to dissolution of the latter. In this case the infiltrated enzymes would have had to be able to penetrate into the wood cell wall. As this was not the case, it was concluded that some kind of highly diffusible, low molecular weight compound must have been produced by the fungus leading to dissolution of the middle lamella (Messner and Srebotnik, 1994).

As similar results were obtained with *Dichomitus squalens*, another selectively delignifying white-rot fungus, but not with *Trametes versicolor*, a simultaneous white-rot fungus, it must be assumed that the selectively delignifying white-rot fungi produce a highly diffusible lignin degrading agent of a molecular mass lower than 5000 kDa. This agent obviously diffuses into the cell wall in the early stages of decay, cleaving bonds in the lignin (or hemicellulose) polymers, thereby facilitating an easier chemical delignification and a decreased energy input in mechanical pulping. Retrospectively, it was fortuitous to select fungi excreting these so far unidentified agents leading to the biopulping effect before delignification at only 2 per cent weight loss.

Due to this new understanding of the biopulping mechanism, the strategy has changed from a long-term process with a high degree of delignification to a relatively short-term process with almost no loss of substrate. This is essential progress as high yield is desirable both in mechanical and chemical pulping. The solubilized lignin received after cooking is used for internal heat and energy generation in chemical pulp mills.

Another approach to biopulping is the use of ascomycetes such as *Ophiostoma piliferum*. This fungus rapidly colonizes the wood chips but does not alter the wood cell wall. Its main effect is the degradation of extractives after colonizing the resin canals and rupturing the ray parenchyma cells and the distruction of pit membranes (Blanchette *et al.*, 1992c). It is assumed that by the latter activity a more even distribution of the cooking chemicals is brought about, resulting in a more uniform pulp production (Wall, unpublished results).

4.3   Lignolytic Enzyme Systems

The consumption of wood or other highly lignified plant tissues by microorganisms requires the excretion of a lignolytic enzyme system. White-rot fungi can be considered to be the most efficient organisms in this respect. So far, four types of lignin degrading enzymes have been isolated from white-rot fungi.

Laccase: A copper-containing oxidase of a molecular mass between 60 and 80 kDa. It catalyzes four one-electron oxidations of mostly phenolic compounds. Free phenoxy radicals are formed as intermediates. Various compounds such as ABTS (Bourbonnais and Paice, 1992), HBT (Call and Mücke, 1997), or 3-HAA (3-hydroxy-anthranilate) (Eggert *et al.*, 1995, 1996) can act as cosubstrates or mediators. While ABTS and HBT are not produced by fungi, 3-HAA was isolated from the white-rot fungus *Pycnoporus cinnabarinus*. It was shown that these low molecular weight compounds also enable the fungus to oxidize non-phenolic lignin and to act at a distance from the fungal hypha.

Manganese peroxidase: Like lignin peroxidase, this was discovered in *P. chrysosporium* and characterized by Kuwahara *et al.* (1984). It is commonly produced by white-rot fungi. Its molecular weight is similar to that of lignin peroxidase, it contains protoporphyrin IX and catalyzes one-electron oxidations of phenolic and non-phenolic compounds. The primary substrate is Mn(II) which is oxidized to Mn(III) and stabilized by forming complexes with organic acids. The Mn(III)-complex must be diffusible in wood and can depolymerize lignin (Wariishi *et al.*, 1991). A new mechanism involving lipid peroxidation, leading to oxidation of the recalcitrant non-phenolic structures in lignin was found by Moen and Hammel (1994) and Bao *et al.* (1994). Recent results pointed to the participation of this mechanism in the depolymerization of lignin by *C. subvermispora* (Srebotnik *et al.*, 1994; Jensen *et al.*, 1996).

Lignin peroxidase has a molar mass of 38 to 43 kD, also contains protoporphyrin IX as a prosthetic group and catalyzes one-electron oxidations of phenolic and non-phenolic compounds generating phenoxy radicals and cation radicals. In the presence of veratryl alcohol, an organic acid, H_2O_2 and oxygen, lignin peroxidase was also found to be able to oxidize Mn(II) (Popp *et al.*, 1990). The enzyme was first detected in the culture fluid of *P. chrysosporium* by Tien and Kirk (1983) and Glenn and Gold (1985) but it is excreted by only a few white-rot fungi.

Other peroxidases: Several other manganese independent peroxidases, also different from lignin peroxidase, have been described in white-rot fungi (Heinzkill and Messner, 1997).

Principally, all lignin degrading enzymes appear to be able to act at a distance from the hypha in the depth of the wood cell wall either via manganese complexes or via other low molecular weight compounds, also called mediators. Other compounds able to penetrate because of their low molecular weight and to oxidize wood components were discovered recently (Enoki *et al.*, 1997; Goodell *et al.*, 1997). The understanding of the underlying mechanism for biopulping is closely related to the discovery of the chemical and biochemical reactions of such low molecular weight compounds.

4.4 The Biopulping Process

Wood is the natural substrate for white-rot fungi; consequently wood chips offer suitable conditions for biopulping fungi but a non-optimized process would be rather slow. It must be speeded up by optimizing parameters such as nutrient supply, quantity and type of inoculum, moisture content of wood chips, and aeration. As all parameters except aeration have to be set at the beginning, scaling-up is of even greater importance for this type of process than for liquid fermentations.

Development of the biopulping process has reached the pilot scale as far as the use of white-rot fungi for mechanical pulping and sulphite pulping is concerned, and has already been tested on a commercial scale with the ascomycete *Ophiostoma piliferum* for kraft pulping.

4.4.1 Inoculation and Nutrient Supply

Like many other basisiomycetes, *C. subvermispora* differs from *O. piliferum* insofar as it only produces clamydospores integrated in the fungal mycelium, but has no conidiospores. From this point of view it might be easier to produce large amounts of fungal inoculum from *O. piliferum*. With *C. subvermispora* the mycelium must be fragmented to increase the number of inoculation points. Nevertheless, clamydospores are resting spores and probably will also guarantee a stable inoculum after drying. Inocula of *O. piliferum* have already been produced in large scale and this fungus is on the market under the trade name Cartapip 97 and is used to decrease the content of extractives on wood chips. Both types of inocula are produced in liquid fermentation at the same time, but solid substrate fermentation may also be a method for inoculum production, especially for *Ceriporiopsis* (Majcherczyk *et al.*, 1996).

In a wood chip pile, conditions such as humidity, available nutrients and temperature are highly favourable to fungal growth and many fungi are able to colonize the chips. One of the most ubiquitous fungi is *Trichoderma*. Some of its species excrete a broad range of compounds, controlling the growth of basidiomycetes (Horvath *et al.*, 1994). In fact, this fungus is commercially used for plant and wood protection (Freitag *et al.*, 1991). As *Ceriporiopsis* is unable to compete with the indigenous microorganisms on wood, the chips have to be decontaminated prior to inoculation. This can be done either by chemicals such as sodium bisulphate (Akhtar *et al.*, 1995a,b) or by a short atmospheric steaming. A steaming period of only 15 seconds was sufficient to give *Ceriporiopsis* the competitiveness needed for an even colonization of the wood chips and to cause the desired biopulping effect (Akhtar *et al.*, 1997). *O. piliferum* is reported to be competitive against the natural wood chip organisms and can be inoculated on contaminated chips. Chen and Schmidt (1996) reported a method to grow *P. chrysosporium* on unsterilized wood chips. According to Pearce *et al.* (1995), some wood decay fungi identified after screening more than 200 strains were found to be able to colonize unsterilized wood chips; this led to high energy savings in mechanical pulping.

The amount of inoculum needed for an evenly distributed and dense growth of *C. subverispora* or other basidiomycetes on the chip surfaces and a dense colonization of the interior of the chips was found to be strongly dependent on the amount of nutrients available to the fungus. When the fungal inoculum was applied to the wood chips suspended in unsterilized corn steep liquor the amount of fungal inoculum could be reduced to 0.25 g/ton (dw/dw) of wood (Akhtar *et al.*, 1996). Corn steep liquor is a cheap, semi-solid by-product of corn milling, and contains mainly protein, lactic acid and sugars. It is used as a feed supplement and also for other fungal fermentations. Experiments with lactose and other sources of organic nitrogen showed comparable results (Akhtar *et al.*, 1997).

4.4.2 *Fermentation*

Considering the amount of wood chips to be treated in a pulp and paper mill, fermentation processes using rotating fermentors would not be feasible. The only acceptable technology will be a static-bed fermentation type as for example in chip silos or even on wood chip piles. Solid-state fermentations are much harder to control than liquid fermentations. They are considered to be gas–liquid–solid mixtures in which an aqueous phase is intimately associated with solid surfaces and is in contact with a gas phase continuous with the external gas environment (Mudget, 1986).

The moisture content of commercial wood chips varies, but is mostly lower than the optimum range for white-rot fungi which also differs between species. Usually, a broad range around 100 per cent (dew) is acceptable. The only method of controlling process parameters such as oxygen–carbon dioxide exchange or heat transfer is via the gas phase. In experiments carried out in 20-litre laboratory fermentors, aeration rates below 0.01 litre/litre/minute were found to have a negative effect on biopulping with *C. subvermispora* (Heimel, 1993). Similar thresholds have been reported for *P. chrysosporium* (Akhtar *et al.*, 1997) and forced aeration was also found to be needed in solid-state fermentations of aspen wood with *Phlebia tremellosa* (Reid, 1989). In *Ceriporiopsis* fermentations the temperature built up from room temperature to 31–33°C within 4 days and remained constant to day 14 (Heimel, 1993). Taking into consideration that in wood chip piles temperatures of 45–50°C are usually reached, and even may increase to the incineration point, this is a surprising result. It can be explained by the fact that in biopulping, after steaming, the wood chips are colonized by *C. subvermispora* as a monoculture, while in wood chip piles thermophilic organisms may succeed mesophilic microorganisms. Obviously *C. subvermispora* is able to stabilize the temperature at its optimum. The difference in the temperatures reached at aeration rates of 0.001, 0.01 and 0.1 volume/volume/minute (vvm) was only 2°C (Heimel, 1993). It demonstrates that the temperature development in biopulping can hardly be controlled by the aeration rate due to the slow heat transfer from the interior of the wood chip to the gas phase of the void volume. It appears that only the oxygen–carbon dioxide exchange can be improved by aeration.

According to biomass determinations based on the ergosterol content of the mycelium of *C. subvermispora*, it was reported that after a period of vigorous growth of 6 days the fungal growth largely decreased until day 14 although the optimum temperature of 33°C was reached in the solid substrate fermentation (Messner *et al.*, 1997). An ergosterol content of 0.7 per cent correlated to a fungal biomass of 5 mg/g wood (dw) (Koller, 1996) (Figure 4.1).

It should be further investigated whether this model based on experiments in 20-litre laboratory fermentors is also applicable to larger volumes. From the experiments it can be concluded that either aerated chip silos or aerated chip piles will be

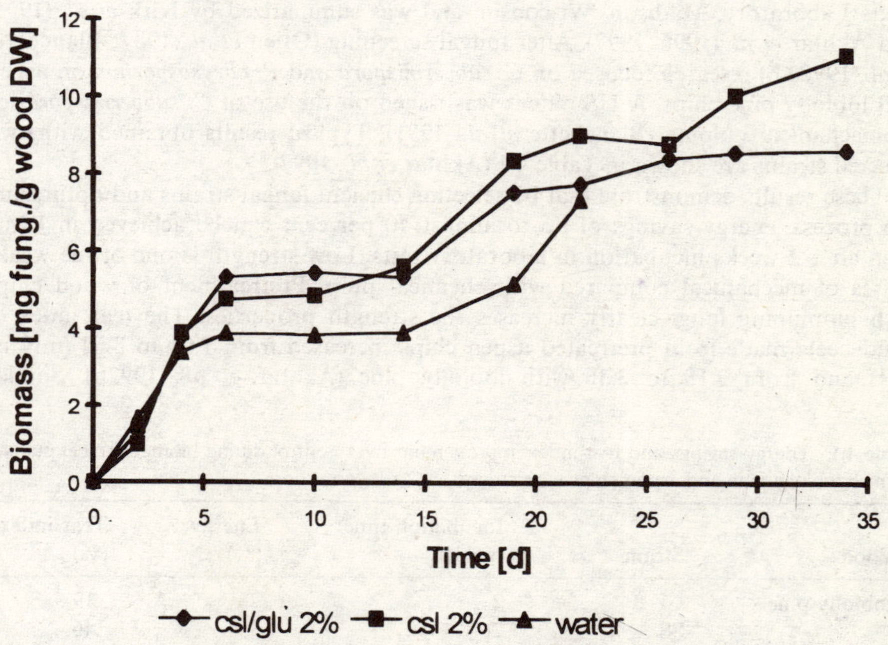

Figure 4.1 Development of biomass of *Ceriporiopsis subvermispora* on spruce wood chips supplemented with 2 per cent corn steep liquor/2 per cent glucose; 2 per cent corn steep liquor; and water, calculated from the ergosterol content of the fungal mycelium

needed to create optimum conditions for *C. subvermispora* or other biopulping fungi. Scale-up of the biopulping process is under way at the Forest Products Laboratory, Madison, Wisconsin, and probably also at some pulp and paper companies.

4.5 Biomechanical Pulping

Refiner mechanical pulp (RMP) is produced by disintegrating chips between rotating metal discs at atmospheric pressure. Refining is carried out in two stages: the first is aimed at fibre separation at the middle lamella after prior softening of the chips by steam while the second step alters the fibre surface for improved fibre bonding in the final paper. Power requirements are 1600–18 000 kWh/ton. In thermomechanical pulping (TMP) the refiners are at 110–130°C and elevated pressure in the first stage to promote fibre liberation at the S_1 cell wall layer, leading to improved fibre bonding compared with RMP. Energy requirements are 1900–2900 kWh/ton, over two-thirds of which is used in the primary pressurized refining step (Biermann, 1993).

It can be concluded from the high values of electrical energy demand that a softening of wood chips by fungal pretreatment can result in great benefits, especially in mechanical pulping. Consequently, most efforts have been invested into this type of application of fungal pretreatment. Fungi which do not lead to a modification of the structure of the wood cell wall, such as *O. pilifreum*, may reduce the resin content but do not bring about energy savings (Fischer *et al.*, 1994). Most of the work on biomechanical pulping was done by the Biopulping Consortium at the Forest Products Laboratory, Madison, Wisconsin, and was summarized by Kirk *et al.* (1993) and Akhtar *et al.* (1996, 1997). After fungal screening (Otjen *et al.*, 1987; Blanchette *et al.*, 1992a,b) research focused on *C. subvermispora* and *P. chrysosporium* on aspen and loblolly pine chips. A US patent was issued on the use of *C. subvermispora* for biomechanical pulping (Blanchette *et al.*, 1991). Typical results obtained with two selected strains are shown in Table 4.1 (Akhtar *et al.*, 1997).

These results demonstrate that by selecting efficient fungal strains and optimizing the process, energy savings of up to almost 40 per cent can be achieved in RMP even after 2 weeks incubation in laboratory tests. Low strength is one of the weaknesses of mechanical compared with chemical pulp. Pretreatment of wood chips with biopulping fungi clearly increases the strength properties. The tear index of handsheets made from pretreated aspen chips increased from 1.01 to 3.62 (mN n^2 g^{-1}) and from 2.18 to 3.36 with loblolly pine (Akhtar *et al.*, 1992b). Similar

Table 4.1 Energy savings and tear index improvement over control during biomechanical pulping of fresh loblolly pine and aspen chips with strains of *C. subvermispora*

Wood	Strain	Incubation time (weeks)	Energy (%)	Tear index (%)
Loblolly pine	CZ-3	2	23	19
	SS-3	2	34	46
	FP-90031	4	42	131
Aspen	SS-3	2	38	0
	FP-90031	4	40	67

improvements were achieved for Norway spruce and birch, but not for eucalyptus wood after fungal treatment (Setliff *et al.,* 1990). It can be concluded that a fully scaled-up industrial process will lead to great benefits for the pulping industry.

Due to chromophore production during fungal pretreatment the brightness was adversely affected and decreased compared with control pulp. Unbleached aspen biopulp reached a brightness level of 51.8 per cent (Elrepho) compared with 62.2 per cent in the control, and 3 per cent H_2O_2 bleached biopulp reached 76.0 per cent compared with 80.0 per cent in the control. Nevertheless, in a two-step bleaching, 78 per cent was reached with the biopulp (Sykes, 1993), demonstrating that bleaching should not create a major problem after biopulping.

Another positive effect of biopulping was identified when waste water from the first refiner passes of aspen chips, treated with either *P. chrysosporium* or *C. subvermispora*, was analyzed for biological oxygen demand (BOD), chemical oxygen demand (COD) and Microtox toxicity. The toxicity of the waste water decreased from 17 to 4 (100/EC50, where EC50 is the median effective concentration), due to the consumption of extractives by the fungus. BOD (g/kg pulp) decreased from 40 to 36 but COD (g/kg pulp) increased from 74 to 100, due to lignin fragments released as a result of fungal pretreatment (Sykes, 1994).

Wood resins cause a number of serious problems in pulp and paper production by creating sticky deposits. Cleaving the ether bonds of triglycerides either by commercial lipases used on pulp (Fischer and Messner, 1992; Fischer *et al.*, 1993) or by lipases produced by fungi during chip colonization (Wendler *et al.*, 1991; Farrell *et al.*, 1994) was found to reduce the pitch problem. When the capacity to decrease the pitch content of loblolly pine chips by *C. subvermispora* during biopulping was compared with that of the commercial pitch control fungus *O. piliferum* both fungi gave the same result – a decrease of approximately 30 per cent – after 4 weeks incubation. After 2 weeks *C. subvermispora* had already degraded 24 per cent compared with 15 per cent for *O. piliferum*, correlating with a 53 per cent decrease in triglycerides (Fischer *et al.*, 1996). These results indicate an additional benefit of biopulping, namely pitch reduction. While no effect of pretreatment with *O. piliferum* on paper strength was found by Fischer *et al.* (1994), an increase in strength parameters was detected by Forde Kohler *et al.* (1996), after pretreatment with Cartapip 97, a commercial product from *O. piliferum*.

4.6   Biochemical Pulping

Various chemical methods exist to break down the chemical structure of lignin and render it soluble in water. Common methods are the sulphite process and the kraft, sulphate or alkaline process, which is most extensively used.

4.6.1   Bio-sulphite Pulping

Sulphite processes use mixtures of sulphurous acid and/or its alkali salts (for example, K^+, Ca^{2+}, Mg^{2+}) to solubilize lignin through the formation of sulphonate functionalities and cleavage of lignin bonds. Only a few paper mills are still using calcium sulphite. In Europe magnesium-based cooking liquors are widely applied as for example in the magnefite process, enabling an efficient chemical and energy recovery.

Biopulping in connection with magnefite pulping was investigated using birch and spruce wood chips and five selected strains of fungi (Messner *et al.*, 1992). The chips were sterilized, supplemented with either synthetic or complex media such as corn steep liquor and inoculated with a suspension of blended mycelium. After 2 and 4 weeks of incubation in a 20-litre static bed laboratory bioreactor aerated with humidified air at 0.01 vvm the chips were cooked. The kappa number of unbleached control pulp was approximately 24 and reached 6.3 after laboratory bleaching in a two-step sequence (EOP–P, where EOP represents alkaline oxygen/peroxide treatment and P a further peroxide bleaching stage). The effect of fungal pretreatment on the pulp properties was monitored by brightness, tear index and tensile index.

Table 4.2 shows that *C. subvermispora* was the best fungus for magnefite biopulping. On spruce chips 30 per cent kappa reduction was gained with this fungus after 2 weeks incubation time (Messner and Srebotnik, 1994). Contrary to *C. subvermispora*, *P. chrysosporium* showed a great selectivity for the wood species as no effect was detected after 2 weeks incubation on spruce chips. Except for *P. chrysosporium*, all fungi tested, known to be selective lignin degraders, showed good results on birch chips. The paper strength properties of handsheets prepared from birch chips, incubated for 4 weeks and measured as tear index and tensile index after 10 and 20 minutes beating time, were reduced by about 10 per cent (Messner and Srebotnik, 1994). These results indicate that after pretreatment of wood chips with selective lignin degrading basidiomycetes, such as *C. subvermispora*, about 30 per cent more lignin can be solubilized in magnefite cooking at a rather low impact on the physical properties of the pulp.

When wood chips supplemented with corn steep liquor or other complex media are incubated with *C. subvermispora*, the chip colour changes to brown after a few days. Unfortunately these chromophoric compounds are not destroyed during cooking, leading to a brightness drop from 62 per cent ISO-brightness to 50 per cent. When this pulp was bleached in an EOP–P bleaching sequence, 4 per cent ISO was still lost compared with the unbleached control of 70 per cent ISO-brightness (Messner *et al.*, 1997). With wood chips supplemented with a synthetic medium the brightness loss was lower and resulted in 1.5 per cent ISO after cooking and 0.8 per cent after bleaching at similar kappa reductions.

Besides optimizing the media or screening for fungal strains that do not decrease the brightness, an approach to avoid the problem of brightness loss is to reduce the cooking time as additional chromophores are created during longer cooking times. Table 4.3 shows that despite a high kappa reduction after magnefite cooking for 315 minutes, no brightness gain was achieved after bleaching due to a more intensive

Table 4.2 Percentage decrease of kappa number of bio-pulped birch chips after magnefite cooking

Fungus	2 weeks	4 weeks
Ceriporiopsis subvermispora	33	48
Phlebia tremollosa		40
Phlebia brevispora		35
Dichomitus squalens		33
Phanerochaete chrysosporium	12	20

Table 4.3 Kappa reduction and brightness loss/gain of spruce wood chips, supplemented with a synthetic medium, pretreated for 2 weeks with *Ceriporiopsis subvermispora* and subsequent magnefite cooking at various times and EOP–P bleaching

Cooking time	Kappa reduction (%)	Brightness (%ISO) loss/gain unbleached	Brightness (%ISO) loss/gain bleached
315 min	−37.7	−1.5	−0.8
240 min	0	+2.3	+3.5

chromophore production at the low kappa values. By reducing the cooking time by 75 minutes, the same kappa level (24) as with the untreated control chips is reached after fungal pretreatment caused by the enzymatic modification of the lignin. Fewer chromophores were produced during cooking, leading to a brightness increase of 3.5 per cent ISO (brightness level 75) after a two-step EOP–P laboratory bleaching sequence. By decreasing the cooking time the capacity of the digester could be increased, leading to a higher productivity, still leading to a small brightness increase. Further work is needed to evaluate the reactivity of fungal pretreated magnefite pulp to various bleaching chemicals and sequences. An Austrian patent has been granted on bio-sulphite pulping (Messner *et al.*, 1995).

4.6.2 *Sodium- and Calcium-based Sulphite Pulping*

The effect of pretreatment of pine chips (*Pinus taeda*) with two strains of *Ceriporiopsis subvermispora* (CZ-3 and L-14807 SS-3) for 2 weeks on kappa number, pulp yield, chemical consumption, brightness and colour, as well as on the effluent parameters has been reported (Scott *et al.*, 1995a, 1996; Akhtar *et al.*, 1997). After sodium bisulphate pulping 27 per cent kappa reduction was reached with both strains at a control value of k 31.2. Calcium-acid sulphite pulped wood chips pretreated with the strain CZ-3 showed 49 per cent kappa reduction compared with 21 per cent kappa reduction for strain SS-3. After sodium sulphite pulping the pulp yield was decreased by 1.7 while after calcium-acid sulphite pulping the yield was comparable with the control. Furthermore, it was found that the consumption of pulping liquor did not increase after calcium-acid sulphite cooking.

Similar bleaching results, as reported above for magnefite pulp, were obtained for calcium-acid sulphite pulp. Due to the fungus induced chromophores the brightness of the unbleached pulp decreased from 54 to 49 per cent brightness. Nevertheless, after 4 per cent hydrogen peroxide bleaching as well as after 1 per cent FAS (formamidine sulphinic acid) bleaching, the same brightness level (80 per cent) was reached with biopulp. The biopulp appeared to be brighter as a result of a lower amount of the yellow component of the reflected light.

As water discharges from pulp and paper mills have been subjected to dramatic regulatory measures and will still have to be decreased in the future, the effect of biopulping on water parameters was assayed (Scott *et al.*, 1995b). While no change was measured in the BOD and COD content of the liquor effluent compared with the control, toxicity was decreased from 17.4 to 7.2 toxicity units, probably due to fungal degradation of extractives, as was also the case with biomechanical pulping.

4.6.3   Conclusions on Bio-sulphite Pulping

The results obtained so far show that sulphite pulping may profit from pretreatment of wood chips with selectively delignifying fungi by a reduced cooking time and a decrease in kappa of the bleached pulp at the same or slightly increased brightness. However, the main objective of chemical pulping is to decrease the amount of bleach chemicals. Chlorine or chlorine dioxide are substituted by oxygen–hydrogen peroxide- or ozone-bleaching sequences because of their adverse environmental effects, but the new totally chlorine-free bleaching sequences are less specific for lignin and lead to pulps with lower strength properties. An increased extraction of lignin during cooking resulting in a lower kappa number, as caused by fungal pretreatment, would be an ideal prerequisite for reducing the amount of bleaching chemicals, resulting in a lower environmental impact and/or a pulp of higher quality. At present the production of chromophores during the modification of the cell wall components by fungi reduces the positive effects of bio-sulphite pulping. More research is needed to overcome this drawback.

Another advantage of fungal pretreatment for sulphite pulping comes from the associated decrease in resin and other extractives thereby making wood species with higher resin content more acceptable for sulphite pulping.

Interesting results were obtained when *C. subvermispora* was used in biobleaching of sulphite pulps for dissolving pulp production (Christov *et al.*, 1996). In contrast to all other experiments described in this chapter, pulp instead of wood chips was inoculated with *C. subvermispora* and incubated for 10 days. Fungal pretreatment of pulp very effectively increased the bleachability of the pulp leading to kappa numbers of around 1.0 compared with 6.7 in the control after an ECF bleaching sequence. Contrary to the results obtained on wood, no adverse effect was observed on brightness. The fungal prebleached pulp reached a brightness level of approximately 80 per cent ISO compared with 56 per cent for the untreated control, but some cellulose degradation occurred. The long incubation time of the pulp and the cellulose loss will probably exclude a technical application for this method, but it is interesting to see that the brightness loss observed with all other biopulping experiments did not occur.

4.6.4   Bio-kraft Pulping

Kraft pulping gives a high strength pulp and is useful for any wood species including high resin wood types. Most of the chemical pulp produced worldwide is kraft pulp. A disadvantage is the difficulty with which the pulp is bleached compared with sulphite pulp (Biermann, 1993). From this point of view a pretreatment with selectively delignifying white-rot fungi to decrease the lignin content after cooking and to render the pulp more accessible for bleaching chemicals would be the ideal combination of methods. Surprisingly, less work has been done on bio-kraft pulping than on bio-sulphite or biomechanical pulping, and has delivered rather inconclusive results. Experiments using *P. chrysosporium* (Oriaran *et al.*, 1990, 1991; Laborsky *et al.*, 1991) on glucose supplemented aspen and red oak chips for 20 and 30 days, respectively, led to 3 and 9 per cent kappa reduction compared with untreated chips. Brightness of the handsheets prepared from unbleached pulp decreased dramatically by 54 and 62 per cent. Comparable with that described for bio-sulphite pulping, pulp with a similar kappa number could be produced at a cooking time reduced by one-third. The pulp produced responded better to refining and had higher tensile and burst indices. As a result of a large screening programme on white-rot fungi from South Africa (Bosman *et al.*, 1993; Wolfaardt *et al.*, 1993) large scale bio-kraft pulping experiments were performed with *S. hirsutum, Pycnoporus sanguineus* and *T. versicolor* following a 9-week treatment. The kappa number decreased by up to 17 per cent but was accompanied by a yield loss and an increase of alkali consumption (Wolfaardt *et al.*, 1996).

From the results reported on bio-kraft pulping it can be concluded that the effect of pretreatment with white-rot fungi on kappa reduction is lower than in bio-sulphite pulping and shows a much higher tendency for colour reduction of the resulting unbleached pulp.

Another approach to biopulping is to use non-wood decay fungi such as the ascomycetes sapstain fungus *O. piliferum*. Similar to the white-rot fungi it penetrates the wood via wood rays and the pit pores of the cell walls and is also able to rupture the pit membranes. While the white-rot fungi are able to modify the wood cell walls after colonization, no attack on lignified cell walls takes place with *O. piliferum* due to its lack of lignolytic enzymes. An important factor in chemical pulping is a uniform liquor penetration of the wood chips. As the liquor penetrates into the lumina of the fibres via pit pores, a prior fungal disruption or dissolution should improve penetration, leading to improved pulping results.

A 2-week treatment of northern Pine softwood chips with *O. piliferum* resulted in a kappa reduction of 9 per cent. Yield, viscosity and strength properties remained constant and the demand of bleach chemicals in the first chlorination step was reduced by 9 per cent at a brightness level of 91.9 per cent (Wall *et al.*, 1994, 1996). On aspen chips treated for 3 weeks, up to 29 per cent kappa reduction was measured. When the bio-kraft pulping effect of *O. piliferum* was compared with that of the white-rot fungus *Phlebia tremellosa*, 5.8 per cent kappa reduction was measured compared with 14.3 per cent (Rocheleau *et al.*, submitted for publication). No brightness loss was created by either of the fungi. The results show that, as with white-rot fungi, improved pulping results can be obtained by increasing penetration of the cooking liquor. Confirmation of this assumption comes from the observation that *O. pilifrum* has no influence on the result in mechanical pulping (Fischer *et al.*, 1994).

4.6.5 Bio-organosolv Pulping

Ferraz *et al.* (1996) investigated the use of white-rot decay as a pretreatment for organosolv delignification of *Eucalyptus grandis* wood and found a threefold increase of the delignification rate after 1 month incubation with *T. versicolor*. Longer incubation times as well as pretreatment with *P. chrysosporium* did not lead to any further improvements in delignification.

4.7 Biopulping of Non-wood Plants

About 10 per cent of the paper produced worldwide is made from non-wood plants such as cotton, straw, canes, grasses and hemp, and paper making from such sources is increasing. The fibres are mostly cooked with sodium hydroxide at a lower temperature and shorter time than wood pulp due to lower lignin content. Straw pulp is similar to hardwood pulp. Fibres may be used for fine papers but also along with secondary fibres in a mixture containing approximately 25–50 per cent non-wood fibres for corrugating medium. The disadvantages of straw for pulping are its high silica content and low drainage rates (Biermann, 1993).

Reed grass (*Phalaris arundinacea*) was treated by Hatakka *et al.* (1996), with selectively delignifying fungi (*P. radiata*, *P. tremellosa*, *Pleurotus ostreatus* and *C. subvermispora*) for 7 and 14 days and cooked for 15 min in NaOH solution with anthraquinone. The highest loss in lignin after 2 weeks cultivation was caused by *C. subvermispora*, leading to kappa 19.7 compared with 21.6 for the control. The fines content was reduced from 9.8 to 7.7 per cent. The viscosity was negatively affected, decreasing by 17 per cent. Consequently, the handsheet properties of a pine/grass fine paper also decreased slightly.

Solid substrate fermentation studies on wheat straw, including ^{14}C-labelled lignin, with different selectively delignifying white-rot fungi strongly suggested lignin degradation by manganese peroxidase mediated by Mn(III) (Martinez, 1997).

Atmospheric refining of jute bast was studied after pretreatment with *C. subvermispora* by Sabharwal *et al.* (1995). The energy consumption in refining was reduced by 33 per cent for fungal treated jute and burst, tensile and tear strengths were enhanced by 39, 22 and 33 per cent, respectively. Similar results were achieved when kenaf bast was biopulped. Similar to that described for the various kinds of biopulping of wood, the brightness increase after a single stage alkaline peroxide bleaching was much less for biopulp, reaching a brightness level of only 57 per cent compared with 70 per cent of the control (Sabharwal *et al.*, 1996).

4.8   Concluding Remarks

The pretreatment of wood chips or non-wood plants with *C. subvermispora* as well as with other selectively delignifying white-rot fungi was found to be beneficial for all kinds of pulp production. Nevertheless, the greatest advantages seem to be related to mechanical pulping. High energy savings, an increase in paper strength, reduced resin content and reduced toxicity of the effluent all point to a great future for biopulping. The benefits for chemical pulping are not so clear at the moment. Although substantially lower kappa values are reached after sulphite cooking, it seems to be critical to overcome the brightness loss in bleaching. More research is needed to study the chemistry of chromophore production, induced by the fungal enzyme system, as well as the effect of different bleaching sequences on biopulp. An interesting approach to the problem could be the high bleaching effect of the same fungus when growing on unbleached pulp instead of wood chips and this seems worth further investigation. The chemical reactions taking place in kraft cooking are even less favourable for wood biopulped with basidiomycetes and increase the trend of brightness loss. However, incubation of wood chips with ascomycetes exerting no enzymatic changes of the wood cell wall lead to some benefits in kraft cooking due to a better penetration of the wood chips by the cooking liquor. The importance of biopulping for straw and other non-wood plants will increase with the growing importance of these raw materials.

The technical development of biopulping has made good progress. With *O. piliferum*, a fully developed industrial process commercially known as CARTAPIP for inoculum production and for inoculation is available. No additional adjustments of the wood chip pile are needed. For *C. subvermispora*, the fungus with potentially greater benefits, the process has been scaled-up at the Forest Products Laboratory in Madison, Wisconsin, to pilot plant level and results are comparable with results achieved in the laboratory (Akhtar, personal communication). It must be decided whether an aerated wood chip pile will be sufficient or whether a silo-type reactor as already used in Scandinavian countries for chip storage will be needed.

The argument of a long incubation time of 2 weeks is sometimes used against biopulping. If it is considered that wood chips are already stored in piles for at least 2 weeks prior to use and that ample space is available at chip yards, biopulping can be regarded as a more controlled procedure, comparable with the present situation.

Theoretically, a pretreatment process for wood chips based purely on enzymes or other biological low molecular weight catalysts can be imagined, but is not feasible at this time. Although research into the decay mechanisms in the wood cell wall has received an important impetus by focusing on low molecular weight compounds, a full understanding of the biochemical mechanisms of biopulping will still take some time.

References

ADAMSKI, Z., GAWECKI, T. and ZIELINSKI, M. H. (1987) In: *Funkcne Integrovane Obhopodarovani Lesoy a Komplexne Vyuztie Dreva*, Medzinarodna Vedecka Konferencia, Zvolen, Czechoslovakia, Akad. Agric. Poznan.

AKHTAR, M., ATTRIDGE, M. C., BLANCHETTE, R. A., MYERS, G. C., WALL, M. B., SYKES, M. S., KONING Jr, J. W., BURGESS, R. R., WEGNER, T. H. and KIRK, T. K. (1992a) The white-rot fungus *Ceriporiopsis subvermispora* saves electrical energy and improves strength properties during biomechanical pulping of wood. In: Kuwahara, M. and Shimada, M., eds, *Biotechnology in the Pulp and Paper Industry, Proceedings of the 5th International Conference on Biotechnology in the Pulp and Paper Industry*, Tokyo: Uni Publishers, pp. 3–8.

AKHTAR, M., ATTRIDGE, M. C., MYERS, G. C., KIRK, T. K. and BLANCHETTE, R. A. (1992b) Biomechanical pulping of loblolly pine with different strains of the white-rot fungus *Ceriopriopsis subvermispora. Tappi J.* **75**, 105–109.

AKHTAR, M., BLANCHETTE, R. A. and BURNES, T. A. (1995a) Using Simons stain to predict energy savings during biomechanical pulping. *Wood Fiber Sci.* **27**, 258–264.

AKHTAR, M., ATTRIGE, M. C., KONING, J. W. and KIRK, T. K. (1995b) Method of pulping wood chips with a fungi using sulfite salt-treated wood chips. US Patent No. 5 460 697.

AKHTAR, M., KIRK, T. K. and BLANCHETTE, R. A. (1996) Biopulping: an overview of consortia research. In: Srebotnik, E. and Messner, K., eds, *Biotechnology in the Pulp and Paper Industry: Recent Advances in Applied and Environmental Research; Proceedings of the Sixth International Conference on Biotechnology in the Pulp and Paper Industry*, Vienna, Austria: Facultas-Univ.-Verl., pp. 187–192.

AKHTAR, M., BLANCHETTE, R. A., MYERS, G. and KIRK, T. K. (1997) An overview of biomechanical pulping research. In: Young, R. A. and Akhtar, M., eds, *Environmentally Friendly Technologies for the Pulp and Paper Industry*, New York: John Wiley & Sons (in press).

ANDER, P. and ERIKSSON, K. (1975) Mekanisk massa fran förröttad flis – en inledande undersökning. *Svensk Papperstidning* **18**, 641–642.

BAO, W., FUKUSHIMA, Y., JENSEN Jr, K. A., MOEN, M. A. and HAMMEL, K. E. (1994) Oxidative degradation of non-phenolic lignin during lipid peroxidation by fungal manganese peroxidase. *FEBS Lett.* **354**, 297–300.

BIERMANN, C. J. (1993) *Essentials of Pulping and Papermaking*, San Diego: Academic Press.

BLANCHETTE, R. A., BURNES, T. A., LEATHAM, G. and EFFLAND, M. J. (1988) Selection of white-rot fungi for biopulping. *Biomass* **15**, 93–101.

BLANCHETTE, R. A., LEATHAM, G. F., ATTRIDGE, M., AKHTAR, M. and MYERS, G. C. (1991) Biomechanical pulping with *Ceriporiopsis subvermispora*. US Patent No. 5 055 159.

BLANCHETTE, R. A., AKHTAR, M. and ATTRIGE, M. C. (1992a) Using Simons stain to evaluate fibre characteristics of biomechanical pulps. *Tappi J.* **75**, 121–124.

BLANCHETTE, R. A., BURNES, T. A., EERDMANS, M. M. and AKTAR, M. (1992b) Evaluating isolates of *Phanerochaete chrysosporium* and *Ceriporiopisis subvermispora* for use in biological pulping processes. *Holzforschung* **46**, 109–115.

BLANCHETTE, R. A., FARRELL, R., BURNES, T. A., WENDLER, P. A., ZIMMERMANN, W., BRUSH, T. and SNYDER, R. A. (1992c) Biological control of pitch in pulp and paper production by *Ophiostoma piliterum*. *Tappi J.* **74**, 102–106.

BLANCHETTE, R. A., KRUEGER, E. W., HAIGHT, J. E., AKHTAR, M. and AKIN, D. E. (1997) Cell wall alterations in loblolly pine wood decayed by the white-rot fungus, *Ceriporiopsis subvermispora*. In: Messner, K., Srebotnik, E. and Fiechter, A., eds, *Low Molecular Weight Compounds in Lignin Degradation*, *J. Biotechnol.* **53**, 203–213.

BOSMAN, J. L., JACOBS, A., MALE, J. R., RABIE, J. C., VAN DER WESTHUIZEN, G. C. A., VENTER, J. S. M. and WOLFAARDT, J. F. (1993) Biopulping potential of South African wood-decay fungi. In: Duarte, J. C., Ferreira, M. C. and Ander, P., eds, *Proceedings, FEMS Symposium, Lignin Biodegradation and Transformation*, Lisboa: Forbitec Editions, pp. 39–40.

BOURBONNAIS, R. and PAICE, M. G. (1992) Demethylation and delignification of kraft pulp by *Trametes versicolor* laccase in the presence of 2,2′-azinobis-(3-ethylbenzthiazoline-6-sulphonate). *Appl. Microbiol. Biotechnol.* **36**, 823–827.

CALL, H.-P. and MÜCKE, I. (1997) History, overview and applications of mediated lignolytic systems, especially laccase-mediator-systems (LIGNOZYM®-Process). In: Messner, K., Srebotnik, E. and Fiechter. A., eds, *Low Molecular Weight Compounds in Lignin Degradation*, *J. Biotechnol.* **53**, 163–202.

CHEN, Y. and SCHMIDT, E. (1996) Improving aspen kraft pulp by a novel, low-technology fungal pretreatment. *Wood Fiber Sci.* **27**, 198–204.

CHRISTOV, L. P., AKHTAR, M. and PRIOR, B. A. (1996) Biobleaching in dissolving pulp production. In: Srebotnik, E. and Messner, K., eds, *Biotechnology in the Pulp and Paper Industry, Recent Advances in Applied and Fundamental Research, Proceedings of the 6th International Conference on Biotechnology in the Pulp and Paper Industry*, Facultas-Univ.-Verl., Wien, pp. 625–628.

COWLING, E. B. and BROWN, W. (1969) Structural features of cellulosic materials in relation to enzymatic hydrolysis. In: Hajni, G. J. and Reese, E. T., eds, *Cellulases and their Application*, Washington, DC: American Chemical Society, pp. 152–187.

DANIEL, G., NILSSON, T. and PETTERSSON, B. (1989) Intra- and extracellular localization of lignin peroxidase during the degradation of solid wood and wood fragments by *Phanerochaete chrysosporium* by using transmission electron microscopy and immunogold labeling. *Appl. Env. Microbiol.* **55**, 871–881.

EGGERT, C., TEMP, U., DEAN, J. F. and ERIKSSON, K.-E. (1995) Laccase-mediated formation of the phenoxazinone derivative, cinnabarinic acid. *FEBS Lett.* **376**, 202–206.

EGGERT, C., TEMP, U., DEAN, J. F. and ERIKSSON, K.-E. (1996) A fungal metabolite mediates degradation of non-phenolic lignin structures and synthetic lignin by laccase. *FEBS Lett.* **391**, 144–148.

ENOKI, A., ITAKURA, S. and TANAKA, H. (1997) The involvement of extracellular substances for reducing molecular oxygen to hydroxyl radical and ferric ion to ferrous iron in wood degradation by wood-decay fungi. In: Messner, K., Srebotnik, E. and Fiechter,

A., eds, *Low Molecular Weight Compounds in Lignin Degradation*, *J. Biotechnol.* Special Issue (in press).

ERIKSSON, K.-E., ANDER, P., HENNINGSSON, B., NILSSON, T. and GOODELL, B. (1976) Method for producing cellulose pulp. US Patent No. 3 962 033.

FARRELL, R. L., BRUSH, T. S., FRITZ, A. R., BLANCHETTE, R. A. and IVERSON, S. (1994) Cartapip: a biological product for control of pitch and resin acid problems in pulp mills. In: *Proceedings, TAPPI Biological Sciences Symposium*, Minneapolis, MN: TAPPI Press, pp. 85–87.

FERRAZ, A., MENDONCA, R., COTRIM, A. R. and DA SILVA, F. T. (1996) The use of white-rot decay as a pretreatment for organosolv delignification of *Eucalyptus grandis* wood. In: Srebotnik, E. and Messner, K., eds, *Biotechnology in the Pulp and Paper Industry: Recent Advances in Applied and Environmental Research; Proceedings of the Sixth International Conference on Biotechnology in the Pulp and Paper Industry*, Vienna, Austria: Facultas-Univ.-Verl., pp. 221–224.

FISCHER, K. and MESSNER, K. (1992) Reducing troublesome pitch in pulp mills by lipolytic enzymes. *Tappi J.* **75**, 10–134.

FISCHER, K., PUCHINGER, L., SCHLOFFER, K., KREINER, W. and MESSNER, K. (1993) Enzymatic pitch reduction of sulphite pulp on pilot scale. *J. Biotechnol.* **27**, 341–348.

FISCHER, K., AKHTAR, M., BLANCHETTE, R. A., BURNES, T. A., MESSNER, K. and KIRK, T. K. (1994) Reduction of resin content in wood chips during experimental biological pulping processes. *Holzforschung* **48**, 285–290.

FISCHER, K., AKHTAR, M., MESSNER, K., BLANCHETTE, R. A. and KIRK, T. K. (1996) Pitch reduction with the white-rot fungus *Ceriporiopsis subvermispora*. In: Srebotnik, E. and Messner, K., eds, *Biotechnology in the Pulp and Paper Industry: Recent Advances in Applied and Environmental Research; Proceedings of the Sixth International Conference on Biotechnology in the Pulp and Paper Industry*, Vienna, Austria: Facultas-Univ.-Verl., pp. 193–198.

FORDE KOHLER, L., DINUS, R., MALCOLM, A., RUDIE, R., FARRELL, R. and BRUSH, T. (1996) Enhancing softwood mechanical pulp properties through chip treatment with *Ophiostoma piliferum*. In: Srebotnik, E. and Messner, K., eds, *Biotechnology in the Pulp and Paper Industry: Recent Advances in Applied and Environmental Research; Proceedings of the Sixth International Conference on Biotechnology in the Pulp and Paper Industry*, Vienna, Austria: Facultas-Univ.-Verl., pp. 225–228.

FREITAG, M., MORRELL, J. J. and BRUCE, A. (1991) Biological protection of wood: status and prospects. *Biodeterioration Abstr.* **5**, 1–13.

GLENN, J. K. and GOLD, M. H. (1985) Purification and characterization of an extracellular Mn(II)-dependent peroxidase from the lignin-degrading basidiomycete *Phanerochaete chrysosoporium*. *Arch. Biochem. Biophys.* **234**, 353–362.

GOODELL, B. J., JELLISON, J., LIU, J. J., DANIEL, G., PASZYNSKI, A., FEKETE, S., KRISHNAMURTHY, S. and XU, G. (1997) Low molecular weight chelators and phenolic compounds isolated from wood decay fungi and their role in the fungal biodegradation of wood. In: Messner, K., Srebotnik, E. and Fiechter, A., eds, *Low Molecular Weight Compounds in Lignin Degradation*, *J. Biotechnol.* **53**, 133–162.

HATAKKA, A., MÄTTÄLÄ, A., HÄRKÖNEN, T. and PAAVILAINEN, L. (1996) Biopulping of gramineous plants by white-rot fungi. In: Srebotnik, E. and Messner, K., eds, *Biotechnology in the Pulp and Paper Industry: Recent Advances in Applied and Environmental Research; Proceedings of the Sixth International Conference on Biotechnology in the Pulp and Paper Industry*, Vienna, Austria: Facultas-Univ.-Verl., pp. 229–232.

HEIMEL, M. (1993) Untersuchung von Weißfäulepilzen auf Einsetzbarkeit im Biopulping Prozeß, unpublished Diplomarbeit, Technische Universität-Wien.

HEINZKILL, M. and MESSNER, K. (1997) The lignolytic system of fungi. In: Anke, T., ed., *Fungal Biotechnology*, 1st edition, Weinheim: Chapman & Hall.

HENNINGSSON, B. H., HENNINGSSON, M. and NILSSON, T. (1972) Defibration of wood by a white-rot fungus. *Royal College of Forestry, Stockholm, Research Notes*, **78**, 1–26.

HORVATH. E. M., BURGEL, J. L. and MESSNER, K. (1994) The production of soluble antifungal metabolites, by the biocontrol fungus *Trichoderma harzianum* in connection with the formation of conidiospores. *Mat. u. Org.* **9**, 1–14.

JENSEN Jr, K. A., BAO, W., KAWAI, S., SREBOTNIK, E. and HAMMEL, K. E. (1996) Manganese-dependent cleavage of nonphenolic lignin structures by *Ceriporiopsis subvermispora* in the absence of lignin peroxidase. *Appl. Env. Microbiol.* **62**, 3679–3686.

KAWASE, K. (1962) Chemical components of wood decayed under natural conditions and their properties. *J. Fac. Agr. Hokkaido Univ.* **52**, 186–245.

KIRK, T. K., BURGESS, R. R. and KONING, Jr, J. W. (1990) Use of fungi in pulping wood: an overview of biopulping research. In: Leatham, G., ed., *Frontiers in Industrial Mycology, Proceedings of Industrial Mycology Symposium, 25–26 June 1990, Madison, WI*, New York: Routledge, Chapman & Hall; 1992, Chapter 5.

KIRK, T. K., KONING, Jr, J. W., BURGESS, R. R., AKHTAR, M., BLANCHETTE, R. A., CAMERON, D. C., CULLEN, D., KERSTEN, P. J., LIGHTFOOT, E. N., MEYERS, G. C., SACHS, I., SYKES, M. and WALL, M. B. (1993) *Biopulping – A Glimpse of the Future?* USDA Forest Service, Research Paper FPL-RP-523, 1–74.

KOLLER, K. (1996) Bestimmung von Ergosterol zur Überwachung von Feststofffermentationen mit Weißfäulepilzen, unpublished Diplomarbeit, Technische Universität-Wien.

KUWAHARA, M., GLENN, J. K., MORGAN, M. A. and GOLD, M. H. (1984) Separation and characterisation of two extracellular H_2O_2-dependent oxidases from lignolytic cultures of *Phanerochaete chrysosporium. FEBS Lett.* **169**, 247–250.

LABORSKY Jr, P., ZHANG, J. and ROYSE, D. J. (1991) Lignin biodegradation of nitrogen supplemented red oak (*Quercus rubra*) wood chips with two strains of *Phanerochaete chrysosporium. Wood Fibre Sci.* **23**, 533–542.

LAWSON Jr, L. R. and STILL, C. N. (1957) The biological decomposition of lignin – literature survey. *Tappi* **40**, 56A–80A.

MAJCHERCZYK, A., BEDAIWY, M., KÜHNE, A., KÖRNER, I., HADAR, Y. and HÜTTERMANN, A. (1996) The production of large amounts of fungal inoculum under unsterile conditions. In: Srebotnik, E. and Messner, K., eds, *Biotechnology in the Pulp and Paper Industry: Recent Advances in Applied and Environmental Research; Proceedings of the Sixth International Conference on Biotechnology in the Pulp and Paper Industry*, Vienna, Austria: Facultas-Univ.-Verl., pp. 199–204.

MARTINEZ, A. T. (1997) Biological deligrification-enzyme mixtures for treating cereal straw and other non-woody material (AIR). In: Eriksson, L., ed., *European Conference on Pulp and Paper Research – The Present and the Future* (in press).

MESSNER, K. and SREBOTNIK, E. (1994) Biopulping: an overview of developments in an environmentally safe paper-making technology. *FEMS Microbiol. Rev.* **13**, 351–364.

MESSNER, K., MASEK, S. and TECHT, G. (1992) Fungal pre-treatment of wood chips for chemical pulping. In: Kuwahara, M. and Shimada M., eds, *Biotechnology in the Pulp and Paper Industry, Proceedings of the 5th International Conference on Biotechnology in the Pulp and Paper Industry*, Tokyo: Uni Publishers, pp. 9–13.

MESSNER, K., SCHIEFERMEIER, M., SREBOTNIK, E. and TECHT, G. (1993) Bio-sulfite pulping: current state of research. In: Duarte, J. C., Ferreira, M. C. and Ander, P., eds, *Proceedings of FEMS Symposium, Lignin Biodegradation and Transformation*, Lisboa: Forbitec Editions, pp. 197–200.

MESSNER, K., TECHT, G., MASEK, S. and SREBOTNIK, E. (1995) Verfahren zur Herstellung von Zellstoff, Osterr. Patent, Nr. 358 589.

MESSNER, K., KOLLER, K., WALL, M. B., AKHTAR, M. and SCOTT, G. (1997) Fungal pre-treatment of wood chips for chemical pulping. In: Young, R. A. and Akhtar, M., eds, *Environmentally Friendly Technologies for the Pulp and Paper Industry*, New York: John Wiley & Sons (in press).

MOEN, M. A. and HAMMEL, K. E. (1994) Lipid peroxidation by the manganese peroxidase of *Phanerochaete chrysosporium* is the basis for phenanthrene oxidation by the intact fungus. *Appl. Environ. Microbiol.* **60**, 1956–1961.

MUDGET, R. E. (1986) Solid state fermentations. In: Demain, A. L. and Solomon, N. A., eds, *Manual of Industrial Microbiology and Biotechnology*, Washington, DC: American Society for Microbiology, pp. 66–83.

ORIARAN, T. PH., LABORSKY, Jr, P. and BLANKEMHORN, P. R. (1990) Kraft pulp and paper making properties of *Phanerochaete chrysosporium* degraded aspen. *Tappi J.* **73**, 147–152.

ORIARAN, T. PH., LABORSKY Jr, P. and BLANKEMHORN, P. R. (1991) Kraft pulp and paper making properties of *Phanerochaete chrysosporium* degraded red oak. *Wood Fiber Sci.* **23**, 316–327.

OTJEN, L., BLANCHETTE, R. A., EFFLAND, M. and LEATHAM, G. (1987) Assessment of 30 white-rot basidiomycetes for selective lignin degradation. *Holzforschung* **41**, 343–349.

PEARCE, M. H., DUNLOP, R. W., FALK, C. J., NORMAN, K. and ROULLO, A. B. (1995) Screening lignin degrading fungi for biomechanical pulping of eucalyptus wood chips. In: *Proceedings 49th Appita Annual General Conference, Hobart, Tasmania, Australia, 2–7 April 1995*, Victoria, Australia: Appita Carlton, pp. 347–351.

PHILIPPI, F. (1893) Die Pilze Chiles, soweit dieselben als Nahrungsmittel gebraucht werden. *Hedwigia* **32**, 115–118.

POPP, J. L., KALYANARAMAN, B. and KIRK, K. (1990) Lignin peroxidase oxidation of Mn^{2+} in the presence of veratryl alcohol, malonic or oxalic acid, and oxygen. *Biochemistry* **29**, 10475–10480.

REID, I. D. (1989) Optimization of solid-state fermentation for selective delignification of aspen wood with *Phlebia tremellosa. Enzyme Microb. Technol.* **11**, 804–809.

REIS, C. J. and LIBBY, C. E. (1960) An experimental study of the effect of *Fomes pini* (Thore) Lloyd on the pulping qualities of pond pine *Pinus serotina* (Michx) cooked by the sulfate process. *Tappi J.* **43**, 489–499.

SABHARWAL, H. S., AKHTAR, M., BLANCHETTE, A. and YOUNG, R. A. (1995) Refiner mechanical and biomechanical pulping of jute. *Holzforschung* **49**, 537–544.

SABHARWAL, H. S., AKHTAR, M., YU, E., D'AGOSTINO, D., YOUNG, R. A. and BLANCHETTE, R. A. (1996) Development of biological pulping processes for nonwoody plants. In: Srebotnik, E. and Messner, K., eds, *Biotechnology in the Pulp and Paper Industry: Recent Advances in Applied and Environmental Research; Proceedings of the Sixth International Conference on Biotechnology in the Pulp and Paper Industry*, Vienna, Austria: Facultas-Univ.-Verl., pp. 233–236.

SCOTT, G. M., LENTZ, M. and AKHTAR, M. (1995a) Fungal pretreatment of wood chips for sulfite pulping. In: *Proceedings of the 1995 TAPPI Pulping Conference; 1–5 October 1995, Chicago, IL*, Atlanta, GA: Tappi Press, Book 1, pp. 355–361.

SCOTT, G. M., AKHTAR, M., SYKES, M., ABUBAKR, S. and LENTZ, M. (1995b) Environmental aspects of biosulfite pulping. In: *Proceedings of 1995 International Environmental Conference; 7–10 May 1995, Atlanta, GA*, Atlanta, GA: Tappi Press, Book 2, pp. 1155–1161.

SCOTT, G. M., AKHTAR, M., LENTZ, M. and ABUBAKR, S. (1996) Bio-sulphite pulping using *Ceriporiopsis subvermispora*. In: Srebotnik, E. and Messner, K., eds, *Biotechnology in the Pulp and Paper Industry: Recent Advances in Applied and Environmental Research; Proceedings of the Sixth International Conference on Biotechnology in the Pulp and Paper Industry*, Vienna, Austria: Facultas-Univ.-Verl., pp. 187–192.

SETLIFF, E. C., MARTON, R., GRANZOW, S. G. and ERIKSSON, K. L. (1990) Biomechanical pulping with white-rot fungi. *Tappi J.* **73**, 141–147.

SREBOTNIK, E. and MESSNER, K. (1994) A simple method that uses differential staining and light microscopy to assess the selectivity Lipf wood delignification by white-rot fungi. *Appl. Environ. Microbiol.* **60**, 1393–1386.

SREBOTNIK, E., MESSNER, K. and FOISNER, R. (1988) Penetrability of white-rot-degraded pine wood by the lignin peroxidase of *Phanerochaete chrysosporium*. *Appl. Environ. Microbiol.* **54**, 2608–2614.

SREBOTNIK, E., JENSEN Jr, K. A. and HAMMEL, K. E. (1994) Fungal degradation of recalcitrant nonphenolic lignin structures without lignin peroxidase. *Proc. Natl. Acad. Sci. USA* **91**, 12794–12797.

STONE, J. E., SCALLAN, A. M., DONEFER, E. and AHLGREN, E. (1969) Digestibility as a simple function of a molecule of similar size to a cellulase enzyme. In: Hajni, G. J. and Reese, E. T., eds, *Cellulases and their Application*, Washington, DC: American Chemical Society, pp. 219–241.

SYKES, M. (1993) Bleaching and brightness stability of aspen biomechanical pulps. *Tappi J.* **76**, 121–126.

SYKES, M. (1994) Environmental compatability of effluents of aspen biomechanical pulps. *Tappi J.* **77**, 160–166.

TIEN, M. and KIRK, T. K. (1983) Lignin-degrading enzyme from the hymenomycete *Phanerochaete chrysosporium* Burds. *Science* **221**, 661–663.

WALL, M. B., BRECKER, J., FRITZ, A., IVERSON, S. and NOEL, Y. (1994) Cartapip treatment of wood chips to improve chemical pulping efficiency. In: *Proceedings of TAPPI J. Biological Sciences Symposium*, pp. 67–76.

WALL, M. B., STAFFORD, G., NOEL, Y., IVERSON, S. and FARRELL, R. L. (1996) Treatment with *Ophiostoma piliferum* improves chemical pulping efficiency. In: Srebotnik, E. and Messner, K., eds, *Biotechnology in the Pulp and Paper Industry: Recent Advances in Applied and Environmental Research; Proceedings of the Sixth International Conference on Biotechnology in the Pulp and Paper Industry*, Vienna, Austria: Facultas-Univ.-Verl. pp. 205–210.

WARIISHI, H., VALLI, K. and GOLD, M. (1991) *In vitro* depolymerization of lignin by manganese peroxidase of *Phanerochaete chrysosporium*. *Biochem. Biophys. Res. Comm.* **176**, 269–275.

WENDLER, P. A., BRUSH, T. S. and FARRELL, R. L. (1991) Biological control of pitch problems on a thermomechanical pulp mill. In: *Proceedings of the 6th International Symposium on Wood and Pulping Chemistry, Melbourne, Australia*, pp. 501–508.

WOLFAARDT, J. F., BOSHOFF, I. E., BOSMAN, J. L., RABIE, J. C. and VAN DER WESTHUIZEN, G. C. A. (1993) Lignin degrading potential of South African wood decay fungi. In: Duarte, J. C., Ferreira, M. C. and Ander, P., eds, *Proceedings of FEMS Symposium, Lignin Biodegradation and Transformation*, Lisboa: Forbitec Editions, pp. 67–69.

WOLFAARDT, J. F., BOSMAN, J. L., JACOBS, A., MALE, J. R. and RABIE C. J. (1996) Bio-kraft pulping of softwood. In: Srebotnik, E. and Messner, K., eds, *Biotechnology in the Pulp and Paper Industry: Recent Advances in Applied and Environmental Research; Proceedings of the Sixth International Conference on Biotechnology in the Pulp and Paper Industry*, Vienna, Austria: Facultas-Univ.-Verl. pp. 211–216.

Enzymes in Pulp Bleaching

LIISA VIIKARI, JOHANNA BUCHERT AND ANNA SUURNÄKKI

5.1 Introduction

The importance of environmental aspects in pulp and paper manufacturing processes has grown dramatically in recent years and has today a central role in marketing of paper products. The pulp and paper industry has already modified the processes to reduce the formation of chlorinated organic compounds and other wastes. Various technologies have been introduced to achieve effluent load levels lower than 0.2 kg of AOX (adsorbable organic halogen) per ton of pulp. This target has already been exceeded in several mills, for example in Scandinavia. The novel methods, allowing replacement of the traditionally used chlorine gas, include extended cooking, oxygen delignification and various oxidative bleaching chemicals. In search of new bleaching methods, enzymatic technologies have also been developed and are now used in the pulp mills.

There are two different approaches for improving the bleachability of pulps by enzymes. By using hemicellulases (xylanases or mannanases) the bleachability of kraft pulps can be increased indirectly, whereas lignin degrading enzymes would result in direct delignification. Although the target lignin is not enzymatically degraded by hemicellulases, the partial enzymatic removal of pulp hemicelluloses improves the chemical extraction of lignin (Viikari *et al.*, 1994). The hemicellulase treatment leads to higher brightness of the pulp and decreased chemical consumption in the bleaching process. This is due to the close association of lignin and hemicelluloses in the pulp fibres. The major enzymes responsible for the positive effect are endo-xylanases. The method was commercialized within five years of its discovery mainly because of the recent advances in xylanase production strains and technologies. Optimization of the industrial production of xylanases has resulted in decreased enzyme prices and improved product properties.

The direct enzymatic degradation of lignin has been the focus of scientists for decades. In spite of active research and basic knowledge gathered, the cell-free degradation of high molecular weight lignin has had limited success. The reaction mechanisms of lignin degrading enzymes, resulting in radical reactions, are difficult to control on the fibre bound, high molecular weight residual lignin. Most experiments with these enzymes have been carried out on isolated model substrates.

Recently, however, very promising results have been obtained using a mediator with the laccase enzyme. The oxidized mediator acts efficiently and is able to degrade lignin specifically. No commercial lignin degrading oxidative enzymes are currently available. It can be expected however, that laccases and peroxidases will soon be introduced to the market, due to their obvious potential.

5.2 Pulping and Bleaching

The main aim in chemical pulping is to remove lignin and to separate the wood fibres from each other in order to render them suitable for the paper making process. The two major processes for carrying out delignification of woody materials are the sulphate (kraft) and sulphite methods (Sjöström, 1993). In the pulping process, the lignified middle lamella located between the wood fibres is solubilized by various chemicals. Today, the predominant pulping method is the kraft process, where the cooking liquors are incinerated and the cooking chemicals recycled. Modern kraft cooking processes are almost entirely closed, that is, they do not release waste waters.

In bleaching, the primary goal is to remove the low amount of residual lignin present in the pulp after cooking, without decreasing the molecular weight of cellulose. Lignin in unbleached pulps represents typically only about 1 per cent of the dry weight. During pulping, however, lignin is chemically modified and condensed, resulting in poorly degradable structures. Cooking and bleaching are separate process phases, differing from each other with respect to the selectivity of the chemicals used. Chlorine is the most selective bleaching agent as it does not impair the quality of the product by depolymerizing cellulose. In the bleaching processes, lignin is sequentially degraded and extracted in several phases. Bleaching sequences are generally composed of at least five phases. Traditionally, the bleaching of chemical pulps has been carried out with elemental chlorine and chlorine dioxide. A typical sequence would consist of a prebleaching stage with chlorine gas (C) and chlorine dioxide (D) in different ratios (for example C/D 80/20) followed by an alkaline extraction (E). The final bleaching has usually been carried out by chlorine dioxide in two phases, with an intermediate alkaline extraction (DED).

Chlorinated organic compounds are mainly formed in bleaching with elemental chlorine. The public concern together with tightened environmental regulations has driven the pulp and paper industry to search for and to utilize alternative bleaching processes. The major aim has been to replace chlorine gas with other chemicals. Reduction in bleach plant effluents can be achieved by reducing the lignin content of pulp prior to bleaching by modified cooking procedures or oxygen delignification (O), or by replacing elemental chlorine with other chemicals such as chlorine dioxide, ozone (Z), oxygen (O), peroxide (P) and/or peroxyacids in bleaching (McDonough, 1995). The alternative bleaching sequences – the oxygen chemical-based totally chlorine free (TCF) and especially the chlorine dioxide-based elemental chlorine free (ECF) sequences – are increasingly used in industrial pulp bleaching.

Compared with elemental chlorine or even with chlorine dioxide, the oxygen-based chemicals are less effective or less selective in reacting with pulp lignin (Sjöström, 1993). They also depolymerize cellulose. To obtain a fully bleached pulp without elemental chlorine, the lignin content of the pulp entering the bleaching process should be as low as possible. Oxygen delignification is commonly used as a prebleaching stage prior to the ECF and TCF bleaching sequences. A reduction of lignin content by about 50 per cent can be achieved using oxygen, with relatively low loss of carbohydrate yield and without impairing the strength properties of pulp. Today, most pulps are produced by the ECF bleaching sequences, although the amount of TCF bleached pulps is increasing.

5.3 Hemicelluloses in Pulps

Hemicelluloses are polysaccharides associated with cellulose and lignin in plants. The two most common hemicelluloses in wood and other plants are xylans and glucomannans. Not only the relative amounts, but also the chemical composition of these two polysaccharides in softwoods and hardwoods vary. The softwood xylan has a backbone of arabino-4-O-methylglucuronoxylan which is composed of D-xylopyranose units connected via β-(1 → 4)-glycosidic linkages. The average molar ratio of arabinose : 4-O-methyl-glucuronic acid : xylose sugar units in softwood xylan is 1.3 : 2 : 10 (Sjöström, 1993). Hardwood xylan contains 4-O-methylglucuronic acid and acetyl side groups. Methylglucuronic acids are linked to the xylan backbone by β-(1 → 2) glucosidic bonds and the acetic acids are esterified at the carbon 2 and/or 3 hydroxyl group. The backbone of softwood glucomannan is composed of β-(1 → 4)-linked D-glucopyranose and D-mannopyranose units, and it is partially substituted by α-galactose and acetyl units. In softwoods two types of glucomannans have been identified which differ in their solubility and molar ratio of gal : glu : man. The low-galactose content fraction has a ratio of 0.1 : 1 : 4 and is generally called glucomannan. This is the main fraction of glucomannan in softwoods. The corresponding ratio in the high-galactose content fraction, galactoglucomannan, is 1 : 1 : 3. Hardwoods also contain a small amount of glucomannan. In addition to xylan and glucomannan, both softwoods and hardwoods contain minor amounts of other hemicelluloses such as galactan and arabinan.

Extensive modifications of hemicelluloses take place during pulping processes. During the conventional kraft cooking, part of the hemicelluloses are first solubilized in the cooking liquor. At the start of the kraft process xylan in wood is partly solubilized by the alkaline cooking liquid and many of the side groups and acetic acid residues are cleaved off. It has recently been observed that the majority of the 4-O-methylglucuronic acid side groups are converted in the early phases of the kraft cook to hexenuronic acid (Teleman *et al.*, 1995). In the later phases of the process when the alkalinity of the cooking liquor decreases, part of the solubilized xylan is relocated onto the cellulose fibres (Yllner and Enström, 1956; Yllner *et al.*, 1957). Although glucomannan is the main hemicellulose in softwood, the bulk of glucomannans are dissolved and degraded during kraft pulping. Reprecipitation of glucomannan has not been reported to take place to the same extent as xylans. Thus, the relative amount of xylan is increased in pine kraft pulp compared with pine wood (Sjöström, 1977). The average amount of hemicelluloses in hardwood (birch) and softwood (pine) wood and pulps are presented in Table 5.1. The amount of xylan in particular varies considerably in different types of kraft pulps. By comparison, in softwood pulps, the amount of glucomannan is practically the same, irrespective of the pulping conditions. In addition to xylan, lignin is also partially readsorbed on the fibres. Lignin has been reported to be linked to hemicelluloses, forming lignin–

Table 5.1 Average content of hemicelluloses in pine and birch wood and pulp

Wood species	Component	Content (% of dry weight)	
		In wood	In pulp
Pine	Xylan	5–11	7–10
	Glucomannan	14–20	10
Birch	Xylan	22–30	20
	Glucomannan	1–4	Trace

carbohydrate complexes (Iversen and Wännström, 1986). Furthermore, hemicelluloses seem to restrict physically the passage of high molecular mass lignin out of the pulp fibre cell wall (Scallan, 1977); thus the removal of hemicelluloses, especially xylan, can be expected to enhance the extractability of residual lignin from pulps.

5.4 Hemicellulases in Bleaching

5.4.1 Hemicellulose-Degrading Enzymes

The two main enzymes which depolymerize the hemicellulose backbone are endoxylanases and endomannanases. Endoxylanases (1,4-β-D-xylan xylanohydrolases, EC 3.2.1.8) catalyze the random hydrolysis of 1,4-β-D-xylosidic linkages in xylans (Figure 5.1). Endomannanases (1,4-β-D-mannan mannanohydrolase, EC 3.2.1.78) catalyze the random hydrolysis of β-D-1,4-mannopyranosyl linkages within the main chain of mannans and various polysaccharides consisting mainly of mannose, such as glucomannans, galactomannans and galactoglucomannans (Figure 5.2). In a complete hydrolysis, small oligosaccharides are further hydrolyzed by β-xylosidase, β-mannosidase and β-glucosidase. β-Xylosidases catalyze the hydrolysis of xylo-oligosaccharides by removing successive xylose residues from the non-reducing termini and, correspondingly, β-mannosidase and β-glucosidase catalyze the hydrolysis of terminal, non-reducing residues in glucomannans. The side-groups are removed by accessory enzymes: α-glucuronidase, α-arabinosidase and α-galactosidase. Esterified side-groups are liberated by acetyl xylan esterase and acetyl galactoglucomannan esterase (Figures 5.1 and 5.2).

Several species of fungi and bacteria are known to produce the whole spectrum of hemicellulose degrading enzymes (Biely, 1985; Coughlan and Hazlewood, 1993). Most of the xylanases characterized are able to hydrolyze xylans from various origins, showing differences only in the spectrum of end products. The main products formed from the hydrolysis of xylans are xylobiose, xylotriose and substituted oligomers of two to five xylosyl residues. The chain length and the structure of the substituted products depend on the mode of action of the individual xylanase. Some xylanases, however, show rather strict substrate specificity. The three-dimensional structures of several low molecular mass xylanases have recently been determined (Törrönen *et al.*, 1994). The structure of the *Trichoderma reesei* pI 9 xylanase is ellipsoidal, having dimensions of about 30 to 40 Å. Unlike most cellulases, it does not contain any separate substrate binding domain. Some bacterial xylanases,

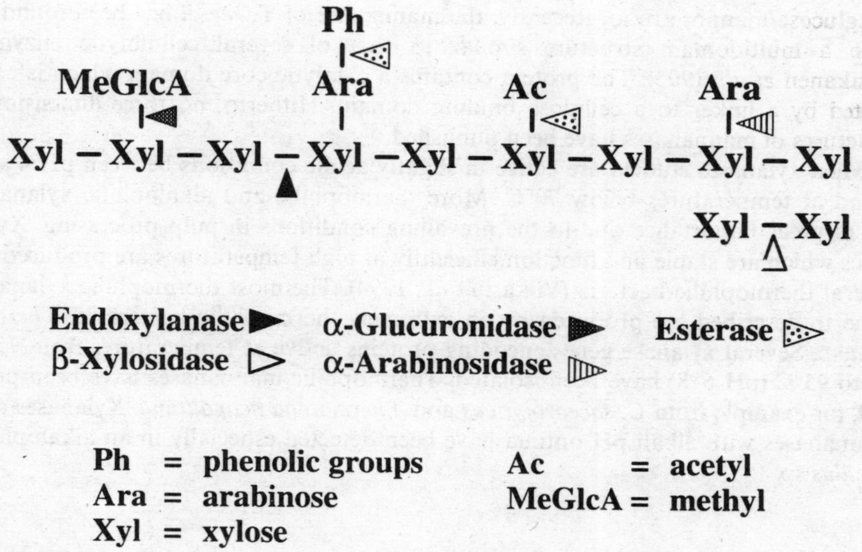

Figure 5.1 Xylan degrading enzymes.

however, have been claimed to contain either a cellulose binding domain (Hazlewood and Gilbert, 1992) or a xylan binding domain (Irwin *et al.*, 1994).

Compared with xylanases, mannanases are a more heterogeneous group of enzymes. The main hydrolysis products from galactomannans and glucomannans are mannobiose, mannotriose and various mixed oligosaccharides. The hydrolysis yield is dependent on the degree of substitution as well as on the distribution of the substituents (McCleary, 1991). The hydrolysis of glucomannans is also affected by

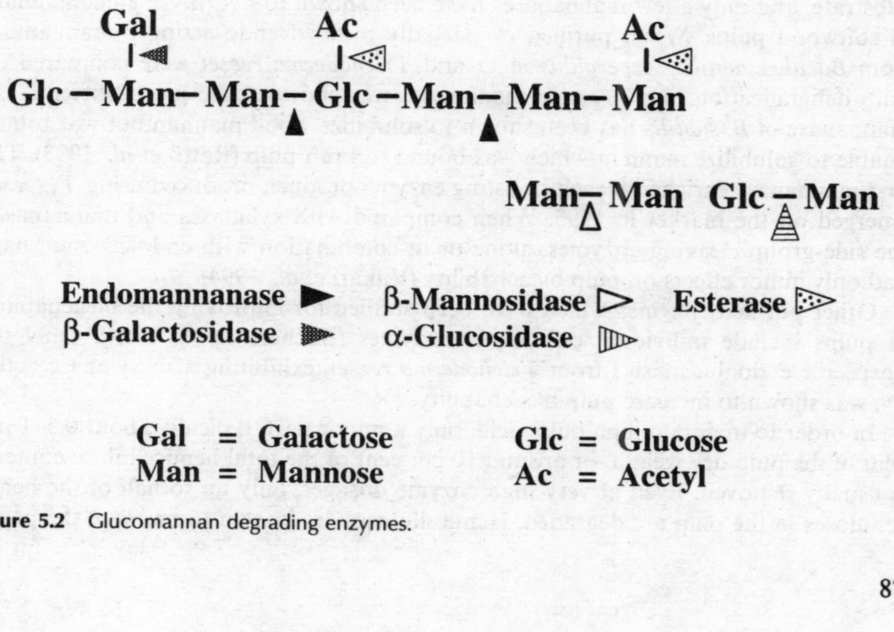

Figure 5.2 Glucomannan degrading enzymes.

the glucose/mannose ratio. Recently, the mannanase of *T. reesei* has been found to have a multidomain structure similar to that of several cellulolytic enzymes (Tenkanen *et al.*, 1995). The protein contains a catalytic core domain which is connected by a linker to a cellulose binding domain. Hitherto, no three-dimensional structures of mannanases have been published.

Most xylanases studied are active in slightly acidic conditions between pH 4 and 6 and at temperatures below 70°C. More thermophilic and alkalophilic xylanases are of great importance due to the prevailing conditions in pulp processing. Xylanases which are stable and function efficiently at high temperatures are produced by several thermophilic bacteria (Viikari *et al.*, 1994). The most thermophilic xylanases hitherto described are produced by an extremely thermophilic bacterium, *Thermotoga* sp. Several xylanase genes encoding proteins active at temperatures from 75°C up to 95°C (pH 6–8) have been isolated. Thermophilic mannanases have been purified, for example from *C. saccarolyticus* and *Thermotoga neapolitana*. Xylanases and mannanases with alkali pH optima have been detected especially in an alkalophilic *Bacillus* sp.

5.4.2 Hydrolysis of Pulp Hemicelluloses

The main enzymes needed to enhance the delignification of both hardwood and softwood kraft pulp have been shown to be endo-β-xylanases (Paice *et al.*, 1988; Buchert *et al.*, 1992; Tenkanen *et al.*, 1992). A positive effect has been achieved with most xylanases studied, independently of the origin of the enzyme. Both fungal and bacterial xylanases have been shown to increase bleachability, and several commercial xylanases are available, varying with respect to their pH and temperature optima (Table 5.2). In practical process conditions, properties of the enzymes such as the pH and temperature optima and stability are of utmost importance. Due to the high temperature and alkalinity of the pulp, the enzymes are generally applied after pH adjustment to about 5–7 and cooling of the pulp to 40–50°C.

Mannanases, on the other hand, appear to be more specific with respect to their substrate, and only a few mannanases have been shown to hydrolyze glucomannans in softwood pulps. When purified or partially purified endo-acting β-mannanases from *Bacillus subtilis*, *Aspergillus niger* and *Trichoderma reesei* were compared on pulp delignification, the *T. reesei* mannanase was shown to be most efficient. The mannanase of *B. subtilis* has been shown to solubilize wood mannan but was totally unable to solubilize mannan which was bound to kraft pulp (Rättö *et al.*, 1993). The first mannanase enriched bleach-boosting enzyme product, produced using *T. reesei*, emerged on the market in 1995. When compared with xylanases and mannanases, the side-group cleaving enzymes, alone or in combination with endoenzymes, have had only minor effects on pulp bleachability (Viikari *et al.*, 1994).

Other purified enzymes which have been studied for improving the bleachability of pulps include individual cellulolytic enzymes (Buchert *et al.*, 1994). Only the unspecific endoglucanase I from *Trichoderma reesei*, exhibiting also xylanase activity, was shown to increase pulp bleachability.

In order to maintain high pulp yield, only a minor part, typically about 0.5–1 per cent of the pulp dry weight, or around 10 per cent of the total hemicellulose content, is usually removed. Even at very high enzyme dosages, only up to half of the hemicelluloses in the pulp are degraded. Hemicelluloses are known to improve the paper

Table 5.2 Commercial bleach-boosting hemicellulases and mannanase

Enzyme	Supplier	Activity/ml	Activity measurement method	pH optimum	Temperature optimum (°C)	Recommended dosage
Irgazyme 40 × 4	Genencor International	4000–4800 U	RBB xylan	6–7	50–60	0.7 U/g 0.5 U/g 1.7 U/g
Irgazyme 40	Genencor International	1000–1200 U	RBB xylan	6–7	50–60	0.7 U/g 0.5 U/g 1.7 U/g
Irgazyme 10A × 4	Genencor International	4000–4800 U	RBB xylan	4–5	50–60	0.7 U/g 0.5 U/g 1.7 U/g
Ecopulp X-200	Primalco	200 000 BXU	XYL/DNS	5–6	55	0.2–0.4 l/ton
Ecopulp X-200/4	Primalco	200 000 BXU	XYL/DNS	3–4.5	45–55	0.2–0.4 l/ton
Ecopulp XM	Primalco	54 000 BXU 36 000 MNU	XYL/DNS MAN/DNS	5–6	50–55	0.5–1 l/ton
Ecopulp T	Primalco	100 000 TXU	XYL/DNS	6–8	55–75	0.3–0.6 l/ton
Cartazyme HS 10	Sandoz	10.10^6 XU		4–5	40–60	0.3–1.0 XU/g
GS35	Iogen	3500 XU	RBB xylan	5.2–7.8	47–58	300 ml/ton
Pulpzyme HB	Novo Nordisk	600 EXU/g	RBB xylan	6–8	40–55	0.3–0.6 EXU/g

RBB, Rhemazol Brilliant Blue xylan, a dyed substrate for xylanase; DNS, measurement reducing sugars; XYL, xylanase activity measured with isolated xylan as substrate; MAN, mannanase activity measured with isolated mannan as substrate

technical properties of the fibres, and thus it is not desirable to remove a major part of hemicelluloses in the enzymatic treatment. The degree of hydrolysis is adjusted by proper dosage of the enzyme amount and by the reaction time. The reaction time is usually also restricted by the mill conditions, that is, the available equipment. Considering the different types of xylans present in pulps – the residual and readsorbed xylans – it would be advantageous to remove specifically only those xylans which hinder the extraction of lignin. However, despite both basic research on substrate specificities of individual xylanases and applied research on different and modified pulps, this question still remains unanswered. The xylanase of *T. reesei* has been observed to hydrolyze xylan from all accessible surfaces of kraft pulps (Suurnäkki *et al.*, 1996 a,b), indicating that the effect of xylanase on bleachability is not only an outer surface phenomenon. The composition of xylan solubilized in limited or extensive treatments has not revealed essential differences, indicating that the structure of xylan is rather similar in all parts of fibres, or that the enzymes are specific to a certain type of xylan.

The effect of enzymes on pulp bleachability has been studied by different methods. In the first phase, enzymes are usually characterized with isolated substrates, which are used in the determination of their activities. These substrates, however, vary extensively with respect to their origin and composition. Furthermore, comparison of enzyme activities is complicated by the utilization of different analysis methods (Bailey *et al.*, 1992). Hence in pulp applications the action of enzymes must be compared on the actual substrate, the pulp. In these tests, the liberation of sugars into the pulp solution has usually been measured either by reducing sugar analysis or by HPLC. In addition, increase in the liberation of lignin-derived compounds after enzymatic treatment (Yang and Eriksson, 1992) or alkaline extraction (Hortling *et al.*, 1994) has been used for evaluation of different enzymes. The most reliable method for comparison of the effects of different enzymes is, however, to bleach the enzymatically treated pulps and to measure the brightness and final lignin content (kappa number).

The action of enzymes is influenced by the electrochemical interactions between the fibres and enzymes (Buchert *et al.*, 1993). The carboxyl groups within the fibre cell wall are mainly responsible for the swelling properties of pulp in water (Scallan, 1983). The surface charge and the swelling of fibres have been reported to affect the action of xylanases. The more negative the surface charge, the less the pulp was hydrolyzed (Buchert *et al.*, 1993). However, the swelling or surface charge may not be the primary factors affecting the hydrolysis. The type of the counter-ions and the degree of the substitution of the carboxyl groups in pulp were found to have a profound effect on the action of xylanases in the pulp matrix. Consequently, metal-free pulps were found to be poorly hydrolyzed by hemicellulases (Buchert and Viikari, 1995). This observed phenomenon of poor hydrolyzability of metal-free pulp is of practical importance in the TCF-bleaching sequences, where the metal removal stage is essential to retain the strength properties of pulp. There seem, however, to be some variations in the specificities of the enzymes, although systematic studies have not been carried out.

5.4.3 Suggested Mechanisms for Improving Bleachability by Hemicellulases

The effect of hemicellulases in bleaching is based on the modification of pulp hemicelluloses, enhancing the removal of lignin in chemical bleaching. It has been proposed that the action of xylanases is due to the partial hydrolysis of reprecipitated xylan (Kantelinen *et al.*, 1993) or to removal of xylan from the lignin–carbohydrate (LC) complexes (Yang and Eriksson, 1992). However, these hypotheses are not mutually exclusive – relocated xylans may contain LC complexes and both mechanisms would allow the enhanced diffusion of entrapped lignin from the fibre wall. Limited removal of pulp xylan is known to increase the leachability of residual lignin from kraft pulps (Hortling *et al.*, 1994) and thus also to increase the pulp bleachability during subsequent bleaching stages. In addition, it has been suggested that the hemicellulase treatment removes chromophoric groups from the pulp (Wong *et al.*, 1996). The suggested mechanisms as well as their consequences are presented in Figure 5.3. The methods used for mechanistic studies have included modified pulping methods, production of model pulps, analysis of degradation products of enzymatic treatments, chemical extractions of lignin and xylans, mechanical peeling, surface composition analysis by ESCA (electron spectroscopy for chemical analysis) and different delignification tests.

It has been suggested that xylanases hydrolyze the reprecipitated xylan lying on the surface of the kraft pulp fibres. Hence enzymatic hydrolysis of the reprecipitated and relocated xylans on the surface of the fibres apparently renders the structure of the fibres more permeable. The increased permeability allows the passage of lignin or lignin–carbohydrate molecules in higher amounts and of higher molecular masses in the subsequent chemical extraction (Kantelinen *et al.*, 1993; Hortling *et al.*, 1994). Both xylan and lignin are dissolved and partially readsorbed on the fibres during pulping. A rather high content of lignin has been observed both in the primary fines – the finest fibre fraction of the pulp – and in the surface material of pine kraft fibres (Laine *et al.*, 1994). Xylanases combined with ESCA have been used to determine the xylan content on the outer surfaces of bleached cellulose fibres. In softwood kraft fibres, removal of xylan by xylanases was found to uncover lignin (Buchert *et al.*, 1996). The relatively low amount of xylan, observed on the outer surfaces of fibres, can obviously easily be removed by xylanases (Suurnäkki *et al.*, 1996c).

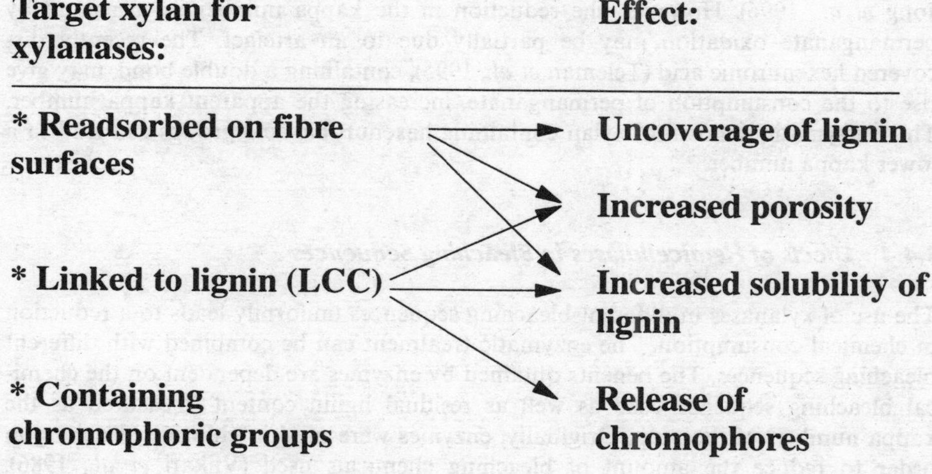

Figure 5.3 Possible mechanisms for improving bleachability of pulps by xylanases

It can thus be expected that removal of xylan improves the extractability of lignin by exposing lignin surfaces. In birch kraft pulp, the primary fines and the fibre surface material were found to be considerably richer in both xylan and lignin than the whole pulp (Suurnäkki *et al.*, 1996a). Therefore, relatively more xylan should be enzymatically removed from the outer surfaces and expose the lignin. The role of reprecipitated xylan in the xylanase-aided bleaching of birch kraft pulps has been confirmed by comparing the effect of xylanase treatment on bleachability of kraft pulps cooked by a batch method and of pulps produced in a flow-through digester and therefore containing only traces of reprecipitated xylan (Kantelinen *et al.*, 1993). On the other hand, xylanase pretreatment has been reported to enhance also the bleachability of softwood pulps produced by novel cooking methods, presumably containing less reprecipitated xylan than conventional softwood kraft pulps. It can thus be expected that the mechanism of xylanase-aided bleaching is not based on hydrolysis of relocated xylan alone.

The partial hydrolysis of xylan or glucomannan may also degrade and improve the extractability of lignin–carbohydrate (LC) complexes. Both softwood and hardwood kraft pulps have been reported to contain LC complexes in which carbohydrates and lignin may be connected to each other by ether or glycosidic linkages (Gellerstedt and Lindfors, 1991). However, no direct evidence for the type of linkage(s) existing between carbohydrates and lignin has yet been presented. Increased solubilization of xylan–lignin complexes both from model pulps and from kraft pulps has been observed by xylanase treatment. According to molecular weight analyses, part of the lignin released during the enzymatic treatment appears to be covalently bound to xylan, whereas most of the lignin may be physically interlinked with xylan in the fibre matrix. The action of xylanases on both reprecipitated and LC–xylan in enhancing bleachability suggests that it is probably not only the type but also the location of the xylan that is important in the mechanism of xylanase-aided bleaching.

It has frequently been observed that xylanase treatment has a slight decreasing effect on the lignin content (measured as the kappa number). This has been explained to be due to removal of lignin fragments or chromophoric structures (de Jong *et al.*, 1996). However, the reduction in the kappa number as measured by permanganate oxidation may be partially due to an artefact. The recently discovered hexenuronic acid (Teleman *et al.*, 1995), containing a double bond, may give rise to the consumption of permanganate, increasing the apparent kappa number. Thus enzymatic removal of xylan containing hexenuronic acid groups can lead to a lower kappa number.

5.4.4 Effects of Hemicellulases in Bleaching Sequences

The use of xylanases in different bleaching sequences uniformly leads to a reduction in chemical consumption. The enzymatic treatment can be combined with different bleaching sequences. The benefits obtained by enzymes are dependent on the chemical bleaching sequence used as well as residual lignin content (measured as the kappa number) of the pulp. Originally, enzymes were used in chlorine bleaching in order to reduce the amount of bleaching chemicals used (Viikari *et al.*, 1986). Enzymes were later combined with various ECF and TCF bleaching sequences to improve the otherwise lower brightness of pulp or to decrease the bleaching costs.

In chlorine bleaching an average reduction of 25 per cent in active chlorine consumption in prebleaching or a reduction of about 15 per cent in total chlorine consumption has been reported both in laboratory scale and in mill trials. As a result, the concentration of chlorinated compounds, measured as AOX, in the bleaching effluent during mill trials was reduced by 15–20 per cent (Viikari *et al.*, 1991). Today xylanases are used both in ECF and TCF sequences. In ECF sequences the enzymatic step is often adopted due to the limiting chlorine dioxide production capacity. The use of enzymes allows bleaching to higher brightness values when chlorine gas is not used. In TCF sequences, the advantage of the enzymatic step is due to improved brightness, maintenance of fibre strength and savings in bleaching costs. The benefits obtained by hemicellulases in different bleaching sequences are summarized in Table 5.3.

The amount of enzyme needed for the bleaching is also a key parameter with respect to both enzyme cost and yield loss and has to be tested in laboratory scale with each individual pulp and bleaching sequence used. Generally, it seems that although hydrolysis (the solubilization of carbohydrates) increases as a function of the enzyme dosage used, no further benefits can be obtained to the bleachability after a certain limit (Buchert *et al.*, 1992). Thus, in order to maximize the positive effect of the enzyme on the pulp kappa number and brightness and simultaneously minimize the yield loss, laboratory scale experiments are required to optimize the enzyme dosage.

5.4.5 *Industrial Use*

The procedures optimized in laboratory scale have easily been scaled up to full industrial scale without the necessity for pilot stages (Viikari *et al.*, 1994). Furthermore, no expensive investments are needed for full scale runs. The only requirement is the addition of pH adjustment facilities and pumps for the enzyme addition to deliver the enzyme solution to the pulp (Koponen, 1991). The enzymatic pretreatment has been shown to be fully compatible with existing industrial equipment, which is a considerable advantage of this method, especially when compared with some other competing technologies. Enzymes are typically mixed with water before being added to the unbleached pulp by a shower bar. Enzymes are allowed to react

Table 5.3 Benefits of xylanase pretreatment in different bleaching sequences

Bleaching sequence	Benefits
Traditional	Reduced Cl_2 consumption
	Reduced AOX
ECF	Reduced ClO_2 consumption
	Reduced AOX
	Increased productivity, when ClO_2 limiting
TCF	Increased brightness
	Reduced chemical consumption
	Retained strength properties

in the high density storage tank for at least 2 hours before the subsequent chemical bleaching steps. Several successful mill trials have been reported and at present a number of mills in Northern America and Scandinavia use enzymes continuously.

Xylanases are sold as concentrated liquids and the amount required per metric ton of pulp is very low, less than a litre. The cost of the enzyme per tonne varies and depends on the dosage required and the supplier. The approximate price in 1996 is less than US$2 per ton. Estimations for the capital cost of enzyme delivery and pH adjustment vary from US$10 000 to 1 000 000 in 1995. The potential economic benefits of enzyme bleaching are significant to the pulp and paper industry. A simple calculation of relative economic benefits in an ECF sequence reveals that the reduction of approximately 5 kg ClO_2/ton of pulp, assuming a chlorine dioxide cost of US$0.70 per kg, leads to savings of about US$2 per ton of pulp in chlorine dioxide costs alone. The costs of oxygen based chemicals (ozone, peroxide) are even higher and the respective savings even more pronounced. Additional savings in alkali can also be expected (Farrell *et al.*, 1996).

5.5 Laccases in Bleaching

The initial studies on the use of enzymes in bleaching were performed with a goal of imitating the wood decaying action of fungi in nature. The largest group of fungi which degrade wood are the Basidiomycetes and the best studied are those which specifically degrade lignin. Research activity using enzymes produced by Basidiomycetes for the laboratory bleaching of pulp followed the discovery of lignin peroxidases (Tien and Kirk, 1983; Gold *et al.*, 1984). It was observed, however, that different mixtures of lignin peroxidases and manganese-dependent peroxidases did not consistently delignify unbleached kraft pulp, although lignin-containing materials appeared to be released into the effluent. In addition to lignin and manganese dependent peroxidases, a prevalent enzyme in many white-rot fungi is laccase – an enzyme which catalyzes the synthesis of lignin in growing plants. Laboratory results have shown that the treatment of softwood kraft pulp with laccase in the presence of a dye (ABTS) causes more than a 25 per cent drop in kappa number, indicating direct delignification (Bourbonnais *et al.*, 1992). Recently a new laccase-mediator concept was published as a technically feasible approach to delignify pulp (Call and Mücke, 1994). In this concept the laccase is combined with a low molecular weight redox mediator resulting in generation of a strongly oxidizing co-mediator which can then specifically degrade lignin. Depending on pulp and the conditions used the laccase-mediator system has been reported to result in kappa reduction of up to 70 per cent in a single step. The laccase-mediator system operates at a pulp consistency of 10–15 per cent, at pH about 4.5 and temperatures 40–56°C. The retention time has been 1–4 hours depending on pulp source and lignin content. The laccase-mediator system is not yet commercialized and requires further improvements in order to be a cost-efficient bleaching system.

5.6 Conclusions

Hemicellulases were the first group of specific enzymes used in large scale in the pulp and paper industry. The method is an example of sustainable technology in the traditional chemical industry with clear environmental benefits, and it is economically attractive. The hemicellulase treatment, together with a chemical extraction, leads to a significant reduction in the residual lignin content of the pulps. The partial hydrolysis of xylan facilitates the extraction of lignin from pulp in higher amounts and with higher molecular weights. However, due to the indirect mode of action, the effect of hemicellulase-aided bleaching is limited. The improved bleachability is mainly based on the action of endo-β-xylanases, a group of enzymes which can be efficiently produced in industrial scale. In addition to lignin modifying enzymes, new commercial hemicellulases with higher pH and temperature optima should improve the applicability of enzymes. Today, several mills are using hemicellulases. Future expectations, however, are focused on the laccase-mediator system. Early results are positive, and it remains to be seen how rapidly further improvement of this method will allow its economic, and hence industrial, application.

References

BAILEY, M. J., BIELY, P. and POUTANEN, K. (1992) Interlaboratory testing of methods for assay of xylanase activity. *J. Biotechnol.* **23**, 257–270.

BIELY, P. (1985) Microbial xylanolytic systems. *Trends Biotechnol.* **3**, 286.

BOURBONNAIS, R., PAICE, M. G. and REID, I. D. (1992) In: Kuwahara, M. and Shimada, M., eds, *Biotechnology in the Pulp and Paper Industry*, Tokyo: Uni Publisher, p. 181.

BUCHERT, J. and VIIKARI, L. (1995) The role of pulp metal profile on enzyme-aided TCF-bleaching. *Paperi ja Puu – Paper and Timber* **77**, 582–587.

BUCHERT, J., RANUA, M., KANTELINEN, A. and VIIKARI, L. (1992) The role of two *Trichoderma reesei* xylanases in bleaching of pine kraft pulp. *Appl. Microbiol. Biotechnol.* **37**, 825–829.

BUCHERT, J., TENKANEN, M., VIIKARI, L. and PITKÄNEN, M. (1993) Role of surface charge and swelling on the action of xylanases on birch kraft pulp. *Tappi J.* **76**, 131–135.

BUCHERT, J., RANUA, M., SIIKA-AHO, M., PERE, J. and VIIKARI, L. (1994) *Trichoderma reesei* cellulases in the bleaching of kraft pulps. *Appl. Microbiol. Biotechnol.* **40**, 941–945.

BUCHERT, J., CARLSSON, G., VIIKARI, L. and STRÖM, G. (1996) Surface characterization of unbleached kraft pulps by enzymatic peeling and ESCA. *Holzforschung* **50**, 69–74.

CALL, H. P. and MÜCKE, I. (1994) State of art of enzyme bleaching and disclosure of a breakthrough system. *Proceedings 1994 Non-Chlorine Bleaching Conference*, USA.

COUGHLAN, M. P. and HAZLEWOOD, G. P. (1993) β-1,4-D-xylan-degrading enzyme systems: biochemistry, molecular biology and applications. *Biotechnol. Appl. Biochem.* **17**, 259.

DE JONG, E., WONG, K.·K. Y., WINDSOR, L. R. and SADDLER, J. N. (1996) The mechanism of xylanase prebleaching of kraft pulp. In Messner, K. and Srebotnik, E., eds, *Biotechnology in Pulp and Paper Industry – Advances in Applied and Fundamental Research*, Vienna: WUA Üniversitätsverlag.

FARRELL, R. L., VIIKARI, L. and SENIOR, D. J. (1996) Enzyme treatments of pulp. In: Jeffries, T. W. and Viikari, L., eds, *Pulp Bleaching: Principles & Practice*, ACS Symp. Ser. 655, Atlanta, GA: Tappi Press, pp. 363–377.

GELLERSTEDT, G. and LINDFORS, E.-L. (1991) On the structure and reactivity of residual lignin in kraft pulp fibres. *Proceedings International Pulp Bleaching Conference*, Stockholm: SPCI, Vol. 1, p. 73.

GOLD, M. H., KUWAHARA, M., CHIU, A. A. and GLENN, J. K. (1984) Purification and characterization of an extracellular hydrogen peroxide requiring diarylpropane oxygenase from the white-rot basidiomycete, *Phanerochaete chrysosporium. Arch. Biochem. Biophys.* **234**, 353–362.

HAZLEWOOD, G. P. and GILBERT, H. J. (1992) The molecular architecture of xylanases from *Pseudomonas fluorescens* subs. *cellulosa.* In: Visser, J., Beldman, G., Kusters-van Someren, M. A. and Voragen, A. G. J., eds, *Xylan and Xylanases*, Amsterdam: Elsevier Science Publishers, p. 259.

HORTLING, B., KORHONEN, M., BUCHERT, J., SUNDQUIST, J. and VIIKARI, L. (1994) The leachability of lignin from kraft pulps after xylanase treatment. *Holzforschung* **48**, 441–446.

IRWIN, D., JUNG, E. D. and WILSON, D. B. (1994) Characterization and sequence of a *Thermomonospora fusca* xylanase. *Appl. Environ. Microbiol.* **60**, 763.

IVERSEN, T. and WÄNNSTRÖM, S. (1986) Lignin-carbohydrate bonds in a residual lignin isolated from pine kraft pulp. *Holzforschung* **40**, 19–22.

KANTELINEN, A., HORTLING, B., SUNDQUIST, J., LINKO, M. and VIIKARI, L. (1993) Proposed mechanism of the enzymatic bleaching of kraft pulp with xylanases. *Holzforschung* **47**, 318–324.

KOPONEN, R. (1991) Enzyme systems prove their potential. *Pulp Pap. Int.* **33**, 20,25.

LAINE, J., STENIUS, P., CARLSSON, G. and STRÖM, G. (1994) Surface characterization of unbleached kraft pulps by means of ESCA. *Cellulose* **1**, 145.

MCCLEARY, B. V. (1991) Comparison of endolytic hydrolases that depolymerize 1,4-β-D-mannan, 1,5-α-L-arabinan and 1,4-β-D-galactan. In: Leatham, G. F. and Himmel, M. E., eds, *Enzymes in Biomass Conversion.* ACS Symp. Ser. 460, Washington, DC: American Chemical Society, p. 437.

MCDONOUGH, T. J. (1995) Recent advances in bleached chemical pulp manufacturing technology. Part 1: Extended delignification, oxygen delignification, enzyme applications, and ECF and TCF bleaching. *Tappi J.* **78**, 55.

PAICE, M., BERNIER, M. and JURASEK, L. (1988) Viscosity enhancing bleaching of hardwood kraft pulp with xylanase from a cloned gene. *Biotechnol. Bioeng.* **32**, 235–239.

RÄTTÖ, M., SIIKA-AHO, M., BUCHERT, J., VALKEAJÄRVI, A. and VIIKARI, L. (1993) Enzymatic hydrolysis of isolated and fibre-bound galactoglucomannans from pine wood and pine kraft pulp. *Appl. Microbiol. Biotechnol.* **40**, 449.

SCALLAN, A. M. (1977) The accomodation of water within pulp fibres. *Proceedings Fibre–Water Interactions in Paper Making*, Oxford, pp. 9–27.

SCALLAN, A. M. (1983) The effect of acidic groups on the swelling of pulps: a review. *Tappi J.* **66**, 73.

SJÖSTRÖM, E. (1977) The behavior of wood polysaccharides during alkaline pulping processes. *Tappi J.* **60**, 151–154.

SJÖSTRÖM, E. (1993) *Wood Chemistry, Fundamentals and Application*, 2nd edition, San Diego, CA: Academic Press.

SUURNÄKKI, A., KANTELINEN, A., BUCHERT, J. and VIIKARI, L. (1994) Enzyme-aided bleaching of industrial softwood kraft pulps. *Tappi J.* **77**, 111–116.

SUURNÄKKI, A., HEIJNESSON, A., BUCHERT, J., VIIKARI, L. and WESTERMARK, U. (1996a) Chemical characterization of the surface layers of unbleached pine and birch kraft pulp fibres. *J. Pulp Paper Sci.* **22**, J43–J47.

SUURNÄKKI, A., HEIJNESSON, A., BUCHERT, J., TENKANEN, M., VIIKARI, L. and WESTERMARK, U. (1996b) Location of xylanase and mannanase action in kraft fibres. *J. Pulp Paper Sci.* **22**, J78–J83.

SUURNÄKKI, A., HEIJNESSON, A., BUCHERT, J., WESTERMARK, U. and VIIKARI, L. (1996c) Effect of pulp surfaces on enzyme-aided bleaching of kraft pulps. *J. Pulp Paper Sci.* **22**, J91–J96.

TELEMAN, A., HARJUNPÄÄ, V., TENKANEN, M., BUCHERT, J., HAUSALO, T., DRA-

KENBERG, T. and VUORINEN, T. (1995) Characterisation of 4-deoxy-β-L-*threo*-hex-4-enopyranosyluronic acid attached to xylan in pine kraft pulp and pulping liquor by NMR spectroscopy. *Carbohydr. Res.* **272**, 55.

TENKANEN, M., BUCHERT, J., PULS, J., POUTANEN, K. and VIIKARI, L. (1992) Two main xylanases of *Trichoderma reesei* and their use in pulp processing. In: Visser, J., Beldman, G., Kusters-van Someren, M. A. and Voragen, A. G. J., eds, *Xylans and Xylanases*, Amsterdam: Elsevier Science Publishers, pp. 547–550.

TENKANEN, M., BUCHERT, J. and VIIKARI, L. (1995) Binding of hemicellulases on isolated polysaccharide substrates. *Enzyme Microb. Technol.* **17**, 499.

TIEN, M. and KIRK, T. K. (1983) Lignin degrading enzyme from the hymenomycete *Phanerochaete chrysosporium* Burds. *Science* **221**, 661.

TÖRRÖNEN, A., HARKKI, A. and ROUVINEN, J. (1994) Three dimensional structure of endo-1,4-β-xylanase II from *Trichoderma reesei*: two conformational states in the active site. *EMBO J.* **13**, 2493.

VIIKARI, L., RANUA, M., KANTELINEN, A., LINKO, M. and SUNDQUIST, J. (1986) Bleaching with enzymes. In: *Biotechnology in the Pulp and Paper Industry, Proceedings 3rd International Conference*, Stockholm, pp. 67–69.

VIIKARI, L., SUNDQUIST, J. and KETTUNEN, J. (1991) Xylanase enzymes promote pulp bleaching. *Paper and Timber* **73**, 384–389.

VIIKARI, L., KANTELINEN, A., SUNDQUIST, J. and LINKO, M. (1994) Xylanases in bleaching: from an idea to the industry. *FEMS Microbiol. Rev.* **13**, 335–350.

WONG, K. K. Y. and SADDLER, J. N. (1992) *Trichoderma* xylanases, their properties and application. *Crit. Rev. Biotechnol.* **12**, 413.

WONG, K. K. Y., CLARKE, P. and NELSON, S. L. (1996) Possible roles of xylan-derived chromophores in xylanase prebleaching of softwood kraft pulp. *ACS Symp. Ser.* **618**, 352.

YANG, J. L. and ERIKSSON, K.-E. L. (1992) Use of hemicellulolytic enzymes as one stage in bleaching of kraft pulps. *Holzforschung* **46**, 481–488.

YLLNER, S. and ENSTRÖM, B. (1956) Studies of the adsorption of xylan on cellulose fibres during the sulphate cook. Part 1. *Svensk Papperstidn.* **59**, 229–232.

YLLNER, S., ÖSTBERG, K. and STOCKMAN, L. (1957) A study of the removal of the constituents of pine wood in the sulphate process using a continuous liquor flow method. *Svensk Papperstidn.* **60**, 795–802.

Anaerobic Treatment of Pulp Mill Effluents

SERGE R. GUIOT AND JEAN-CLAUDE FRIGON

6.1 Introduction

Pulp and paper mill effluents pose a unique challenge to treatment plant operators who have to deliver discharges to the environment with negligible impact. The pulp and paper industry generates large volumes of effluents which contain a broad variety of compounds. These compounds are added, formed or dissolved during the process of manufacturing pulp. The volume of discharged water per ton of manufactured pulp varies over a broad range and depends on the pulping process used, the operating practices, the equipment and the production rate. In the pulp and paper industries, internal measures have been undertaken to reduce wastewater discharges (e.g., dry debarking, prolonged cooks, countercurrent washing, chemical charges, substitution of chlorine with chlorine dioxide, oxygen delignification, and hydrogen peroxide bleaching) (McCubbin, 1984).

The manufacture of pulp and paper includes sequentially part or all of the following stages: wood preparation, chemical or mechanical pulping, pulp bleaching and paper making (newsprint, paper or paperboard). Today, Kraft mills still account for a large share of the total pulp production (newsprint and bleached pulp). The Kraft process is now being challenged by thermomechanical (TMP) and chemithermomechanical pulping (CTMP), which are more competitive (reduced water usage; higher pulp yield) (Murray and Richardson, 1993). In general, the pulping and bleaching processes are the main sources of wastewater load in the pulp and paper industry. Minor loads are generated in the paper machines.

Despite the anaerobic process having some limitations (e.g., lower chemical oxygen demand (COD) specific removal rate than aerobic systems, longer initial system start-up, slow biomass acclimation to more difficult compounds, longer recovery after an upset, and sensitivity to toxic compounds), interest in anaerobic systems for treating some pulp and paper wastewaters has risen steadily since the 1980s. In addition to the production of a gaseous fuel (biogas), the low sludge yield and the low energy and chemical requirements of the anaerobic treatment significantly reduce the operating costs. Additionally, anaerobic systems have proven to be more effective in dehalogenation reactions compared with activated sludge

(Salkinoja-Salonen *et al.*, 1984). The potential for application of anaerobic technology in the pulp and paper industry has increased as the waste streams have become more concentrated due to efficient water recycling (Lettinga *et al.*, 1991).

Some pulp mill effluents are easily amenable to anaerobic treatment: condensates from chemical and semi-chemical pulping; sulphite spent liquor; effluents from mechanical and secondary fibre pulping; and white waters (Rintala and Puhakka, 1994). The great majority of the full-scale anaerobic treatment plants treat these effluents with ease. A smaller number of full-scale operations have been applied to difficult pulp and paper effluents, such as those from chemical and semi-chemical pulping, CTMP, bleaching and debarking.

6.2 Composition and Anaerobic Treatability of Various Pulp and Paper Industry Effluents

In the next sections, the characteristics of effluents, their potential for anaerobic degradation and the performance range of their anaerobic treatment are discussed based on laboratory, pilot studies, and full-scale experience on segregated effluents.

6.2.1 Debarking Effluent

The major components of the debarking effluent are polymeric tannins (30–55 per cent), non-tannic and tannic phenol monomers (10–20 per cent), simple carbohydrates (30–40 per cent) and resin compounds (5 per cent) which include: long-chain fatty acids (LCFA), resin acids, apolar phenols, and volatile terpenes. Phenols and carbohydrates form the majority of the easily anaerobically biodegradable fraction of the effluent (40–60 per cent). However, most of the bark tannins are condensed oligomers linked by C-C bonds with molecular weight (MW) ranging from 500 to 3000 daltons, which are not biodegradable. They are toxic to methanogenic bacteria. The soluble bark COD concentrations corresponding to 50 per cent inhibition of methanogenic activity (50% IC_{MA}) range from 880 to 1930 mg COD L^{-1}. The 50% IC_{MA} of bark mixed tannic oligomers ranges between 240 and 500 mg L^{-1} (Field *et al.*, 1988; Sierra-Alvarez *et al.*, 1994). Possible pretreatments prior to anaerobic treatment of debarking effluent include dilution of the effluent, polymerization by high-pH autoxidation of the low MW tannins into non-toxic high MW (in excess of 3000 daltons) tannins and humic-like compounds, and fungal treatment. Autoxidation, precipitation by calcium of the tannic humic polymers, and anaerobic treatment may reduce the initial soluble COD by 79 per cent, half of which is converted into methane (Field *et al.*, 1990a,b).

6.2.2 Mechanical and Thermomechanical Pulping Effluent

Mechanical pulping (MP) produces an easily biodegradable effluent since its main constituents are carbohydrates (80–90 per cent), with lower amounts of volatile fatty acids (VFA) and extractives (10–20 per cent). Thermomechanical pulping (TMP) wastewater differs from the MP effluent in that the heating of the wood chips, prior to the mechanical pulping, solubilizes lignin. This lignin can represent as much as 40 per cent of the effluent organics (Järvinen *et al.*, 1980; Sierra-Alvarez *et al.*, 1991). Carbohydrates contribute to another 40 per cent, with the balance being VFA and alcohols. It was shown that the acidic form of TMP effluent could be successfully degraded (68–87 per cent); however, methanogenic inhibition was observed in alkaline TMP effluents (Sierra-Alvarez *et al.*, 1991). The COD removal in laboratory or pilot scales could be as high as 60–70 per cent (Rintala and Vuoriranta, 1988; Schnell *et al.*, 1990) at high organic loading rates (OLR) (12–31 kg COD m^{-3} d^{-1}). Salkinoja-Salonen *et al.* (1985) have reported 80–90 per cent biochemical oxygen demand (BOD) and COD removal with a fluidized bed reactor at an OLR of 3–4 kg COD m^{-3} d^{-1}. A full-scale upflow anaerobic sludge blanket (UASB) reactor was able to remove 30–40 per cent and 45–55 per cent of the COD and BOD respectively, as well as all fish toxicity, although a large fibre accumulation in the reactor reduced sludge activity (MacLean *et al.*, 1990).

6.2.3 Chemi-thermomechanical Pulping Effluent

The CTMP process is merely the TMP process with the addition of sodium sulphite. The average CTMP effluent contains lignin (30–40 per cent), organic acids (35–40 per cent) and polysaccharides (10–15 per cent). High concentrations of fatty and resin acids have been found in CTMP effluents (Walden and Howard, 1981; Liu *et al.*, 1993). Lower COD removal is achieved from anaerobic treatment of CTMP effluent compared with that of the TMP effluent. This is because CTMP effluent contains more lignin and lignosulphonates which are not anaerobically biodegradable. In addition, compounds known to inhibit methanogenesis, such as sulphate, sulphite, and resin and fatty acids are present in CTMP effluent (Anderson *et al.*, 1987; Pichon *et al.*, 1988; McCarthy *et al.*, 1990). Nevertheless, COD and BOD removal performances of up to 55 and 77 per cent respectively, were shown at OLR varying between 4 and 22 kg COD m^{-3} d^{-1} (Wilson *et al.*, 1987).

6.2.4 Kraft Pulping Effluent

In the Kraft pulping process, wood chips are cooked at 160–180°C with an alkaline white liquor that contains sodium hydroxide and sodium sulphide. This chemical recipe cleaves the ether bonds of the lignins (Rintala and Puhakka, 1994). After separation from the fibres, the spent black liquor is evaporated to a high concentration and then combusted to recover energy and chemicals (Kringstad and Lindström, 1984). In some processes weak black liquors are produced and are disposed of. Black liquor is mainly composed of lignin (29–41 per cent), saccharidic acids (19–28 per cent), other acids (12–16 per cent) and wood extractives (4–7 per cent) (McCubbin, 1984). The black liquor causes a 50% IC_{MA} at concentrations of 2.1–2.6 g COD L^{-1}. This inhibition is the result of wood resin compounds and lignin derivatives (Sierra-Alvarez *et al.*, 1994). Without pretreatment but with proper dilution, black liquor could be anaerobically treated in continuous systems at OLR of 6 kg COD m^{-3} d^{-1} with 50 per cent COD and 95–98 per cent BOD removal. The COD removal could further be improved to 85 per cent after peroxide pretreatment (Zuxuan *et al.*, 1983).

An important group of Kraft process wastewaters consists of foul condensates from the wood chips digester vent and from the black liquor evaporation (KEC). The KEC is usually amenable for anaerobic treatment as its organic compounds are mainly acetate (49–80 per cent), methanol (6–39 per cent), ethanol (6–13 per cent), and furfural (0–17 per cent). The minor constituents are guaicol, acetone and phenol. COD removals of 70–88 per cent can be attained at OLRs of 12–15 kg COD m^{-3} d^{-1} (Pipyn *et al.*, 1987; Qiu *et al.*, 1988). Thermophilic treatment could achieve 83–92 per cent of COD removal for loading rates of 12–38 kg COD m^{-3} d^{-1} (Yamaguchi *et al.*, 1990).

6.2.5   Sulphite Pulping Effluent

In contrast to the Kraft process, the sulphite process solubilizes lignins as lignosulphonic acids at elevated temperature with an acidic solution of calcium, magnesium or sodium sulphite (Kringstad and Lindström, 1984). The chemicals in the sulphite spent liquors are not commonly recovered in contrast to Kraft liquors (Rintala and Puhakka, 1994). Sulphite pulping effluent toxicity is believed to be caused by phenolic compounds. Additionally the effluents are deficient in nutrients (Table 6.1). In fact, only 20 per cent of the effluent COD is anaerobically biodegradable. Ninety-five per cent of the biodegradable COD was effectively removed when the sulphite spent liquor was diluted by 60 per cent with water (Chave *et al.*, 1988).

The sulphite evaporator condensate (SEC) obtained after chemical recovery is composed principally of acetic acid (20–90 per cent), methanol (6–23 per cent) and furfural (4–15 per cent), with smaller concentrations of formaldehyde, formic acid, acetaldehyde and methylglyoxal (Aivasidis, 1985; Ferguson and Benjamin, 1985; Gunnarsson and Rosén, 1985). Toxicity in SEC is caused by sulphur compounds, mainly sulphite which is often present in high concentrations. Many anaerobic degradability studies have been performed with SEC at the laboratory level and pilot-scale and have shown removal efficiencies of 50–93 per cent of the COD and 84–97 per cent of the BOD at OLR of 1–16 kg COD m^{-3} d^{-1} with different types

Table 6.1   Nitrogen and phosphorous balances in various pulp mill wastewaters

Effluent	Nitrogen (% of BOD)	Phosphorus (% of BOD)	References
Debarking	1.9–3.0	0.3–0.5	Virkola and Honkanen (1985)
Kraft pulping	1.3	0.3	Salkinoja-Salonen *et al.* (1985)
KEC	1.5–6.0	0.01–0.09	Norrman and Narbuvold (1984); Pipyn *et al.* (1987); Lee *et al.* (1989)
Sulphite pulping	1.1	0.1	Salkinoja-Salonen *et al.* (1985)
NSSC spent liquor	0.4–1.0	0.06–0.08	Cocci *et al.* (1985); Lee *et al.* (1989)
MP	2.9	0.7	Salkinoja-Salonen *et al.* (1985)
TMP	0.4–1.1	0.08–0.4	Frostell (1984b); Lee *et al.* (1989)
Paper and paperboard	0.5–2.2	0.09–0.24	Salkinoja-Salonen *et al.* (1985) Luonsi *et al.* (1986); Rintala and Vuoriranta (1988)
Minimal ratio	2	0.4	Pohland (1992)

of reactors (Frostell, 1984a; Ferguson and Benjamin, 1985; Gunnarsson and Rosén, 1985; Kroiss *et al.*, 1985; Salkinoja-Salonen *et al.*, 1985). Exceptional performance was obtained with a fixed bed loop reactor working at 100 and 206 kg COD m^{-3} d^{-1} with a COD removal of 78–84 per cent (Aivasidis, 1985).

6.2.6 Semi-chemical Pulping Effluent

In the neutral sulphite semi-chemical (NSSC) pulping process, the wood chips are cooked with a neutral sodium sulphite solution, combined with a mechanical process. The NSSC spent liquor produced has been shown to be anaerobically biodegradable when combined with other effluents of the mill (Hall *et al.*, 1986; Wilson *et al.*, 1987). When not diluted, the spent liquor is inhibitory to methanogenic activity due to the presence of tannins (Habets *et al.*, 1988). Lack of trace elements and nutrients also can cause low methanogenic activity (Table 6.1). Evaporation of the NSSC spent liquor for chemical recovery also generates a condensate rich in BOD which is mainly composed of acetate. This condensate was efficiently biodegraded (70 per cent BOD removal) at a low OLR of 1.1 kg COD m^{-3} d^{-1} (Cocci *et al.*, 1985).

6.2.7 Non-chlorine Bleaching Effluent

Bleaching of MP and TMP pulps is usually performed with dithionite, dithionate or alkaline hydrogen peroxide (H$_2$O$_2$). Recently, Kraft pulp bleaching has been carried out experimentally in a chlorine-free process (Ristolainen *et al.*, 1996). Dithionite bleaching produces a low BOD effluent (McCubbin, 1984) while alkaline peroxide bleaching effluent contains more biodegradable organics, consisting of 60 per cent carbohydrates and 40 per cent acids (acetic, formic) and methanol (Järvinen *et al.*, 1980). Driessen and Wasenius (1994) treated peroxide bleached TMP effluent with a pilot-scale UASB reactor, obtaining 55–60 per cent COD removal at an OLR of 20 kg COD m^{-3} d^{-1}. However, an acidogenic pretreatment was necessary to remove H$_2$O$_2$ which was inhibitory to the methanogenesis. Multi-stage treatment systems to remove the H$_2$O$_2$ toxicity were also tested to treat CTMP effluent and COD removals of up to 60 per cent at OLRs of 4–6 kg COD m^{-3} d^{-1} were obtained (Anderson *et al.*, 1987).

6.2.8 Chlorine Bleaching Effluent

Chlorine bleaching is performed in order to remove the residual 5–10 per cent of lignin in the pulp. Approximately 1 kg of extractives, 20 kg of polysaccharides, and 50 kg of lignin are dissolved from 1 ton of softwood pulp. Chlorine can react with all of these organic wastes to produce structurally diverse organochlorine compounds, from high MW chlorolignins to simple chloroaromatic monomers (Murray and Richardson, 1993). The easily degradable organic fraction of the bleaching effluent consists mainly of methanol and hemicelluloses. The bleaching effluent represents an important fraction (50–60 per cent) of the total BOD load from a mill (Virkola and Honkanen, 1985). Anaerobic toxicity of chlorine bleaching effluent was mainly attributed to its relatively high content of low MW chlorinated organics and wood extractives such as resin acids, volatile terpenes, triterpenes and hydroxystilbene (Rintala and Puhakka, 1994; Sierra-Alvarez *et al.*, 1994). The effluent can nevertheless be treated after dilution with extraction stage effluents (Ferguson and Dalentoft, 1991; La Fond and Ferguson, 1991) or condensates (Qiu *et al.*, 1988; Welander *et al.*, 1988). Long-term adaptation of the anaerobic biomass was also beneficial (Parker *et al.*, 1993). As reviewed by Rintala and Puhakka (1994), many studies have been conducted on Kraft bleach effluent (KBE) anaerobic treatment, showing 20–50 per cent COD removal. Non-diluted KBE were treated anaerobically with a COD removal performance never exceeding 40 per cent (Salkinoja-Salonen *et al.*, 1985; Poggi-Varaldo *et al.*, 1994).

6.2.9 *Paper Making Effluent*

The main dissolved components of paper making effluents (white water) are easily biodegradable. They include cellulose, hemicellulose and starch. In addition to these carbohydrates, the paperboard and corrugated medium wastewater also contains a significant fraction of lignin and extractives. Additives to the paper making (alum, biocides, latex, alkyl ketene dimers) are not inhibitory to the methanogens at the residual concentrations usually found in the effluents (Jopson *et al.*, 1986).

These effluents are often used at full-scale plants to dilute other effluents to non-toxic concentrations (Rintala and Puhakka, 1994). Full-scale treatment of a paper machine effluent could reach a COD removal of over 80 per cent at OLR up to 28 kg COD m^{-3} d^{-1} (Velasco *et al.*, 1986). COD removal of a paperboard mill effluent was 60 per cent at OLRs of 5–11 kg COD m^{-3} d^{-1} (Eroglu *et al.*, 1994) while that of a corrugated medium effluent was 60–80 per cent despite a low OLR of 2.5 kg COD m^{-3} d^{-1} (Norrman and Narbuvold, 1984).

6.3 Anaerobic Degradation and Toxicity of the Different Groups of Chemical Constituents of Pulp and Paper Effluents

As discussed above, the wastewaters of the forest industry contain a large variety of compounds, from high to low MW, and both toxic and non-toxic to anaerobic bacteria. This section provides an overview of the various classes of compounds commonly encountered in the effluents, in terms of their anaerobic toxicity and their potential for anaerobic degradation. Some compounds are more resistant or totally recalcitrant to anaerobic biodegradation, while others are readily anaerobically biodegradable. The high MW carbohydrates, such as cellulose and hemicellulose, are easily hydrolyzed by exo-enzymes that are currently secreted by fermentative bacteria (the primary fermenters of the anaerobic consortium). The monosaccharides (e.g., glucose, arabinose, xylose, mannose, galactose, methylglucuronic acid), which are either already present in the pulping effluent or produced by the above enzyme hydrolysis during the water treatment, are fermented by the acidogenic bacteria into

VFAs, CO_2 and H_2. Fermentative bacteria also easily degrade the non-volatile monocarboxylic (e.g., glycolic, lactic, glyceric and deoxytetronic) and dicarboxylic (e.g., oxalic, malonic, succinic, tartronic and malic) acids, present in the bleaching effluents, into acetate, or a mixture of acetate, propionate and/or butyrate, plus H_2 and CO_2. Propionate and longer-chain VFAs, ethanol, and some aromatic compounds (e.g., benzoate) are degraded into acetate, CO_2 and H_2, by the obligate H_2-producing acetogens (OHPA), another important intermediate group in the anaerobic consortium (McInerney and Byrant, 1980). Unsaturated LCFAs, after hydrogenation, and saturated LCFAs are β-oxidized by OHPAs into acetate and H_2 (plus CO_2 with C-odd acids) (McInerney, 1988). In sulphate-rich environments, VFAs and LCFAs can also de degraded by sulphate reducing bacteria (SRB) with H_2S instead of H_2 as the reduced end-product (Colleran *et al.*, 1995). Furfural ($C_5H_4O_2$), a heterocyclic aromatic compound present in SEC at concentrations ranging from 10 to 1280 mg L^{-1}, is mineralized into CH_4 and CO_2 by consortia including SRBs, hydrogenotrophic and aceticlastic methanogens. In the absence of sulphate, the reducing equivalents may be scavenged by hydrogenotrophic methanogens via the interspecies transfer of H_2 (Sahm *et al.*, 1992). Methanol may be reduced into CH_4 by some aceticlastic methanogens (e.g. *Methanosarcina* spp.) or into acetate and sometimes butyrate by a minor group referred to as the H_2-consuming acetogens. Formic acid is reduced to CH_4 by the hydrogenotrophic methanogens while acetate is cleaved to CH_4 and CO_2 by the aceticlastic methanogens (McInerney and Bryant, 1980).

6.3.1 *Lignin and Lignin-derived Compounds*

The distribution of dissolved lignin fragments in pulp and paper effluents covers a broad range of MWs, from 100 to several tens of thousands of daltons. Only lignin monomers and oligomers (up to three units, i.e. MW approx. 600 daltons) are anaerobically mineralized, whereas higher MW lignin is recalcitrant to anaerobic degradation (Colberg, 1988; Field *et al.*, 1988). The share of polymeric and oligomeric lignin derivatives in the wastewater COD mainly determines the degree of its recalcitrance towards anaerobic treatment (Sierra-Alvarez and Lettinga, 1991b).

Fermentative and acetogenic bacteria, or SRBs, are required for the conversion of the aromatic compounds to the methane precursors, acetate, H_2 and CO_2 (Chen *et al.*, 1988; Colberg, 1988; Häggblom and Young, 1995). The anaerobic consortia convert the oligomers first into intermediate aromatic monomers by cleaving the β-aryl-ether bond, the most common intermonomeric linkage in lignin (Colberg and Young, 1985). Lignin-derived monomeric acids (e.g., coumaric, caffeic, ferulic, cinnamic, syringic and vanillic), and phenolics (e.g., guaiacol, cresol, resorcinol, catechol and hydroquinone) are degraded through different steps which may include decarboxylation, hydroxylation, dehydroxylation, demethylation, and/or demethoxylation reactions (Grbic-Galic, 1985; Berry *et al.*, 1987). Then, degradation of the final aromatic intermediates, essentially phenol or benzoate, proceeds through complete reduction of the aromatic ring into a substituted cyclohexane. Ring fission yields aliphatic acids that are β-oxidized into acetate, H_2 and CO_2 (Colberg, 1988). The reduction of the aromatic ring is exergonic; the ring fission and further β-oxidation are endergonic at standard conditions. The reaction is thus feasible if the partial pressure of H_2 is kept low by methanogens (Holliger *et al.*, 1988).

MW was found to be an important factor determining the inhibitory effect of lignin derivatives towards methanogenic bacteria. The MW fraction over 10 000 daltons is almost completely free of anaerobic toxicity. In contrast, lignin monomeric derivatives are toxic. However the low MW lignin derivatives differ considerably in their inhibitory potential, which is essentially related to the functional groups substituted on the aromatic ring. The anaerobic toxicity decreases in the following substituent order: aldehyde > apolar alkyl > alcohol > carboxylic (Sierra-Alvarez *et al.*, 1991; Sierra-Alvarez and Lettinga, 1991b).

6.3.2 Terpenes

Terpenoid hydrocarbons (terpinene, pinene, limonene, *p*-cymene, squalene) are commonly found in pulp and paper wastewaters, namely in the condensates and the bleaching effluents. They are recalcitrant to anaerobic biodegradation, even after prolonged periods of incubation. The initial hydroxylation of the unsubstituted hydrocarbons is impeded in the absence of oxygen, due to the lack of polar substitution (Schink, 1985). Monoterpenes cause methanogenic inhibition at low concentrations (50% IC_{MA} 40–350 mg L^{-1}). Squalene, although highly hydrophobic, is not inhibitory, even at high concentrations (1300 mg L^{-1}) (Sierra-Alvarez and Lettinga, 1990).

6.3.3 Resin Acids

Resin acids are monocarboxylic tricyclic acids found in pulp and paper mill effluents in concentrations commonly ranging from 2 to 40 mg L^{-1} (Kennedy *et al.*, 1992; Liu *et al.*, 1993). Concentrations can however reach up to several hundreds of mg per litre in more concentrated waters of pulping processes such as CTMP (McFarlane and O'Kelly, 1988; Habets and de Vegt, 1991) because they may form micelles in the liquid phase and also adsorb strongly onto the fine fibres of the wastewaters (Richardson *et al.*, 1991; Hall and Liver, 1996). There are eight resin acids that are commonly monitored in pulp mill effluents: dehydroabietic, abietic, palustric, isopimaric, levopimaric, pimaric, neoabietic and sandaracopimaric (McFarlane and Clark, 1988; Qiu *et al.*, 1988; Bisaillon *et al.*, 1991; Kennedy *et al.*, 1992; Zender *et al.*, 1994).

The literature does not yield a clear consensus on the fate of the resin acids or on their inhibitory effect on anaerobic sludge. In many case studies, anaerobic treatment of resin acid-containing effluents is successful in eliminating COD, although to a lesser extent than with aerobic treatment (Wilson *et al.*, 1987; Winslow, 1989; Schnell *et al.*, 1990; Rintala and Lepisto, 1992). However, many batch studies indicate that resin acids are probably responsible for decreased efficiencies during the anaerobic treatment of pulp and paper effluents (Field *et al.*, 1988; McCarthy *et al.*,

1990; Sierra-Alvarez and Lettinga, 1990; Sierra-Alvarez *et al.*, 1991; Kennedy *et al.*, 1992). The aceticlastic methanogens are the most inhibited group of bacteria, while the activity of acetogens and hydrogenotroph methanogens seem to be less inhibited or even unaffected (Patel *et al.*, 1991; Patoine *et al.*, 1997).

The incomplete removal of resin acids following anaerobic treatment is widely reported: from 44 to 63 per cent in anaerobic lagoons or in UASB reactors (Schnell *et al.*, 1990; Stuthridge *et al.*, 1991). It has not been ascertained whether this removal is achieved by means of biodegradation or by adsorption. In fact, the limited solubilities and the hydrophobic character of resin acids suggests that removal may be governed largely by partitioning onto biomass (Hall and Liver, 1996). At some point of the cumulative resin acid loading, the adsorption and biodegration capacities of the biomass are surpassed, which results in the release of a resin acid concentration in the effluent similar to or higher than that of the influent. A comparison of published studies suggests that the accumulation of resin acids on the sludge has to surpass 100 mg/g volatile suspended solid (VSS) for inhibition to occur. This also explains why highly concentrated effluent would not be inhibitory on a short-term basis, while other diluted effluent would still undergo a delayed inhibition in the long term.

6.3.4 Chlorinated Organic Compounds

The adsorbable organic halogens (AOX), which include all the chlorinated organics which can adsorb onto activated carbon, may range from less than 10 to 100 mg Cl L^{-1} in bleach Kraft waters (Jokela *et al.*, 1993; Parker *et al.*, 1993; Ferguson, 1994). It is commonly thought that the high MW chlorolignins account for about 80 per cent of the AOX of the bleach effluents (Fitzsimons *et al.*, 1990). However, a more recent study has shown that over 65 per cent of AOX were of rather low MW (<1000 daltons) (Jokela *et al.*, 1993). The apparent high MW of chlorolignins is related more to intermolecular associations than to formations with covalent bonds.

The high MW chlorolignins are thought to be generally non-toxic as their large size precludes their penetration or transport across the cell membranes. Chlorinated monoaromatics (e.g. chlorinated phenols, guaiacols, veratroles, syringols, resorcinols and catechols) are highly toxic to methanogens and their toxicity increases with increasing Cl-number as well as apolarity (Sierra-Alvarez and Lettinga, 1991a).

AOX have been removed at levels of 35–65 per cent in anaerobic systems (Lettinga *et al.*, 1991; Parker *et al.*, 1993; Rintala and Puhakka, 1994). The rate and extent of AOX removal decreased with increasing MW (Fitzsimons *et al.*, 1990). For most chloroaromatics, the reductive dehalogenation (the replacement of chlorine by hydrogen) proceeds in a stepwise fashion until all halogens are removed, without it being necessary that the aromatic ring be cleaved (Horowitz *et al.*, 1983). Reductive dehalogenation usually occurs first at the *ortho*-position, next at the *para*-position, then at the *meta*-position (Cozza and Woods, 1992). Under optimum conditions, acclimated consortia can completely dechlorinate the chlorophenols to phenol, which is then mineralized to CH_4 plus CO_2. However in other cases, partial dechlorination is reported, and the accumulation of di- and mono-chlorophenols is observed. Less chlorinated compounds are usually less toxic, less likely to bioaccumulate and more amenable to subsequent aerobic degradation (Woods *et al.*, 1989; Hendriksen and Ahring, 1992). An almost endless list of halogenated aromatics have been reported to be reductively dehalogenated by unidentified bacteria in environmental samples, mixed cultures and enriched consortia (Mohn and Tiedje, 1992). Only a few pure cultures of anaerobic bacteria which dehalogenate chloro-aromatics have been isolated: *Desulfomonile tiedjei* DCB-1 (Shelton and Tiedje, 1984; DeWeerd *et al.*, 1990); *Clostridium* strain DCB-2 (Madsen and Licht, 1992); strain 2CP-1 (Cole *et al.*, 1994); *Desulfitobacterium dehalogenans* JW/IU-DC1 (Utkin *et al.*, 1994); and *Desulfitobacterium* strain PCP-1 (Beaudet *et al.*, 1996). The latter is capable of dechlorination at all positions. Enrichments can be performed as they all, except strain DCB-2, conserve energy from reductive dehalogenation for growth (halorespiration) (Holliger and Schumacher, 1994). This selective advantage may facilitate their exploitation for depollution and remediation.

6.3.5 Sulphur Compounds

Pulp and paper manufacturing uses sulphur in various forms: sulphate or sulphite in chemical pulping; sulphite in CTMP; and dithionite ($S_2O_4{}^{2-}$) or dithionate ($S_2O_6^{2-}$) in bleaching of mechanical pulp. Organic sulphur compounds (methyl mercaptan, methyl sulphide, dimethyl disulphide and difurfuryl disulphide) are present in chemical pulping condensates. The concentrations at which sulphate and sulphite are reported to inhibit methanogenic bacteria vary greatly: 1600–10 000 and 25–1440 mg L^{-1}, respectively (Khan and Trottier, 1978; Eis *et al.*, 1983; Karhadkar *et al.*, 1987; Rinzema and Lettinga, 1988; Puhakka *et al.*, 1990).

Sulphate and dithionate in the pulp and paper effluents, except in the spent liquors, are present at non-inhibitory concentrations (Rintala and Puhakka, 1994). However, sulphate is used as an electron acceptor by SRBs which produces hydrogen sulphide. The non-ionized H_2S is toxic to methanogens and acetogens. Methanogens become inhibited by 50 per cent at a non-ionized H_2S concentration of only 50 mg L^{-1}, and completely at 200–250 mg L^{-1}. Acetogens seem to be similarly sensitive to H_2S while hydrogenotrophic methanogens are less sensitive. Anaerobic treatment of effluents with a COD/SO_4^{2-} ratio in excess of 10, particularly at influent COD concentration exceeding 10 g L^{-1}, is likely to result in H_2S concentrations that are toxic (Pichon *et al.*, 1988; Colleran *et al.*, 1995).

Several methods can be used to alleviate potential H_2S toxicity from anaerobic treatment of pulping effluents: addition of molybdenum (1.0 mM) which specifically inhibits SRBs; sulphide precipitation by iron salts (Stephenson *et al.*, 1993); filtration of the biogas onto steelwool, packed silica, peat moss, or dry activated sludge; or biogas scrubbing within alkaline and/or ferric ion solutions and recirculation of the H_2S-free biogas through the reactor (Lee *et al.*, 1989). Another strategy is to segregate the sulphate reduction and the methanogenesis. The first stage is basically an acidogenic/sulphidogenic anaerobic reactor. The H_2S is stripped out from this reactor by thorough recirculation of the washed biogas (Maree *et al.*, 1987; Särner *et al.*, 1988). Both H_2S-containing biogas and liquid can also be treated biologically in an aerobic reactor immobilizing sulphur-oxidizing bacteria (*Thiothrix*, *Thiobacillus* and *Beggiatoa* spp.) (Buisman *et al.*, 1991; Lanting and Shah, 1992). Typically two-thirds of the sulphides are converted into elemental sulphur, the balance consisting of sulphate and thiosulphate.

6.4 Prospects

The presence of numerous toxicants in forest industry wastewaters is a key factor which has led the industry to opt overwhelming for aerobic effluent treatment in spite of its onerous operating costs. Anaerobic toxicity in wastewaters, however, does not necessarily preclude anaerobic treatment. A variety of techniques can be employed to diminish the inhibition potential, such as dilution and detoxification pretreatments.

Alkaline oxidative polymerization (tannic oligomers) (Field *et al.*, 1990a), precipitation with aluminium, iron or calcium salts (tannic polymers, LCFA, resin acids) (Welander, 1988), destructive oxidation with H_2O_2 or ozone (Field *et al.*, 1990b) and fungal lignolytic enzymes are among the techniques proposed for the pretreatment of the pulp and paper effluents. As chemical oxidation processes are prohibitively expensive, the powerful enzymatic oxidative systems of white-rot fungi deserve attention. Ligninases and manganese peroxidases are able to depolymerize lignins (Livernoche *et al.*, 1981) and to degrade many recalcitrant aromatics (Reddy, 1995). Such fungal pretreatment allowed for increases by up to 45 and 85 per cent in COD and colour removal, respectively, in subsequent anaerobic treatment of bleach effluents or spent liquors (Prasad and Joyce, 1993; Feijoo *et al.*, 1995).

Finally, aerobic/anaerobic coupling biotreatment in a single unit could be considered for some cases. The concept of aerating an anaerobic reactor may seem counterintuitive. However, a limited supply of oxygen to a UASB reactor minimally supports aerobic metabolism in an otherwise anaerobic environment, where most of the substrate carbon is still converted into methane (Guiot *et al.*, 1993, 1995; Shen and Guiot, 1996). The coupled reactor showed better overall performance in resin acid removal and lower amounts were adsorbed onto the biomass (8–12 mg resin acids/g VSS in the coupled reactor, compared with 68 mg in the control UASB) (Stephenson *et al.*, 1996). Supporting both anaerobes for low cost organics removal and aerobes for resin acid tolerance and degradation in the same reactor can extend the effectiveness of either treatment system on its own.

The importance of optimum conditions for anaerobic populations should be re-emphasized. These conditions include: supply of nutrients, inasmuch as inorganic nitrogen and phosphorus are often deficient in pulp and paper waters (Table 6.1), and of trace metals (selenium, iron, cobalt, nickel and molybdenum); and establishment, by long solid retention time and selective stress (Aivasidis, 1985), of a population acclimated to toxicants.

6.5 Conclusions

Pulp and paper mill effluents pose a particular challenge to anaerobic technologies, which would, however, merit renewed consideration if the problems linked to anaerobic toxicity of some compounds could be overcome. Furthermore, traditional anaerobic treatment flowsheets might need to be modified to provide detoxification prior to methane production. Segregation and optimal selective regrouping of effluents; physical, chemical or biological pretreatments; and simultaneous aerobic/anaerobic biotreatments might be avenues to explore. Aerobic post-treatment remains inevitable, for detoxification and/or effluent polishing (Zitomer and Speece, 1993). While the investment cost of a sequential anaerobic/aerobic treatment train remains in the same range as an activated sludge plant alone, the operational cost of the former system would be an order of magnitude lower than the latter, if the energy recovery from biogas is considered as a saving (Habets and Knelissen, 1985; Luonsi *et al.*, 1986; Vochten *et al.*, 1988).

Acknowledgement

The authors are grateful to L. Petti for careful proofreading of the manuscript. NRC Paper No. 34556.

References

AIVASIDIS, A. (1985) Anaerobic treatment of sulfite evaporator condensate in a fixed bed loop reactor. *Wat. Sci. Technol.* **17**, 207–221.

ANDERSON, P.-E., GUNNARSSON, G., OLSSON, R., WELANDER, T. and WIKSTROM, A. (1987) Anaerobic treatment of CTMP effluent. *Pulp Paper Can.* **88**, T223–T236.

BEAUDET, R., VILLEMUR, R., BOUCHARD, B., McSWEEN, G., LÉPINE, F. and BISAILLON, J. G. (1996) Anaerobic dechlorination of pentachlorophenol by *Desulfitobacterium* strain PCP-1 (Abstract Q-172). In: *Proceedings of the 96th General Meeting of the American Society for Microbiology*, New Orleans, Louisiana, 19–23 May, Washington DC: AMS, p. 415.

BERRY, D. F., FRANCIS, A. J. and BOLLAG, J.-M. (1987) Microbial metabolism of homocyclic and heterocyclic aromatic compounds under anaerobic conditions. *Microbiol. Rev.* **51**, 43–59.

BISAILLON, J. G., PERRON, J., PAQUET, M., PAQUETTE, G., LÉPINE, F. and BEAUDET, R. (1991) Effet d'un traitement anaérobie sur la charge organique et la toxicité d'un effluent thermomécanique. *Sci. Techniques Eau* **24**, 137–142.

BUISMAN, C. J. N., LETTINGA, G., PAASSCHENS, C. W. M. and HABETS, L. H. A. (1991) Biotechnological sulphide removal from effluents. *Wat. Sci. Technol.* **24**, 347–356.

CHAVE, E., FUENTES, J. L. and GOMA, G. (1988) Anaerobic treatment of spent sulphite liquors. In: Tilche, A. and Rozzi, A., eds, *Proceedings of the 5th International Symposium on Anaerobic Digestion*, Bologna, Italy, 20–25 May, Bologna: Monduzzi, pp. 635–638.

CHEN, W., OHMIYA, K., SHIMIZU, S. and KAWAKAMI, H. (1988) Isolation and characterization of an anaerobic dehydrodivanillin-degrading bacterium. *Appl. Environ. Microbiol.* **54**, 1254–1257.

COCCI, A. A., BROWN, G. J., LANDINE, R. C. and PRONG, C. F. (1985) Anaerobic-aerobic treatment of NSSC pulp mill effluent – a major biogas energy and pollution abatement method. *Pulp Paper Can.* **86**, T32–T35.

COLBERG, P. J. (1988) Anaerobic microbial degradation of cellulose, lignin, oligolignols, and monoaromatic lignin derivatives. In: Zehnder, A. J. B., ed., *Biology of Anaerobic Microorganisms*, New York: John Wiley & Sons, pp. 333–372.

COLBERG, P. J. and YOUNG, L. Y. (1985) Aromatic and volatile acid intermediates observed during anaerobic metabolism of lignin-derived oligomers. *Appl. Environ. Microbiol.* **49**, 350–358.

COLE, J. R., CASCARELLI, A. L., MOHN, W. W. and TIEDJE, J. M. (1994) Isolation and characterization of a novel bacterium growing via reductive dehalogenation of 2-chlorophenol. *Appl. Environ. Microbiol.* **60**, 3536–3542.

COLLERAN, E., FINNEGAN, S. and LENS, P. (1995) Anaerobic treatment of sulphate-containing waste streams. *Antonie van Leeuwenhoek*, **67**, 29–46.

COZZA, C. L. and WOODS, S. L. (1992) Reductive dechlorination pathways for substituted benzenes: a correlation with electronic properties. *Biodegradation* **2**, 265–278.

DeWEERD, K. A., MANDELCO, L., TANNER, R. S., WOESE, C. R. and SUFLITA, J. M. (1990) *Desulfomonile tiedjei* gen. nov. and sp. nov., a novel anaerobic, dehalogenating, sulfate-reducing bacterium. *Arch. Microbiol.* **154**, 23–30.

DRIESSEN, W. J. B. M. and WASENIUS, C. O. (1994) Combined anaerobic/aerobic treatment of peroxide bleached TMP mill effluent. *Wat. Sci. Technol.* **29**, 381–389.

EIS, B. J., FERGUSON., J. F. and BENJAMIN, M. M. (1983) The fate and effect of bisulfate in anaerobic treatment. *J. Water Pollut. Control Fed.* **55**, 1355–1365.

EROGLU, V., OZTÜRK, I., UBAY, G., DEMIR, I. and KORKURT, E. N. (1994) Feasibility of anaerobic pre-treatment for the effluents from hardboard and laminated board industry. *Wat. Sci. Technol.* **29**, 391–397.

FEIJOO, G., VIDAL, G., MOREIRA, M. T., MÉNDEZ, R. and LEMA, J. M. (1995) Degradation of high molecular weight compounds of Kraft pulp mill effluents by a combined treatment with fungi and bacteria. *Biotechnol. Lett.* **17**, 1261–1266.

FERGUSON, J. F. (1994) Anaerobic and aerobic treatment for AOX removal. *Wat. Sci. Technol.* **29**, 149–162.

FERGUSON, J. F. and BENJAMIN, M. M. (1985) Studies of anaerobic treatment of sulfite process wastes. *Wat. Sci. Technol.* **17**, 113–121.

FERGUSON, J. F. and DALENTOFT, E. (1991) Investigation of anaerobic removal and degradation of organic chlorine from Kraft bleaching wastewaters. *Wat. Sci. Technol.* **24**, 241–250.

FIELD, J. A., LEYENDECKERS, M. J. H., SIERRA-ALVAREZ, R., LETTINGA, G. and HABETS, L. H. A. (1988) The methanogenic toxicity of bark tannins and the anaerobic biodegradability of water soluble bark matter. *Wat. Sci. Technol.* **20**, 219–240.

FIELD, J. A., LETTINGA, G. and HABETS, L. H. A. (1990a) Oxidative detoxification of aqueous bark extracts. Part I: Autoxidation. *J. Chem. Tech. Biotechnol.* **49**, 15–33.

FIELD, J. A., SIERRA-ALVAREZ, R., LETTINGA, G. and HABETS, L. H. A. (1990b) Oxidative detoxification of aqueous bark extracts. Part II: Alternative methods. *J. Chem. Tech. Biotechnol.* **49**, 35–53.

FITZSIMONS, R., EK, M. and ERIKSSON, K. E. L. (1990) Anaerobic dechlorination/degradation of chlorinated organic compounds of different molecular masses in bleach plant effluents. *Environ. Sci. Technol.* **24**, 1744–1748.

FROSTELL, B. (1984a) Anaerobic/aerobic pilot-scale treatment of a sulphite evaporator condensate. *Pulp Paper Can.* **85**, T57–T62.

FROSTELL, B. (1984b) Full scale anaerobic treatment of a pulp and paper industry wastewater. In Bell, J. M., ed., *Proceedings of the 39th Industrial Waste Conference*, Purdue University, West Lafayette, IN, 8–10, May, Boston: Butterworth, pp. 687–696.

GRBIC-GALIC, D. (1985) Fermentative and oxidative transformation of ferulate by a facultatively anaerobic bacterium isolated from sewage sludge. *Appl. Environ. Microbiol.* **50**, 1052–1057.

GUIOT, S. R., FRIGON, J. C., DARRAH, B., LANDRY, M. F. and MACARIE, H. (1993) Coupled aerobic and anaerobic treatment of toxic wastewater. *Med. Fac. Landbouww. Univ. Gent* **58**, 1761–1769.

GUIOT, S. R., KUANG, X., BEAULIEU, C., CORRIVEAU, A. and HAWARI, J. A. (1995) Anaerobic and aerobic/anaerobic treatment for tetrachloroethylene (PCE). In: Hinchee, R., Leeson, A. and Semprini, L., eds, *Bioremediation of Chlorinated Solvents*, Vol. 3(4), Columbus, OH: Battelle Press, pp. 191–198.

GUNNARSSON, L. and ROSÉN, B. (1985) Anaerobic treatment of sulphite evaporator condensate in a pilot plant of novel design. *Wat. Sci. Technol.* **17**, 271–279.

HABETS, L. H. A. and DE VEGT, A. L. (1991) Anaerobic treatment of bleached TMP and CTMP effluent in the Biopaq system. *Water Sci. Technol.* **24**, 331–345.

HABETS, L. H. A. and KNELISSEN, J. H. (1985) Application of the UASB-reactor for anaerobic treatment of paper and board mill effluent. *Wat. Sci. Technol.* **17**, 61–76.

HABETS, L. H. A., TIELBAARD, M. H., FERGUSON, A. M. D., PRONG, C. F. and CHMELAUSKAS, A. J. (1988) On site high rate UASB anaerobic demonstration plant treatment of NSSC wastewater. *Wat. Sci. Technol.* **20**, 87–97.

HÄGGBLOM, M. M. and YOUNG, L. Y. (1995) Anaerobic degradation of halogenated phenols by sulfate-reducing consortia. *Appl. Environ. Microbiol.* **61**, 1546–1550.

HALL, E. R. and LIVER, S. F. (1996) Interactions of resin acids with aerobic and anaerobic biomass – II. Partitioning on biosolids. *Water Res.* **30**, 672–678.

HALL, E. R., ROBSON, P. D., PRONG, C. F. and CHMELAUSKAS, A. J. (1986) Anaerobic pilot studies of NSSC wastewater treatment. *Tappi J.* **69**, 79–84.

HENDRIKSEN, H. V. and AHRING, B. K. (1992) Metabolism and kinetics of pentachlorophenol transformation in anaerobic granular sludge. *Appl. Microbiol. Biotechnol.* **37**, 662–666.

HOLLIGER, C. and SCHUMACHER, W. (1994) Reductive dehalogenation as a respiratory process. *Ant. van Leeuwenhoek* **66**, 239–246.

HOLLIGER, C., STAMS, A. J. M. and ZEHNDER, A. J. B. (1988) Anaerobic degradation of recalcitrant compounds. In: Hall, E. R. and Hobson, P. N., eds, *Proceedings of the 5th International Symposium on Anaerobic Digestion*, Bologna, Italy, 22–26 May, Oxford: Pergamon Press, pp. 211–224.

HOROWITZ, A., SUFLITA, J. M. and TIEDJE, J. M. (1983) Reductive dehalogenations of halobenzoates by anaerobic lake sediment microorganisms. *Appl. Environ. Microbiol.* **45**, 1459–1465.

JÄRVINEN, R., VAHTILA, M., MANNSTRÖM, B. and SUNDHOLM, J. (1980) Reduced environmental load in TMP. *Pulp Paper Can.* **81**, 39–43.

JOKELA, J. K., LAINE, M., EK, M. and SALKINOJA-SALONEN, M. S. (1993) Effect of biological treatment on halogenated organics in bleached Kraft pulp mill effluents studied by molecular weight distribution analysis. *Environ. Sci. Technol.* **27**, 547–557.

JOPSON, R. N., WEBB, L. J. and BOARD, N. P. (1986) Effect of papermaking additives on anaerobic effluent treatment. In: *Proceedings of the PIRA Paper and Board Division Seminar on Cost-effective Treatment of Papermill Effluents Using Anaerobic Technologies*, January 14–15, Leatherhead, UK: PIRA, pp. 1–11.

KARHADKAR, P. P., AUDIC, J. M., FAUP, G. M. and KHANNA, P. (1987) Sulfide and sulfate inhibition of methanogenesis. *Wat. Res.* **21**, 1061–1066.

KENNEDY, K. J., McCARTHY, P. J. and DROSTE, R. L. (1992) Batch and continuous anaerobic toxicity of resin acids from chemithermo-mechanical pulp wastewater. *J. Ferment. Bioeng.* **73**, 206–212.

KHAN, A. W. and TROTTIER, T. M. (1978) Effect of sulfur compounds on anaerobic degradation of cellulose to methane by mixed cultures obtained from sewage sludge. *Appl. Environ. Microbiol.* **35**, 1027–1034.

KRINGSTAD, K. P. and LINDSTRÖM, K. (1984) Spent liquors from pulp bleaching. *Environ. Sci. Technol.* **18**, 236A–248A.

KROISS, H., SVARDAL, K. and FLECKSEDER, H. (1985) Anaerobic treatment of sulfite pulp mill effluents. *Wat. Sci. Technol.* **17**, 145–156.

LA FOND, R. A. and FERGUSON, J. F. (1991) Anaerobic and aerobic biological treatment processes for removal of chlorinated organics from kraft bleaching wastes. In: *Proceedings of the TAPPI Environmental Conference*, San Antonio, TX, 7–10 April, Atlanta: TAPPI Press, pp. 797–812.

LANTING, J. and SHAH, A. S. (1992) Biological removal of hydrogen sulfide from biogas.

In: Wukasch, R. F., ed., *Proceedings of the 46th Purdue Industrial Waste Conference*, Purdue University, West-Lafayette, IN, Chelsea, MI: Lewis Publishers, pp. 709–714.

LEE, J. W. J. PETERSON, D. L. and STICKNEY, A. R. (1989) Anaerobic treatment of pulp- and paper-mill wastewaters. *Environ. Prog.* **8**, 73–87.

LETTINGA, G., FIELD, J. A., SIERRA-ALVAREZ, R., VAN LIER, J. B. and RINTALA, J. (1991) Future perspectives for the anaerobic treatment of forest industry wastewaters. *Wat. Sci. Technol.* **24**, 91–102.

LIU, H. W., LO, S. N. and LAVALLÉE, H. C. (1993) Study of the performance and kinetics of aerobic biological treatment of a CTMP effluent. *Pulp Paper Can.* **94**, T502–T507.

LIVERNOCHE, D., JURASEK, L., DESROCHERS, M. and VELIKY, I. A. (1981) Decolorization of a kraft mill effluent with fungal mycelium immobilized in calcium alginate gel. *Biotechnol. Lett.* **3**, 701–706.

LUONSI, A., VUORIRANTA, P. and HYNNINEN, P. (1986) Reduction of pulp and paper industry effluent loading. *Wat. Sci. Technol.* **18**, 109–125.

MACLEAN, B., DE VEGT, A. and VAN DRIEL, E. (1990) Full-scale anaerobic/aerobic treatment of TMP/BCTMP effluent at Quesnel River Pulp. In: *Proceedings of the Tappi Environmental Conference*, Seattle, WA, 9–10, April, Atlanta: TAPPI Press, pp. 647–661.

MADSEN, T. and LICHT, D. (1992) Isolation and characterization of an anaerobic chlorophenol-transforming bacterium. *Appl. Environ. Microbiol.* **58**, 2874–2878.

MAREE, J. P., GERBER, A. and HILL, E. (1987) An integrated process for biological treatment of sulfate-containing industrial effluents. *J. Wat. Pollut. Cont. Fed.* **59**, 1069–1074.

MCCARTHY, P. J., KENNEDY, K. J. and DROSTE, R. L. (1990) Role of resin acids in the anaerobic toxicity of chemithermo-mechanical pulp wastewater. *Wat. Res.* **24**, 1401–1405.

MCCUBBIN, N. (1984) Techniques de base de l'industrie des pâtes et papiers, et ses pratiques de protection environnementale, Report No. SPE 6-EP-83-1F, Environment Canada.

MCFARLANE, P. and O'KELLY, N. (1988) Treatment of CTMP effluents in New Zealand. In: *Proceedings of the Technology Transfer Workshop on Treatment of CTMP Effluents*, Burlington, Ontario, Canada, July, Burlington, Ontario: Wastewater Technology Center.

MCFARLANE, P. N. and CLARK, T. A. (1988) Metabolism of resin acids in anaerobic systems. *Wat. Sci. Technol.* **20**, 273–276.

MCINERNEY, M. J. (1988) Anaerobic hydrolysis and fermentation of fats and proteins. In: Zehnder, A. J. B., ed., *Biology of Anaerobic Microorganisms*, New York: John Wiley & Sons, pp. 373–415.

MCINERNEY, M. J. and BRYANT, M. P. (1980) Basic principles of bioconversions in anaerobic digestion and methanogenesis. In: Sofer, S. and Zaborsky, O., eds, *Biomass Conversion Processes for Energy and Fuels*, New York: Plenum, pp. 277–296.

MOHN, W. W. and TIEDJE, J. M. (1992) Microbial reductive dehalogenation. *Microbiol. Rev.* **56**, 482–507.

MURRAY, W. D. and RICHARDSON, M. (1993) Development of biological and process technologies for the reduction and degradation of pulp mill wastes that pose a threat to human health. *Crit. Rev. Environ. Sci. Technol.* **23**, 157–194.

NORRMAN, J. and NARBUVOLD, R. (1984) Anaerobic treatability of waste waters from pulp and paper industries. *Biotech. Adv.* **2** 329–345.

PARKER, W. J., HALL, E. R. and FARQUHAR, G. J. (1993) Assessment of design and operating parameters for high rate anaerobic dechlorination of segregated Kraft mill bleach plant effluents. *Wat. Environ. Res.* **65**, 264–270.

PATEL, G. B., AGNEW, B. J. and DICAIRE, C. J. (1991) Inhibition of pure cultures of methanogens by benzene ring compounds. *Appl. Environ. Microbiol.* **57**, 2969–2974.

PATOINE, A., MANUEL, M. F., HAWARI, J. A. and GUIOT, S. R. (1997) Toxicity reduction and removal of dehydroabietic and abietic acids in a continuous anaerobic reactor. *Wat. Res.* **31**, 825–831.

PICHON, M., ROUGHER, J. and JUNET, E. (1988) Anaerobic treatment of sulphur-containing effluents. *Wat. Sci. Technol.* **20**, 133–141.

PIPYN, P., EECKHAUT, M., OMBREGT, J. P., TERAOKA, H., KISHIMOTO, H., SAKAMOTO, H., KURODA, J. and MASSUE, Y. (1987) Anaerobic treatment of Kraft pulp mill condensate. In: *Proceedings of the TAPPI Environmental Conference*, Portland, OR, 27–29 April, Atlanta: TAPPI Press, pp. 173–178.

POGGI-VARALDO, H. M., RINDERKNECHT, N. and RIOS-LEAL, E. (1994) Anaerobic pretreatment and detoxification of pulp mill wastewaters. In: Delisle, C. E. and Bouchard, M. A., eds, *Compte-rendus du 7ᵉ Symposium International sur le Traitement des Eaux Usées*, Montréal, Canada, 15–17 November, Montréal: Université de Montréal, pp. 46–63.

POHLAND, F. G. (1992) Anaerobic treatment: fundamental concepts, applications, and new horizons. In: Malina, J. F. J. and Pohland, F. G., eds, *Design of Anaerobic Processes for the Treatment of Industrial and Municipal Wastes*, Vol. 7, Lancaster, PA: Technomic Publishing Co., pp. 1–40.

PRASAD, D. Y. and JOYCE, T. W. (1993) Sequential treatment of E1 stage kraft bleach plant effluent. *Bioresource Technol.* **44**, 141–147.

PUHAKKA, J. A., SALKINOJA-SALONEN, M., FERGUSON, J. F. and BENJAMIN, M. M. (1990) Carbon flow in acetotrophic enrichment cultures from pulp mill effluent treatment. *Wat. Res.* **24**, 515–519.

QIU, R., FERGUSON, J. F. and BENJAMIN, M. M. (1988) Sequential anaerobic and aerobic treatment of Kraft pulping wastes. *Wat. Sci. Technol.* **20**, 107–120.

REDDY, C. A. (1995) The potential for white-rot fungi in the treatment of pollutants. *Curr. Opin. Biotechnol.* **6**, 320–328.

RICHARDSON, D. A., ANDRAS, E. and KENNEDY, K. J. (1991) Anaerobic toxicity of fines in chemi-thermomechanical pulp wastewaters: a batch assay-reactor study comparison. *Wat. Sci. Technol.* **24**, 103–112.

RINTALA, J. A. and LEPISTO, S. S. (1992) Anaerobic treatment of thermomechanical pulping whitewater at 35–70°C. *Wat. Res.* **26**, 1297–1305.

RINTALA, J. A. and PUHAKKA, J. A. (1994) Anaerobic treatment in pulp- and paper-mill waste management: a review. *Biores. Technol.* **47**, 1–18.

RINTALA, J. and VUORIRANTA, P. (1988) Anaerobic-aerobic treatment of thermomechanical pulping effluents. *TAPPI J.* **71**, 201–207.

RINZEMA, A. and LETTINGA, G. (1988) Anaerobic treatment of sulfate containing wastewater. In: Wise, D. L., ed., *Biotreatment Systems*, Vol. 3, Boca Raton: CRC Press, pp. 65–109.

RISTOLAINEN, M., ALÉN, R. and KNUUTINEN, J. (1996) Characterization of TCF effluents from Kraft pulp bleaching. 1. Fractionation of hardwood lignin-derived material by GPC and UF. *Holzforschung* **50**, 91–96.

SAHM, H., NEY, U. and SCHOBERTH, S. M. (1992) Anaerobic treatment of wastewater from the pulp and paper industry. In: Ladisch, M. R. and Bose, A., eds, *Proceedings of the 9th International Biotechnology Symposium and Exhibition*, Crystal City, VA, USA, 16–21 August, Washington DC: American Chemical Society, pp. 431–434.

SALKINOJA-SALONEN, M. S., VALO, R., APAJALAHTI, J., HAKULINEN, R., SILAKOSKI, L. and JAAKOLA, T. (1984) Biodegradation of chlorophenolic compounds in wastes from wood-processing industry. In: Klug, M. J. and Reddy, C. A., eds, *Current Perspectives in Microbial Ecology*, Washington DC: American Society for Microbiology, pp. 668–676.

SALKINOJA-SALONEN, M. S., HAKULINEN, R., SILAKOSKI, L., APAJALAHTI, J., BACKSTRÖM, V. and NURMIAHO-LASSILA E.-L. (1985) Fluidized bed technology in the anaerobic treatment of forest industry wastewaters. *Wat. Sci. Technol.* **17**, 77–88.

SÄRNER, E., HULTMAN, B. G. and BERGLUMD, A. E. (1988) Anaerobic treatment using new technology for controlling H$_2$S toxicity. *Tappi J.* **71**, 41–45.

SCHINK, B. (1985) Degradation of unsaturated hydrocarbons by methanogenic enrichment cultures. *FEMS Microbiol. Ecol.* **31**, 69–77.

SCHNELL, A., DORICA, J., HO, C., ASHIKAWA, M., MUNNOCH, G. and HALL, E. R. (1990) Anaerobic and aerobic pilot-scale effluent detoxification studies at an integrated newsprint mill. *Pulp Paper Can.* **91**, 75–80.

SHELTON, D. R. and TIEDJE, J. M. (1984) Isolation and partial characterization of bacteria in an anaerobic consortium that mineralizes 3-chlorobenzoic acid. *Appl. Environ. Microbiol.* **48**, 840–848.

SHEN, C. F. and GUIOT, S. R. (1996) Long-term impact of dissolved oxygen on the activity of anaerobic granular sludge. *Biotech. Bioeng.* **49**, 611–620.

SIERRA-ALVAREZ, R. and LETTINGA, G. (1990) The methanogenic toxicity of wood resin constituents. *Biol. Wastes* **33**, 211–226.

SIERRA-ALVAREZ, R. and LETTINGA, G. (1991a) The effect of aromatic structure on the inhibition of acetoclastic methanogensis in granular sludge. *Appl. Microbiol. Biotechnol.* **34**, 544–550.

SIERRA-ALVAREZ, R. and LETTINGA, G. (1991b) The methanogenic toxicity of wastewater lignins and lignin related compounds. *J. Chem. Tech. Biotechnol.* **50**, 443–455.

SIERRA-ALVAREZ, R., KORTEKAAS, S., VAN EEKERT, M. and LETTINGA, G. (1991) The anaerobic biodegradability and methanogenic toxicity of pulping wastewaters. *Wat. Sci. Technol.* **24**, 113–125.

SIERRA-ALVAREZ, R., FIELD, J. A., KORTEKAAS, S. and LETTINGA, G. (1994) Overview of the anaerobic toxicity caused by organic forest industry wastewater pollutants. *Wat. Sci. Technol.* **29**, 353–364.

STEPHENSON, R. J., BRANION, R. M. R. and PINDER, K. L. (1993) Sulphur management strategies in anaerobic treatment of a BCTMP/TMP effluent. *Wat. Poll. Res. J. Can.* **28**, 635–664.

STEPHENSON, R. J., PATOINE, A., FRIGON, J. C. and GUIOT, S. R. (1996) Coupled aerobic/anaerobic treatment of resin acids. In: Hall, E., ed., *Proceedings of the 5th IAWQ Symposium on Forest Industry Wastewaters*, Vancouver, BC, Canada, 10–13 June.

STUTHRIDGE, T. R., CAMPIN, D. N., LANGDON, A. G., MACKIE, K. L., McFARLANE, P. N. and WILKINS, A. L. (1991) Treatability of bleached kraft pulp and paper mill wastewaters in a New-Zealand aerated lagoon treatment system. *Wat. Sci. Technol.* **24**, 309–317.

UTKIN, I., WOESE, C. and WIEGEL, J. (1994) Isolation and characterization of *Desulfitobacterium dehalogenans* gen. nov., sp. nov., an anaerobic bacterium which reductively dechlorinates chlorophenolic compounds. *Int. J. Syst. Bacteriol.* **44**, 612–619.

VELASCO, A. A., SÄRNER, E. and BONKOSKI, W. A. (1986) Full scale anaerobic-aerobic biological treatment of a semichemical pulping waste-water. In: *Proceedings of the TAPPI Environmental Conference*, New Orleans, LA, 21–23 April, Atlanta: TAPPI Press, pp. 197–206.

VIRKOLA, N. E. and HONKANEN, K. (1985) Wastewater characteristics. *Wat. Sci. Technol.* **17**, 1–28.

VOCHTEN, P., SCHOWANEK, S., SCHOWANEK, W. and VERSTRAETE, W. (1988) Aerobic vs anaerobic wastewater treatment. In: Hall, E. R. and Hobson, P. N., eds, *Proceedings of the 5th Int. Symposium on Anaerobic Digestion*, Bologna, Italy, 20–25 May, Oxford: Pergamon Press, pp. 91–104.

WALDEN, C. C. and HOWARD, T. E. (1981) Toxicity of pulp and paper mill effluents – a review. *Pulp Paper Can.* **82**, T143–T147.

WELANDER, T. (1988) An anaerobic process for treatment of CTMP effluent. *Wat. Sci. Technol.* **20**, 143–147.

WELANDER, T., MALMQVIST, A. and YU, P. (1988) Anaerobic treatment of toxic forest industry wastewaters. In: Hall, E. R. and Hobson, P. N., eds, *Proceedings of the 5th*

International Symposium on Anaerobic Digestion, Bologna, Italy, 20–25 May, Oxford: Pergamon Press, pp. 267–274.

WILSON, R. W., MURPHY, K. L. and FRENETTE, E. G. (1987) Aerobic and anaerobic treatment of NSSC and CTMP effluent. *Pulp Paper Can.* **88**, 31–35.

WINSLOW, F. B. (1989) Operating experiences with an anaerobic-aerobic treatment system. In: *Proceedings of the Tappi Environmental Conference*, Orlando, FL, 17–19 April, Atlanta: TAPPI Press, pp. 465–471.

WOODS, S. L., FERGUSON, J. F. and BENJAMIN, M. M. (1989) Characterization of chlorophenol and chloromethoxybenzene biodegradation during anaerobic treatment. *Environ. Sci. Technol.* **23**, 62–68.

YAMAGUCHI, M., TANIMOTO, Y., OGAWA, S., MINAMI, K., OKAMURA, K., NARI-TOMI, T. and HAKE, J. (1990) Thermophilic methane fermentation of evaporator condensate from a kraft pulp mill. In: *Proceedings of the TAPPI Environmental Conference*, Seattle, WA, 9–10 April, Atlanta: TAPPI Press, pp. 631–639.

ZENDER, J. A., STUTHRIDGE, T. R., LANGDON, A. G., WILKINS, A. L., MACKIE, K. L. and MCFARLANE, P. N. (1994) Removal and transformation of resin acids during secondary treatment at a New Zealand bleached Kraft pulp and paper mill. *Wat. Sci. Technol.* **29**, 105–121.

ZITOMER, D. H. and SPEECE, R. E. (1993) Sequential environments for enhanced biotransformation of aqueous contaminants. *Environ. Sci. Technol.* **27**, 227–244.

ZUXUAN, W., LIXON, Y. and NIANGUNO, L. (1983) Disposal of paper black liquor and furfuraldehyde sewage through anaerobic digestion process. In: van der Brink, W. J., ed., *Proceedings of the Symposium on Anaerobic Wastewater Treatment*, Noordwijkerhout, The Netherlands, 23–25 November, The Hague: TNO Corp. Communications, pp. 369–382.

Bioremediation of Soils Contaminated with Organic Wood Preservatives

ABDOLHAMID BORAZJANI AND SUSAN V. DIEHL

7.1 Introduction

Creosote and pentachlorophenol (PCP) are two of the major and highly effective organic wood preservatives. Creosote consists mostly of aromatic single to multiple ring compounds termed polycyclic aromatic hydrocarbons (PAH). Over 200 different chemical components have been identified in creosote; however, it is generally agreed that creosote contains several thousand different compounds which are identifiable with gas chromatography/mass spectroscopy (GC/MS). At least 16 PAHs found in creosote have been categorized as priority pollutants by the Environmental Protection Agency (EPA). Creosote is a blend of various coal tar distillates with specific physical characteristics that meet standards of the American Wood Preservers' Association (AWPA). Creosote has been used worldwide for more than 150 years. During the early part of the twentieth century, there was a significant expansion and increase in the number of creosote treating plants for production of treated wood crossties for the expanding railroad industry in the USA. This demand for crosstie material peaked at about 70 million ties in 1920 and slowly declined to 50 million in 1925 and 30 million in 1930 (Webb, 1987). Since then the use of creosote has been declining. The annual USA consumption in 1993 was estimated at 73.7 million gallons (Micklewright, 1994).

PCP was developed by the Monsanto Chemical Company in 1935. The typical composition of commercial technical grade PCP is 85–90 per cent pure PCP, 4–8 per cent tetrachlorophenol, 2–6 per cent other chlorinated phenols, and 0.1 per cent dioxin and furans (Crosby, 1981). PCP must be dissolved in petroleum oil before being impregnated into the wood because of its low solubility in water. In the USA during 1993, 20.7 million lb of PCP was consumed for pressure-treating wood products (Micklewright, 1994).

Although wood products treated with creosote and PCP do not represent a threat to the environment during use, past improper disposal methods and accidental spillage have resulted in extensive contamination of surface and subsurface soils and aquifers. Sites with this type of contamination are considered hazardous and require remedial action. The Land Disposal Restrictions under the US Resource and Conservation Recovery Act (RCRA) limit materials which can be disposed of by landfill, impoundment or land application in the USA. Soils contaminated with creosote and PCP require treatment to best demonstrated available technology (BDAT) levels prior to landfilling of the soils. In many cases the only two alternatives are either incineration or bioremediation. Incineration is highly effective but quite expensive and has some potential environmental disadvantages. In most large-scale soil contamination problems, bioremediation seems to be the best alternative.

Bioremediation is the process by which hazardous organic materials are biologically degraded, usually to innocuous materials such as carbon dioxide, methane, water, inorganic salts, biomass, and/or by-products that are less complex and less toxic than the parent compound (Anderson, 1995). The process can be carried out through either aerobic or anaerobic pathways by indigenous heterotrophs or by specially selected organisms and can be applied either *in situ* or *ex situ*. This chapter reviews the aerobic application of different bioremediation techniques, including land treatment (landfarming), bioventing, biopiling and bioslurry treatment, for the clean-up of creosote and PCP contaminated soils.

The concept of bioremediation can be traced back centuries to such processes as composting of organic wastes for soil conditioners and mulch (Thomas *et al.*, 1992). As early as the 1940s, industrial wastes were treated by bioremediation when the petroleum industry disposed of refinery wastes by land treatment, which included biological, physical and chemical processes. Management of wastes by land application was unregulated at that time (Anderson, 1995). Several field studies during the 1970s examined the disposal and degradation of agricultural pesticides. Goulding (1974) applied 2,4-D manufacturing wastes (50–250 ppm) to Oregon desert soils and determined that 2,4-D residues were not detectable after 540 days, whereas dichlorophenol, another manufacturing waste product, was persistent. Daughton and Hsieh (1977) added a mixed microbial culture, adapted to grow on parathion, to soil containing 5000 and 10 000 ppm of formulated parathion and found that inoculated soil samples degraded 86 and 21 per cent of the parathion, respectively, within three weeks, whereas non-inoculated soils showed no significant parathion degradation. Biological treatment of soil contaminated with organic wood preservatives became technically acceptable during the 1980s after several successful EPA sponsored demonstration studies (Loehr, 1989; McGinnis *et al.*, 1989; Sims *et al.*, 1989; McGinnis *et al.*, 1991).

7.2 Land Treatment

Land treatment is the use of biological, chemical and physical soil processes to transform waste. Although wastes could be added to soil for treatment, land treatment techniques are usually applied to soil that is already contaminated. RCRA defines land treatment as 'the hazardous waste management technology pertaining to application and/or incorporation of waste into the upper layer of the soil in order to degrade, transform, or immobilize hazardous constituents contained in the applied waste'. Landfarming is one of the most common land treatment technologies. It is a relatively simple technology and one of the most economical means of disposing of wastes. The polluted soils could be moved to a location better suited for control of the landfarming process (*ex situ*) or treated in place (*in situ*). In *ex-situ* cases, a liner may be needed to stop potential water migration problems, elevated earthen barriers may be needed to control run-on and run-off of rainfall or irriga-

Table 7.1 Bioremediation of creosote, tetrachlorophenol and PCP from Landfarm 7, International Paper site in Joplin, MO, for 1995

Date	PAHs[a] (mg/kg)	PCP (mg/kg)	Tetrachlorophenol (mg/kg)	Toxicity[b] (EC_{50})
January	1356	506	21.4	1.09
December	660	71	2.2	76.65
Degradation	51%	86%	90%	

[a] Average of 24 PAHs including acenaphthene, pyrene, benzo(a)pyrene, indeno(1,2,3-cd) pyrene, dibenzo(a,h)anthracene, dibenzofuran, *p*-cresol, *o,m*-cresol, 2-methylnaphthalene, 1-methylnaphthalene, naphthalene, fluorene, anthracene, phenanthrene, pyrene, fluoranthrene, benzo(a)anthracene, chrysene, benzo(b)fluoranthene, benzo(k)fluoranthene, benzo(g,h,i)perylene, and carbazole

[b] Toxicity of leachate determined by Microtox toxicity assay. Leachate prepared by sonicating 5 g soil in 15 ml water for 10 min. EC_{50} is the effective concentration that kills 50 per cent of the test organism. The lower the EC_{50}, the more toxic the sample

tion, and leachate collection systems may be needed to retain any water that passes through the soil.

In most *ex-situ* landfarming units, treatment beds are lined with high density liners with heat welded seams and clean sand is placed on top of the liner. The sand provides protection for the liner and proper drainage for water as it leaches through contaminated soils. Perforated drainage pipes are placed laterally on top of the synthetic liner in the sand bed for collection of soil leachate. The treatment beds can be completely covered by a plastic film, greenhouse or roof. An overhead spray irrigation system contained within the greenhouse provides moisture control and a means of distributing nutrients and microbial inocula (as needed) to the soil treatment. In this system, contaminated leachate draining from the soil is transported by the drain pipes, collected in a gravity-flow lined sump, and pumped to an on-site bioreactor for treatment. This construction has successfully treated soils contaminated with creosote, petroleum hydrocarbons and chlorinated pesticides (Borow, 1989).

A more technical version of this system is currently in operation at an International Paper, Inc. wood treating plant in Joplin, Missouri. At this site, several roofed landfarm units (approx. 70 000 m^2) have been constructed to treat soil contaminated with PCP and creosote constituents (Figure 7.1). The bed is lined with a high density polyethylene (HDPE) liner with 30 cm of clean sand on top of the liner and a series of 5-cm perforated PVC pipes 90 cm apart in the sand. Contaminated soil (1.8 m) is located on top of the leachate collection systems. Aerobic microbial degradation of PAHs and PCP occurs in the top 30–35 cm of this soil. The remaining 1.45–1.5 m of soil, because of lack of oxygen, remains inactive (Loome, 1995). In order to contain the treatment soil, earthen barriers are constructed of sufficient height to contain the treatment soil and prevent any water run-off from the treatment area. The landfarm roofs are located 7.6 m above the soil with a sprinkler system positioned directly under the roof in order to provide enough height for movement of tilling machinery. The roof prevents soil saturation from rain. The sprinkler system regulates soil moisture at 40–60 per cent of field capacity in order to maintain an active microbial population. Nutrients in the form of turkey manure are added to the active layer of each landfarm to provide a suitable microbial environment. The active layer of each landfarm is tilled twice weekly to enhance the bioremediation of contaminants by introducing oxygen and distributing nutrients and moisture. These units have been in full operation since January 1995. Several thousand cubic metres of PCP and creosote contaminated soil from the active layer of these landfarms have been cleaned up. Data for 1995 for the eight landfarm units in landfarm 7 at this company site are given in Table 7.1. Decontaminated soils are collected in a decontamination storage area outside the landfarm but remain covered by the landfarm roof. These storage areas are also protected by the HDPE liner and contained by an earthen barrier (Figure 7.1). The water collected from these systems is treated in a bioreactor before being discharged (Loome, 1995).

McGinnis *et al.* (1991) conducted a comprehensive field demonstration project at a site in Wiggins, Mississippi, which contained soils contaminated with PCP and creosote. Nine 24 × 5-m cells were constructed in an area cleared of surface debris. An earthen barrier (1.7 m) around the site controlled run-on and run-off water. To monitor the migration potential of PCP and creosote, six 30-cm, square glass brick pan lysimeters were buried at random locations within each cell at three depths (0.4, 0.9 and 1.3 m). Pan lysimeters or free-drainage lysimeters collect soil moisture by gravity drainage and are often used below HDPE liners (Nielsen, 1991). Each cell was sloped (0.5 per cent) to a drainage sump connected to a 15-cm drainage line that drained into a 75 000-litre storage tank. The drainage water was pumped from this tank through an activated carbon filter before being discharged. One monitoring well was placed about 20 m up-gradient (groundwater hydraulic gradient) from the facility, and two wells were placed about 30 and 50 m down-gradient from the facility. Each cell had an independently controlled irrigation system that maintained

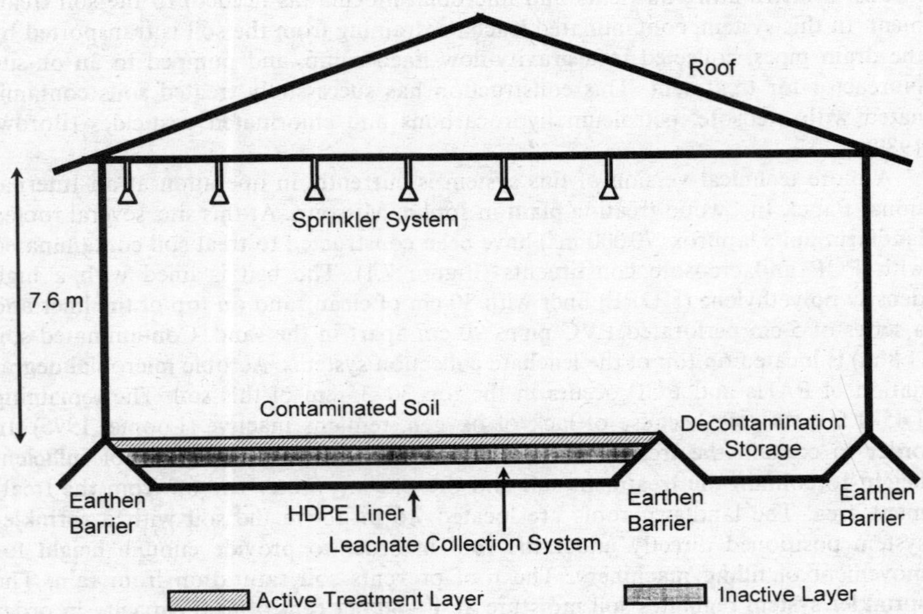

Figure 7.1 Side view of roofed landfarm unit

the soil moisture conditions needed for microbial activity and controlled wind dispersal of the soil. Chicken house bedding (0.75 m^3) containing a mixture of sawdust and chicken manure was rototilled into each cell and provided 3.3 per cent of mixture (weight) in the zone of incorporation.

The cells were loaded with recycled waste treating solutions from the creosote and PCP treating cylinders at the site. To maximize uniformity of application, the wastes were applied in a criss-cross pattern to a 22 × 4.6 m area in each cell. After loading, the cells were rototilled to mix the waste material into the soil. Three cells were loaded with creosote containing wastes, three were loaded with PCP containing wastes, and three cells without waste served as controls. The loading rates were as high as 3954 ppm creosote and 463 ppm PCP. The cells were loaded three times over a 10-month period. The cells were tilled about every 2 to 4 weeks during the 30-month study. PAHs and PCP were significantly degraded within 2 years to 53 ppm PAHs (average of all creosote cells) and 10.5 ppm PCP (average for all cells). No PAHs or PCP leached through the soil into the monitoring wells. Based on the amount of PAHs and PCP recovered from the lysimeters, investigators concluded that the movement of these compounds in the soil pore liquid was negligible. They also concluded that PAHs and PCP in soil can be transformed at useful rates and that soil treatment of creosote and PCP wood treating wastes appears to be a viable management alternative. Land treatment of these compounds could be applied in many wood-treating site scenarios.

In-situ landfarming systems also involve the controlled manipulation and management of soil microbial, physical, and chemical processes in order to achieve degradation and detoxification of wood preserving organic wastes without physically removing the contaminated soil (Sims *et al.*, 1989). Successful application of *in-situ* treatment requires good site characterization and an understanding of site, soil and waste interactions. Soil management techniques for *in-situ* landfarming do not differ from *ex-situ* techniques and include the addition of nutrients (mainly nitrogen and phosphorus), tilling, irrigation, pH adjustment, and addition of bulking materials and specific non-indigenous microorganisms. Stimulation of factors that influence indigenous microbial activity is recommended over augmentation with non-indigenous microorganisms because augmentation has significant technical, economic and regulatory constraints (Sims *et al.*, 1989). *In-situ* landfarming is best suited for sites containing heavy clay soils and very deep groundwater tables. The clay layer provides an impermeable barrier to the downward migration of organic contaminants. Two *in-situ* landfarming units have been in full operation for several years at a south Mississippi location. To monitor the movement of creosote constituents, several lysimeters were buried at different locations in each landfarm unit. Significant degradation of PAHs has occurred since operation began, and all but one lysimeters have shown no migration of the chemicals to below the active zone.

7.3 Biopiling

Soil pile treatment is an *ex-situ* process where soil is excavated and piled on a liner, mixed with amendments, and ventilated to promote biological oxidation of organic contaminants (Anderson, 1995). This technique is similar to *ex-situ* land treatment, except that the soil is not tilled. Oxygen is added through piped air or water. The soil pile system may be totally enclosed if control of air emissions is necessary

(Anderson, 1995), and has been used extensively for treating soils with volatile constituents because vacuum aeration systems can readily collect and treat gas emissions. Two types of soil pile systems are water-based and aerated systems (Anderson, 1995). In the water-based system, the contaminated soil is spread on a liner, and a constant flow of water containing nutrients and/or inoculum is added to the piles. Leachate that drains to a sump is sent back to the piles by the water system. The water-based system is limited in size because of oxygen transfer and water movement (Anderson, 1995). For the air soil pile system, soil is mixed with nutrients and/or inoculum prior to construction of the piles. Pipes are placed in the soil piles at different depths and spacing depending on the permeability of the soil and height of the piles. The air pipes connected to a vacuum pump or blower provide oxygen to the piles. The exhaust air can be collected and treated if necessary. The air system is more versatile, can be larger, and can handle higher contaminant concentrations compared with the water-based system (Anderson, 1995).

Two examples of successful application of the air based system are being demonstrated at two inactive wood treating sites in southeast USA. At Lake City, Florida, PCP contaminated soil piles 50 m long, 20 m wide and 2.5 m deep were constructed. Five per cent sawdust was added as a bulking agent and chicken manure was added to provide nutrients. Piping (10-cm diameter flexible perforated plastic pipe) in the piles was connected to a 2 hp electric blower so that no place in the pile was more than 1 m from an air source. One unvented soil pile served as a control (McGinnis *et al.*, 1992a). Respiration rates indicated that PCP concentrations ranging between 30 and 300 ppm were not inhibitory to microbial metabolism (McGinnis *et al.*, 1992a). The PCP concentration in one soil pile was reduced from 443 ppm to 11 ppm within 242 days. In the unvented control, the PCP concentration was reduced from 484 to 61 ppm within 339 days (McGinnis *et al.*, 1992b).

In a 1-year biopiling study of creosote contaminated soil at a site in south Mississippi, twelve $4 \times 3 \times 1$ m soil piles were constructed in a randomized complete block design (Figure 7.2) with three replicates per treatment (Hurt, 1996). The site was initially cleared of vegetation and sloped for capture of run-off water. An earthen barrier was constructed to prevent flooding and release of contaminated water. Treatments were unvented controls, vented piles, vented piles with 1 per cent kenaf core added (by volume), and vented piles with 1 per cent kenaf core inoculated with *Cladosporium* sp. added. Kenaf is a specialty fibre crop with excellent oil absorbency properties. The nine vented piles were constructed with five 3-m sections of 10-cm diameter perforated PVC vent pipes that were capped on one end. The piles were built in layers with an initial layer of 30 cm of contaminated soil. Three pipes were oriented horizontally on top of this bed, equidistant from the sides and from each other. An additional 35-cm layer of soil was placed on top of these pipes, and two vent pipes were situated above the gaps of the bottom pipe layer. The top tier of vent pipes was then covered with 35 cm of contaminated soil. Ten-centimetre diameter flexible tubing was then attached to the exposed end of each vent pipe. The flexible tubing led to a 10-cm diameter PVC manifold connected to a 0.5 hp electric blower. Positive pressure of 0.5 psi was constantly maintained in all of the piping, decreasing with distance from a blower. This manifold was constructed so that one blower served each block. Sprinklers were placed on each pile for water delivery. The water was supplied from a small sump pond located downhill of the study area. In this way a semi-closed system was present, water was recycled, and the use of chlorinated potable water was avoided. Of the 16 PAHs monitored in this study,

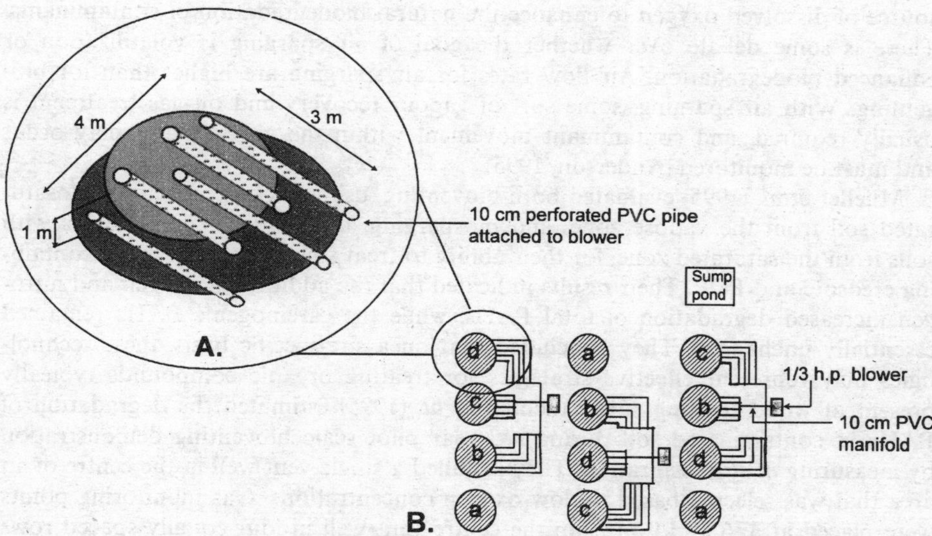

Figure 7.2 Schematic of biopile treatment study for creosote contaminated soil. (A) Enlarged view of a single vented pile; (B) Randomized complete block design for study area. a, unvented control piles; b, vented piles; c, vented piles plus kenaf; d, vented piles plus kenaf and fungus

only six exhibited discernable reductions with the largest reductions occurring by day 90. After 360 days, total PAHs were reduced by 3853 mg/kg in control piles, 3313 mg/kg in vented piles, 4186 mg/kg in kenaf piles, and 4060 mg/kg in fungal piles (Hurt, 1996). The lack of degradation after day 90 was probably due to the loss of co-metabolic induction, insufficient aeration and low bioavailability of contaminants. Biopiling proved to be an inadequate technique for enhancing bioremediation of PAH contaminated soil in this study.

7.4 Bioventing and Biosparging

Bioventing is considered an *in-situ* remediation process which uses a low flow of induced air to increase oxygen levels in order to stimulate aerobic biodegradation of organic contaminants in unsaturated soil and groundwater, but minimizing volatilization of the contaminants. The indigenous aerobic microbial communities often present in subsurface contaminated soil are usually capable of converting organic contaminants, such as PAHs, chlorinated phenols, and petroleum hydrocarbons, to less hazardous substances, but the addition of oxygen, limiting nutrients, or adjustment of certain physical parameters are required to stimulate the needed microbial activities. Bioventing can be established with or without nutrient addition. Bioventing has advantage over other types of remediation in that it is highly economical: it reduces excavation and drilling, the need for off-gas treatment caused by low air flow rates, and the time required for remediation. This technique can be influenced negatively by low soil moisture and low temperatures.

Air sparging or biosparging is the introduction of air into the soil below the water table to volatilize dissolved and adsorbed contaminants and to serve as a source of dissolved oxygen to enhance the natural biodegradation of contaminants. There is some debate over whether the goal of air sparging is volatilization or enhanced biodegradation. Air flow rates for air sparging are higher than for bioventing. With air sparging, some sort of vapour recovery and off-gas treatment is usually required, and contaminant movement within the groundwater may occur and must be monitored (Anderson, 1995).

Mueller *et al.* (1995) evaluated both bioventing, using packed columns of unsaturated soil from the vadose zone, and biosparging, using intact core samples with soils from the saturated zone, for their ability to treat soil and groundwater containing creosote and PCP. Their results indicated that the addition of oxygen and nitrogen increased degradation of total PAHs, while the carcinogenic PAHs remained essentially unchanged. They concluded that on a site-specific basis, these technologies may represent effective strategies for treating organic compounds typically present at wood treating sites. Alleman *et al.* (1995) estimated the degradation of PAHs in contaminated soil during a 3-year pilot scale bioventing demonstration by measuring *in-situ* respiration. They installed a single vent well in the centre of an area that was selected based on low oxygen concentrations. Gas monitoring points were placed at 3, 6 and 9 m from the centre vent well in four equally spaced rows extending towards the corners of the plot. Soil contamination ranged from 1000 to 19 000 mg PAH/kg soil. A non-vented control plot was located 46 m from the treatment plot. Data indicated a 13.4 per cent and 17.3 per cent degradation of total selected PAHs for the first and second year respectively. They concluded that venting increased the microbial activity, although not all of the respiration can be attributed conclusively to PAH metabolism.

7.5 Bioslurry Reactors

Slurry-phase bioremediation treats contaminated soil, sediment or sludges as an aqueous soil suspension in bioreactors or lined lagoons. Suspensions ranging from 30–50 per cent dry solids by weight are typical, although 20–30 per cent are not uncommon (Anderson, 1995). Slurry-phase remediation is based on the movement of contaminants into the aqueous phase where they are more susceptible to microbial degradation (Jerger and Woodhull, 1995). Large debris (>0.64 cm diameter) must first be removed from the contaminated soils. Soils are then mixed with nutrient-amended water, air is added to maintain adequate oxygen levels, and mixing is required to keep the solids in suspension. The systems maximize the transfer of contaminants into the aqueous phase and provide adequate contact time between the contaminant and the microorganisms (Anderson, 1995). A typical design of a slurry phase biotreatment system is shown in Figure 7.3.

Jerger and Woodhull (1995) treated 9600 m^3 of soil highly contaminated with PAHs in a 680 000-litre slurry bioreactor at an abandoned creosote wood preserving site in Canton, Mississippi. Total PAH concentrations ranged from 8000 to 10 000 mg/kg. Preparation of the soil for the slurry reactors included screening the contaminated soil through a power screen, loading the soil into the slurry mix tank, adding water to form a slurry, rescreening the slurry through a shaker screen, mixing the slurry, nutrients and conditioning chemicals, and transfer to the bioreactors. Aeration maintained the dissolved oxygen levels at >2 mg/kg. The pH, temperature, dissolved oxygen concentrations and other biological parameters were

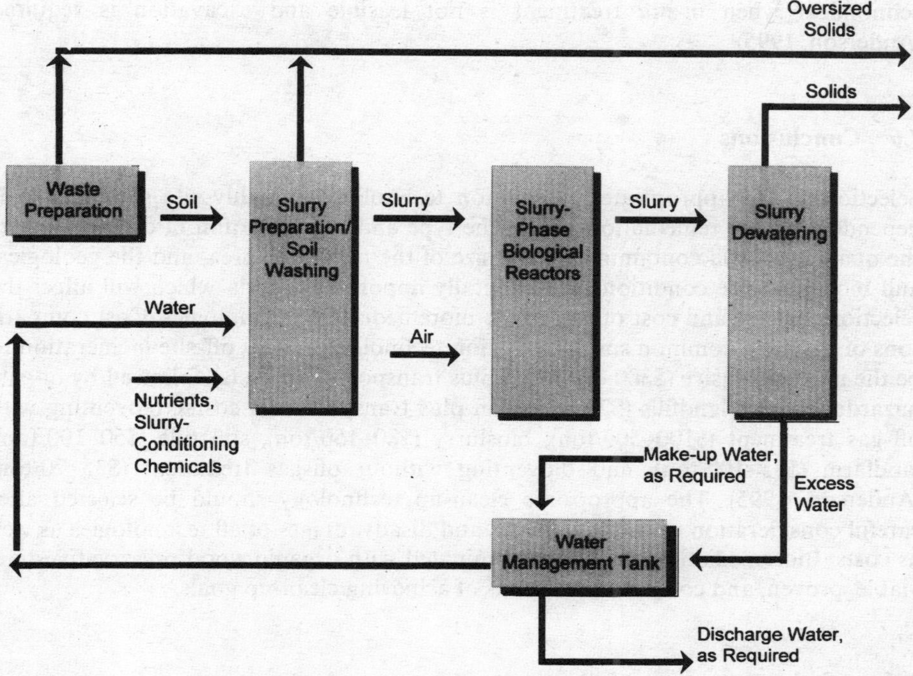

Figure 7.3 Full-scale slurry bioremediation process. Redrawn with permission from Woodhull and Jerger (1994). Reprinted by permission of John Wiley & Sons, Inc.

measured daily. An operation time of 8–12 days per batch was required to meet criteria. Risk based treatment criteria were 950 mg/kg dry weight for total PAHs and 180 mg/kg for benzo(a)pyrene equivalent carcinogenic PAHs. After this time, the slurry was transferred to a dewatering unit, the soil was dried, and the water was sent back to the reactors. A 95 per cent reduction of total PAHs was achieved within 5–10 days of treatment. Three variables that significantly influenced costs were contaminant concentration, residence time and slurry solids concentration.

In a second bioslurry reactor demonstration study, Brown *et al.* (1995) combined chemical and biological oxidation to treat soil contaminated with PAHs. One reactor received a daily addition of salicylate and succinate to enhance the biodegradation of the less recalcitrant PAHs. Effluent from this reactor was fed into a second reactor where Fenton's reagent ($Fe^{2+} + H_2O_2$) was added to accelerate the oxidation of 4- to 6-ring PAHs. A third reactor received the effluent from the second, the pH was adjusted, and nutrients and oxygen were added for biological polishing. Overall, there was an 85 per cent removal of total PAHs and a 66 per cent removal of carcinogenic PAHs, with at least half of the reduction occurring in the first reactor. Slurry reactors can often treat the more recalcitrant contaminants and higher contaminant concentrations more effectively than other types of bioremediation technologies because the environmental conditions can be tightly regulated and the percentage of the contaminant solids can be decreased to less toxic levels. The major cost for slurry remediation is soil movement; it may thus only be economical when *in-situ* treatment is not feasible and excavation is required (Anderson, 1995).

7.6 Conclusions

Selection of the appropriate remediation technology is highly site specific and is dependent on the remediation goals. The type and concentration of contamination, the quantity of soil contaminated, the size of the treatment area, and the geological and biological site conditions are all vitally important criteria which will affect the selection, success and cost of any of the bioremediation technologies. Cost comparisons of the most common soil remediation technologies show off-site incineration to be the most expensive ($300–1200/ton plus transportation costs), followed by off-site hazardous waste landfills ($200–300/ton plus transportation costs), bioventing with off-gas treatment ($100–500/ton), bioslurry ($80–150/ton), soil pile ($50–100/ton), landfarm ($35–100/ton), and bioventing without off-gas treatment ($35–70/ton) (Anderson, 1995). The appropriate clean-up technology should be selected after careful consideration of the advantages and disadvantages of all technologies as well as costs. Bioremediation of soils contaminated with organic wood preservatives is a viable, proven, and cost-effective means of achieving clean-up goals.

Acknowledgments

The authors would like to acknowledge Mr Wayne Ryland, Mr Dick Russell and Mr Jim Loome of International Paper Company for use of their data from the Joplin landfarm. Approved for publication as Article No. FPA-062-0596 of the Forest and Wildlife Research Center, Mississippi State University.

References

ALLEMAN, B. C., HINCHEE, R. E., BRENNER, R. C. and McCAULEY, P. T. (1995) Bioventing PAH contamination at the Reilly tar site. In: Hinchee, R. E., Miller, R. N. and Johnson, P. C., eds, *In Situ Aeration: Air Sparging, Bioventing, and Related Remediation Processes*, Columbus, OH: Battelle Press, pp. 473–482.

ANDERSON, W. C. (1995) *Innovative Site Remediation Technology, Vol. 1, Bioremediation*, Annapolis: American Academy of Environmental Engineers.

BOROW, H. S. (1989) Biological cleanup of extensive pesticide contamination in soil and groundwater. In: *Proceedings, Hazardous Materials Control Research Institute's Second National Conference*, 22–29 November, Washington, DC, pp. 51–56.

BROWN, K. L., DAVILA, B., SANSEVERINO, J., THOMAS, M., LANG, C., HAGUE, K. and SMITH, T. (1995) Chemical and biological oxidation of slurry-phase polycyclic aromatic hydrocarbons. In: Hinchee, R. E., Skeen, R. S. and Sayles, G. D., eds, *Biological Unit Processes for Hazardous Waste Treatment*, Columbus, OH: Battelle Press, pp. 113–127.

CROSBY, D. G. (1981) Environmental chemistry of pentachlorophenol. *Pure Appl. Chem.* **53**, 1051–1080.

DAUGHTON, C. G. and HSIEH, D. P. H. (1977) Accelerated parathion degradation in soil by inoculation with parathion-utilizing bacteria. *Bull. Environ. Contam. Toxicol.* **18**, 48.

GOULDING, R. L. (1974) *Waste Pesticide Management: Final Narrative Report*, July 1969–June 1972, Oregon State University, Corvallis Report to EPA.

HURT, K. (1996) Biopile Treatment of Creosote Contaminated Soil at a closed Wood-treating Facility, PhD Dissertation, Mississippi State University.

JERGER, D. E. and WOODHULL, P. M. (1995) Economics of a commercial slurry-phase biological treatment process. In: Hinchee, R. E., Skeen, R. S. and Sayles, G. D., eds, *Biological Unit Processes for Hazardous Waste Treatment*, Columbus, OH: Battelle Press, pp. 105–111.

LOEHR, R. (1989) *Treatability Potential for EPA Listed Hazardous Wastes in Soil*, EPA/600-2/89/011, Ada, OK: Robert S. Kerr Environmental Research Laboratory.

LOOME, J. (1995) Environmental Supervisor, International Paper Co., Joplin, MO, (personal communication).

MCGINNIS, D., DUPONT, R. R. and EVERHART, K. (1992a) Determination of respiration rates in soil piles to evaluate aeration efficiency and biological activity, Presentation at 85th Annual Meeting and Exhibition of the Air and Waste Management Association, Kansas City, MO.

MCGINNIS, D., Dupont, R. R., ST LAURENT, G. and EVERHART, E. (1992b) Evaluation of the effectiveness of soil venting to enhance the degradation of pentachlorophenol in soil. In: *Proceedings, Emerging Technologies in Hazardous Waste Management*, ACS, Atlanta, GA, pp. 500–503.

MCGINNIS, G. D., BORAZJANI, A., MCFARLAND, L. K., POPE, D. F. and STROBEL, D. A. (1989) *Characterization and Laboratory Studies for Creosote and Pentachlorophenol Sludges and Contaminated Soils*, EPA/600/2-88/055, Ada, OK: Robert S. Kerr Environmental Research Laboratory.

MCGINNIS, G. D., BORAZJANI, A., POPE, D. F., STROBEL, D. A. and MCFARLAND, L. K. (1991) *On-site Treatment of Creosote and PCP Sludges and Contaminated Soil*, EPA1600/2-91/019, Washington, DC.

MICKLEWRIGHT, J. T. (1994) A report to the wood-preserving industry in the United States. In: *Wood Preservation Statistics*, Woodstock, MD: American Wood-Preservers' Association.

MUELLER, J. G., TISCHUK, M. D., BROURMAN, M. D. and VAN DE STEEG, G. E. (1995) *In situ* bioremediation strategies for organic wood preservatives. In: Hinchee, R. E., Miller, R. N. and Johnson, P. C., eds, *In situ Aeration: Air Sparging, Bioventing, and Related Remediation Processes*, Columbus, OH: Battelle Press, pp. 571–580.

NIELSEN, D. M. (1991) *Practical Handbook of Ground-water Monitoring*, Chelsea, MI: Lewis Publishers.

SIMS, J. L, SIMS, R. C. and MATHEWS, J. E. (1989) *Bioremediation of Contaminated Surface Soil*, EPA/600/9-89/073, Ada, OK: Robert S. Kerr Environmental Research Laboratory.

THOMAS, J. M., WARD, C. H., RAYMOND, R. L., WILSON, J. T. and LOEHR, R. C. (1992) Bioremediation. In: Lederberg, J., ed., *Encyclopedia of Microbiology*, Vol. 1, San Diego: Academic Press, p. 369.

WEBB, D. A. (1987) Creosote–recent development and environmental considerations. *AWPA Proc.* **83**, 11–18.

WOODHULL, P. M. and JERGER, D. E. (1994) Bioremediation using a commercial slurry-phase biological treatment system: site-specific applications and costs. *Remediation* **4**, 353–362.

Bioremediation of Wood Treated with Preservatives Using White-Rot Fungi

ANDRZEJ MAJCHERCZYK AND ALOYS HÜTTERMANN

8.1 Introduction

With regard to its physical properties, especially to its durability, wood is a very special natural product. As long as trees which produce timber continue to grow, it is necessary that their boles be resistant to bacterial or fungal decay. This may not always be the case, but in principle, most trees are blessed with longevity, the oldest known specimens having endured for more than three thousand years. The carbon cycle would however be interrupted if after the death of the tree the wood had a slow rate of decay; therefore, natural degradation begins soon after the death of the tree with white-rot fungi being the most prominent organisms which prey upon the tree carcasses.

Preservation of wood from decay has accompanied its usage for building houses, ships, wheels, ploughs and other tools. Methods to conserve wood by daubing it with natural oils were already comparatively well developed 4000 years ago by the Egyptians. Asyrian and Chinese documentations describe the use of timber from certain trees for better durability (Moll, 1920). Usage of durable wood for construction was well known around 1000 BC in Greece and the records of Pliny the Elder (Rackham, 1945) in early Rome confirm the already established technology of wood and its preservation. Treating the surface with salts, burning it to induce protecting tars, and the application of hydrophobic toxic oils were methods used to protect timber until the eighteenth century (Hösli, 1982). Although these methods make it possible to preserve construction timber for several years against fungi and insects, a cheap source of wood preservatives became necessary with increasing wood usage. It was not until the nineteenth century with the development of coal chemistry that such products became available. The first substance which offered good general protection was coal tar creosote; however, it could only be used on exterior surfaces. Since the beginning of the last century, poisonous salts such as mercury chloride, as well as salts of arsenic, copper, zinc and iron have proved useful for protecting wood used on the exterior and interior of buildings (Graham, 1973). In addition, chlorinated tar components, for example, chloronaphthalene and chlorophenols, have been developed, these are much more specific in controlling fungi and less poisonous than the metal salts.

In the last 50 years numerous chemicals have been tested as wood protectants but only a very few have found general usage. As well as creosote – still the most important wood preservative – other organic compounds, for example pentachlorophenol, naphthenic acid, copper-8-quinolinolate, bis(tributyltin)oxide, 3-iodo-2-propynyl butyl carbamate, 4,5-dichloro-2-N-octyl-4-isothiazolone-3-one, dialkyldimethylammonium chloride and propiconazole have commonly been used for wood conservation (Zabel and Morrell, 1992). Among waterborne preservatives only chromated copper arsenate (CCA), ammoniacal copper zinc arsenate (ACZA), chromated zinc chloride (CZC), acid copper chromate (ACC), a few organometal compounds, fluorides, and boron compounds are used to any extent. The application of any given type of preservative is determined by the required durability of wood, price, indoor or outdoor usage, contact with soil or sea water and the desired characteristics of the wood surface after treatment. Additionally, various insecticides such as hexachlorocyclohexane (γ-HCH, Lindane), DDT, chloro-cyclodiene compounds (e.g. dieldrin), pyrethroids and carbamates have been added to many commercial preservatives (Richardson, 1993).

All the preservatives used have one thing in common: they all make the disposal of wood more difficult. In many countries it is considered to be a highly problematic waste which cannot be placed in a normal municipal landfill. The costs for the disposal of wood treated with preservatives are therefore increasing: up to \$500 for one ton of contaminated wood now has to be paid for its thermal disposal in special furnaces. This method of disposal cannot solve the problem due to the limited capacities of furnaces, and construction of new installations in countries with a high population density is not viewed favourably. On the other hand, wood as a natural product should be treated as a part of the terrestrial carbon cycle: if possible the contamination should be remedied by natural means and the residue returned to natural environmental cycles.

8.2 Bioremediation – a Strategy Adopted from the Natural Carbon Cycle

Bioremediation is becoming more and more widely accepted as a less expensive alternative to physical and chemical means of degradation of organic pollutants. It applies the reactions and mechanisms of the natural carbon cycle in the degradation of chemicals which are foreign to nature. Two different major strategies are utilized for the mineralization of organic compounds in the carbon cycle: the classical biochemical pathway and radical chemistry.

The biochemical strategy which a cell uses to mineralize organic compounds with the ultimate goal of converting as much of its energy into a usable form, i.e. ATP, is strictly limited by certain physiological restraints. The energy changes accompanying the reactions in the cell are limited; otherwise, the enthalpy of the reaction would injure the cell structure. These limitations affect the cellular chemistry so that only a few types of reaction can take place in many single steps; conversion of even moderately complex chemicals like glucose to CO_2, water and energy requires about 30 single reactions, each of them catalyzed by a different enzyme. This small-step strategy yields the maximal available energy of the metabolized molecule for the cell; however, the disadvantage is that highly specific catalysts, i.e. enzymes, are very expensive to make. It is obvious that this strategy of mineralization of organic compounds is only profitable for cells if the compound to be metabolized is present at a sufficiently high concentration; otherwise, the costs of supplying the catalysts would be higher than the gains from metabolizing the substrate. This method of mineralizing organic compounds in a given organism must thus be restricted to a few compounds which occur with a certain reliability in its ecological niche. In addition, the catalysts used are so specific that they only react at a reasonable rate with one single compound. As these two considerations imply, a given population of (micro)organisms is usually limited in its metabolic activity to a certain spectrum of organic chemicals. Any additional compound will not be mineralized by this type of metabolic strategy unless new organisms are introduced into the system or the enzyme spectrum of the organisms is changed by mutation.

The natural alternative to this rather limited degradation potential is radical chemistry. The most prominent natural compound whose synthesis and degradation operates via a radical mechanism is lignin, the substance which gives the lignified plant cell wall its pressure stability and which is quantitatively the second most abundant natural compound in the biosphere. The lignin molecule contains no bonds which can be cleaved hydrolytically under conditions prevalent within cells. All lignin degrading enzymes have some features in common: they are extracellular and the first reaction they catalyze is the elimination of an electron from a substrate, which can either be lignin itself, a metal ion, or a low molecular aromatic compound such as veratryl alcohol. After this initial step very complex reactions follow, which involve both depolymerizations and polymerizations and eventually lead to the complete mineralization of the lignin molecule. It has to be assumed that only part of this mineralization actually takes place inside the cell and that the most important reactions must take place outside the cell. Due to the low specificity of the enzymes involved, they can also be utilized in the remediation of aromatic compounds present in the biosphere.

8.3   Biodegradation of Wood Preservatives

The usage of wood preservatives during the last 50 years has been accompanied by intensive studies investigating their microbial degradation. The purpose of these investigations has usually been to determine the detoxification mechanisms in the protected wood and to build a basis for the improvement of preservative compositions. A full review of all works on this subject cannot be given here and only a few of the early publications will be mentioned. The first studies concentrated on coal tar creosote, as one of the oldest, most powerful and also most toxic wood preservatives giving a long-term protection effect. The microbial breakdown of many creosote-like components was described as early as 1928 by Gray and Thornton. Since then various creosote tolerant bacteria and fungi, mostly *Pseudomonas* and *Streptomyces*, have been isolated from treated wood. Some microorganisms, such as *Hormodendrum resinae*, isolated from the creosote treated timber, can utilize coal tar and creosote as an N-source and C-source (Marsden, 1951, 1954) and most of them can degrade phenols and up to three-ring polycyclic aromatic hydrocarbons (PAH) (Drisko and O'Neill, 1966; Kerner-Gang, 1975). Removal of the phenolic fraction of creosote, which is especially important for its preservative qualities and toxic effect on wood-rotting fungi (Da Costa *et al.*, 1969), allows other organisms to colonize and subsequently degrade wood.

The biodegradation of another commonly used group of preservatives, chloro-phenols – especially the most active of these, pentachlorophenol (PCP) – was the object of intensive research during their widespread application in the early 1960s (Stranks and Hulme, 1975). Enzymatic detoxification of chlorophenols by oxidative enzymes such as laccase, tyrosinase and peroxidase was reported by Lyr (1962, 1963a) and was considered to be a key mechanism in the degradation of these com-pounds in timber by fungi. Chlorophenols were dechlorinated to a large extent by this treatment and also partially polymerized. Detoxification of PCP and its adsorp-tion to the fungal mycelium was confirmed directly for the white-rot fungus *Trametes versicolor* (Lyr, 1963b). Many PCP tolerant fungi, mostly Ascomycetes or Fungi Imperfecti, were isolated from treated wood and were shown to be able to remove this preservative, converting it to a less toxic pentachloroanisole (Cserjesi and Johnson, 1972). PCP was extensively removed by treatment with tolerant lower fungi in a first step, and then the wood was degraded by Basidiomycetes in a second treatment, thus explaining the natural degradation of PCP treated timber (Duncan and Deverall, 1964). Furthermore, brown-rot fungi such as *Coniophora puteana* were found to be PCP tolerant (Unligil, 1968) and in the ensuing years many white-rot fungi were found capable of degrading this compound in treated wood (e.g., Lamar and Dietrich, 1992; Majcherczyk and Hüttermann, 1993). Not surprisingly, in con-trast to wood penetrating fungi, PCP-degrading bacteria which had been isolated from contaminated soil failed to remove PCP from wood (McBain *et al.*, 1995).

Comparatively little is known about the biodegradation of other organic wood preservatives probably due to their lack of widespread usage. Organoiodine wood preservatives, for example, were found to be detoxified by brown- and especially white-rot fungi (Lee *et al.*, 1992).

Inorganic preservatives, especially heavy metals, are the most persistent in the environment and cannot be effectively removed by biological consortia. This does not mean however that these compounds protect wood indefinitely. Many fungi and bacteria were found to be very resistant to mixtures of copper, arsenate, chromium and fluorine in treated wood. Most were not able to degrade wood components but they were able to detoxify and also partially remove the preservatives. Metal ions were found to be methylated or precipitated by extracellular organic acids. In many cases salts of zinc, copper, chrome or arsenate were remobilized from wood and removed by leaching or bound to the fungal hyphae (Stranks and Hulme, 1975). A complementary action involving preservative-tolerant, metal-removing fungi and intolerant, wood-degrading fungi was suggested (Madhosingh, 1961). Cu, Zn and As tolerant soft-rot microfungi which were able to degrade cellulose were isolated (Henningson and Nilsson, 1975) and a complementary action of soft-rot fungi and bacteria was also reported (Singh *et al.*, 1992). Biological detoxification of timber and subsequent wood deterioration would return treated wood to the natural carbon cycle but the heavy metals applied for preservation must be considered as a permanent fixture in the environment.

8.4 The Use of White-Rot Fungi for Bioremediation of Wood

With regard to the bioremediation of treated wood it follows that the degradation of wood preservatives is limited to those organisms which are able to penetrate wood and grow in such an environment. Since wood preservatives which were used some 50 years ago are not of natural origin but are usually composed of a widely varying number of compounds, most being of aromatic nature, it is obvious that classical biochemical metabolization is not likely to occur with these substances. The alternative metabolic pathway, radical chemistry, seems to be more promising for their remediation. This idea has also been confirmed in the cited studies with regard to the natural detoxification of wood preservatives.

The most efficient producers of extracellular enzymes known thus far, which unspecifically oxidize aromatic compounds via the elimination of an electron or a hydrogen atom, are white-rot fungi. They generate such a high redox-potential that this metabolic approach was called 'enzymatic combustion' (Kirk and Farrell, 1987), which means that they 'burn down' all available aromatic compounds present in the proximity of the mycelia without receiving any immediate return from this action. Numerous experiments have shown that white-rot fungi in liquid culture are able to degrade mixtures of even very high condensed PAH, including the carcinogen benzo(a)pyrene; the first study reporting this was published in 1985 (Bumpus *et al.*, 1985). The spectrum of substances which has successfully been degraded (mostly under laboratory conditions) includes virtually all important xenobiotica (e.g. Higson, 1991; Morgan *et al.*, 1991; Shah *et al.*, 1992; Paszczynski and Crawford, 1995). From the data obtained worldwide to date, it is obvious that the low substrate specificity of enzymes from white-rot fungi might be utilized for the remediation of recalcitrant aromatic xenobiotic compounds used as wood preservatives.

8.4.1 Remediation of Creosote Treated Wood

Creosote, which can be obtained in various distillation ranges, is a very complex mixture of a few hundred compounds, mainly polycyclic aromatic hydrocarbons,

Table 8.1 Main components of a typical creosote oil used for impregnation of railway ties

Acenaphthene	Ethylnaphthalene (isomers)
Acridine	Fluoranthene
Anthracene	Fluorene
Benz(a)anthracene	Indeno(c,d)pyrene
Benzo(a)pyrene	Isoquinoline
Benzo(b + k)fluoranthene	Methyldibenzofurane (isomers)
Benzo(b)thiophene	Methylfluorene (isomers)
Benzo(e)pyrene	Methyl-1H-indene (isomers)
Benzo(h)quinoline	Methylnaphthalene (isomers)
Benzoperylene	Methylphenanthrene (isomers)
Carbazole	Methylpyrene (isomers)
Chrysene	Naphthalene
Cyclopenta(def)phenanthrene	Phenanthrene
Dibenz(a,h)anthracene	Propenylbenzene
Dibenzofurane	Propynylbenzene
Dibenzothiophene	Propenylnaphthalene (isomers)
Dimethyl-1,1'-biphenyl (isomers)	Pyrene
Dimethylbenzene (isomers)	Trimethylnaphthalene (isomers)
Dimethylnaphthalene (isomers)	2,4,5-Trimethylbenzaldehyde

which may vary depending on the manufacturer (Table 8.1). If creosote treated wood is incubated with white-rot fungi, some degradation of all components is obtained in a relatively short time (Figure 8.1). Not all fungi are able to degrade creosote under these conditions to the same extent, and their ability to remove creosote can vary significantly between different strains. The typical distribution of the radioactivity of labelled PAHs after treatment with a white-rot fungus is presented in Figure 8.2. Degradation products of single components are only partially comparable with those obtained with complex PAH mixtures in pure, liquid cultures of fungi. The complexity of the reaction products creates an additional analytical barrier and the generally low mineralization of PAH is therefore not a guarantee of reliable information on the success of the remediation. If an attempt is made to reuse such bioremediated wood instead of further degrading it, the analytical monitoring of the removal of single components should be accompanied by determination of the total toxicity after treatment with fungi. A combination of both analytical methods and toxicological testing seems to be more reliable and describes the real purpose of the process: the detoxification of contaminated wood.

8.4.2 DDT, Lindane and PCP

Most preservative products are mixtures of fungicides and insecticides. In many cases an additional preservative is used to enhance the protection of that part of the timber which is immersed in water or has a direct contact with soil; it is also

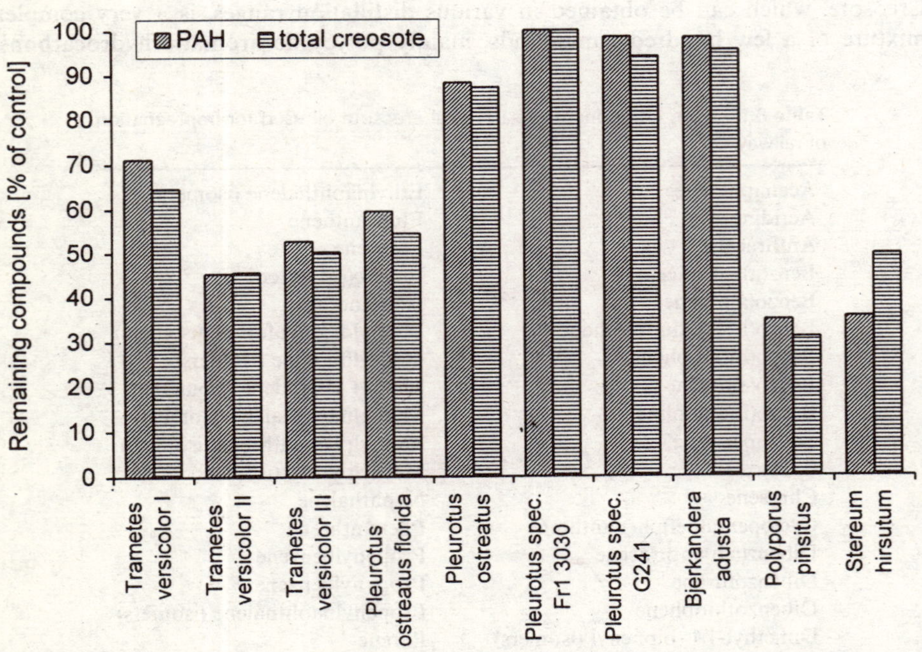

Figure 8.1 Degradation of total extractable creosote components, especially polycyclic aromatic hydrocarbons (PAH), after 3-week treatment of railway ties with white-rot fungi

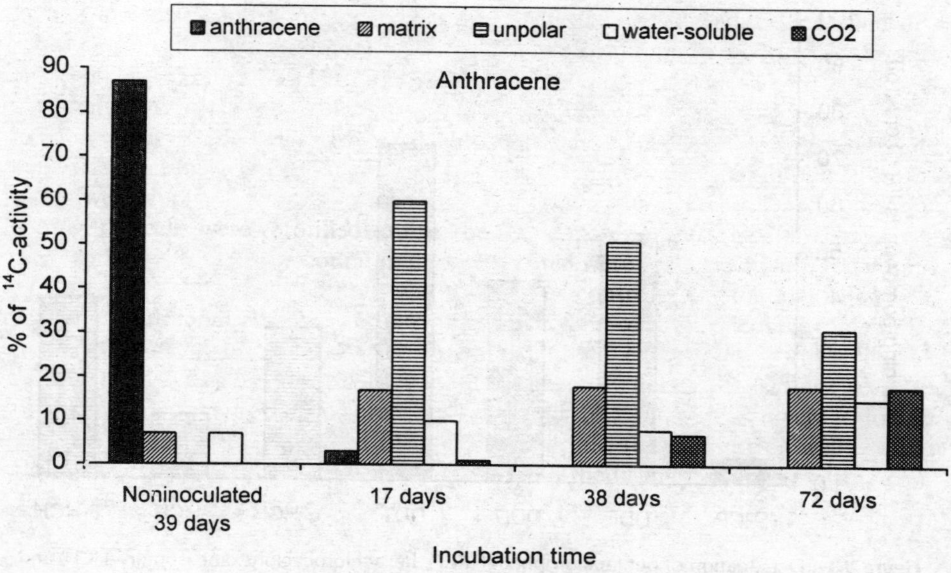

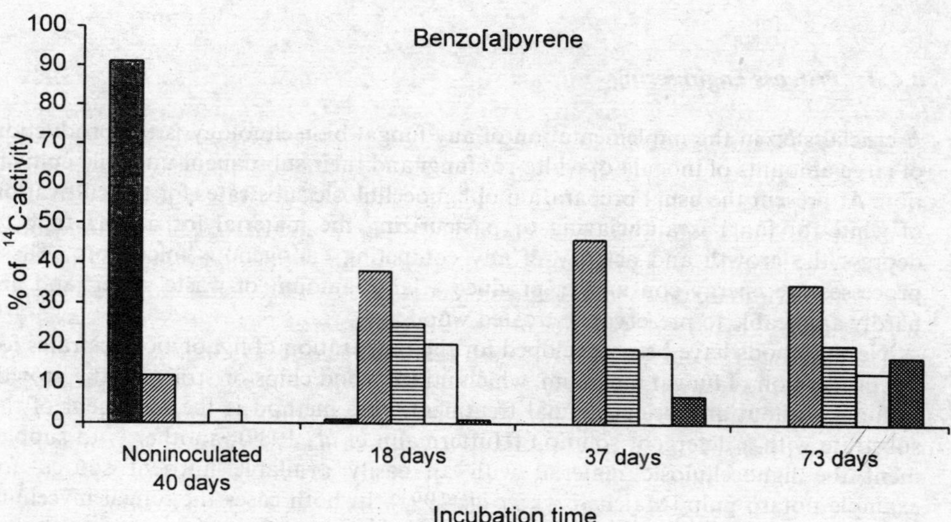

Figure 8.2 Distribution of ^{14}C-activity during degradation of ^{14}C-labelled anthracene and benzo(a)pyrene in wood by *Pleurotus ostreatus* (Zeddel, 1994)

common practice to repeat conservation after a few years. A typical wood preservative widely used in the 1950s and 1960s in Germany was composed of 1% pentachlorophenol (PCP) and 1% γ-hexachlorcyclohexane (γ-HCH, lindane). Both classes of substance and also insecticides such as DDT have been degraded by remediation of contaminated wood with white-rot fungi (Figure 8.3), thereby supporting the universal potential of the oxidative 'burn out'.

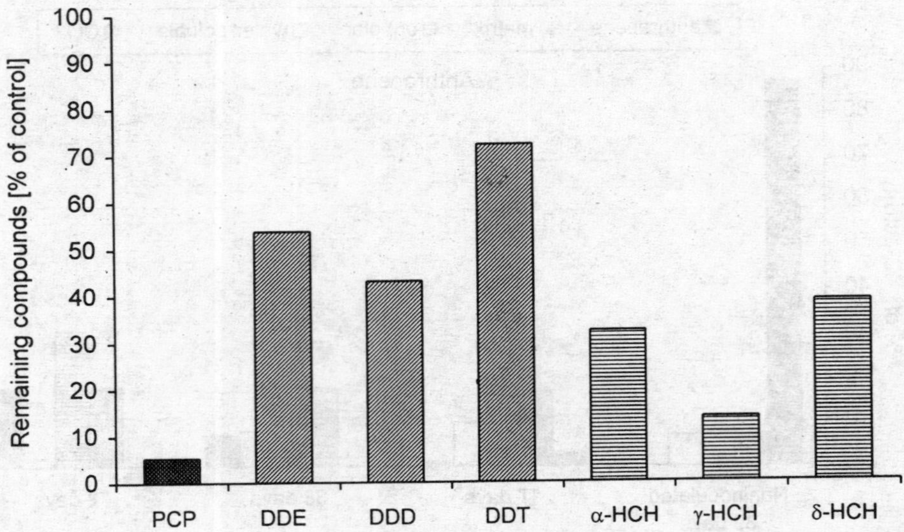

Figure 8.3 Degradation of pentachlorophenol (PCP), hexachlorocyclohexane isomers (HCH) and insecticides (DDT, DDD, DDE) after 4-week treatment of preserved wood with *Pleurotus ostreatus*

8.4.3 Process Engineering

A crucial step in the implementation of any fungal biotechnology is the production of large amounts of inocula of white-rot fungi and their subsequent unsterile cultivation. At present the usual preparation of lignocellulosic substrates for the cultivation of white-rot fungi is autoclaving or pasteurizing the material for a long time to depress the growth and activity of any competing endogenous microflora. These processes are energy consuming, produce a large amount of waste water, and are hardly applicable to preservative treated wood.

New methods have been developed for the preparation of lignolytic substrates for the production of fungal inoculum, which utilize wood chips or straw for the growth of fungi without any prior thermal treatment. One method is the treatment of the substrate with a detergent solution (Hüttermann *et al.*, 1989); another is to supplement the lignocellulosic material with an easily available nutrient source, for example potato pulp (Majcherczyk *et al.*, 1991). In both cases the fungal mycelium of typical wood-degrading fungi such as *Pleurotus*, or *Trametes* or many other organisms, is able to grow on the treated substrate and to overcome competition problems. One advantage of these methods, especially with the addition of potato pulp, is the resulting high activity of the ligninolytic enzymes of the fungi as indicated by the preferential degradation of lignin (Majcherczyk *et al.*, 1996). A feasible technical process of wood remediation could work according to the following scheme (Figure 8.4):

- The wood is first shred into small pieces.
- The wood chips are mixed with water, potato pulp and fungal inoculum.
- The mixture is put into a container and aerated.

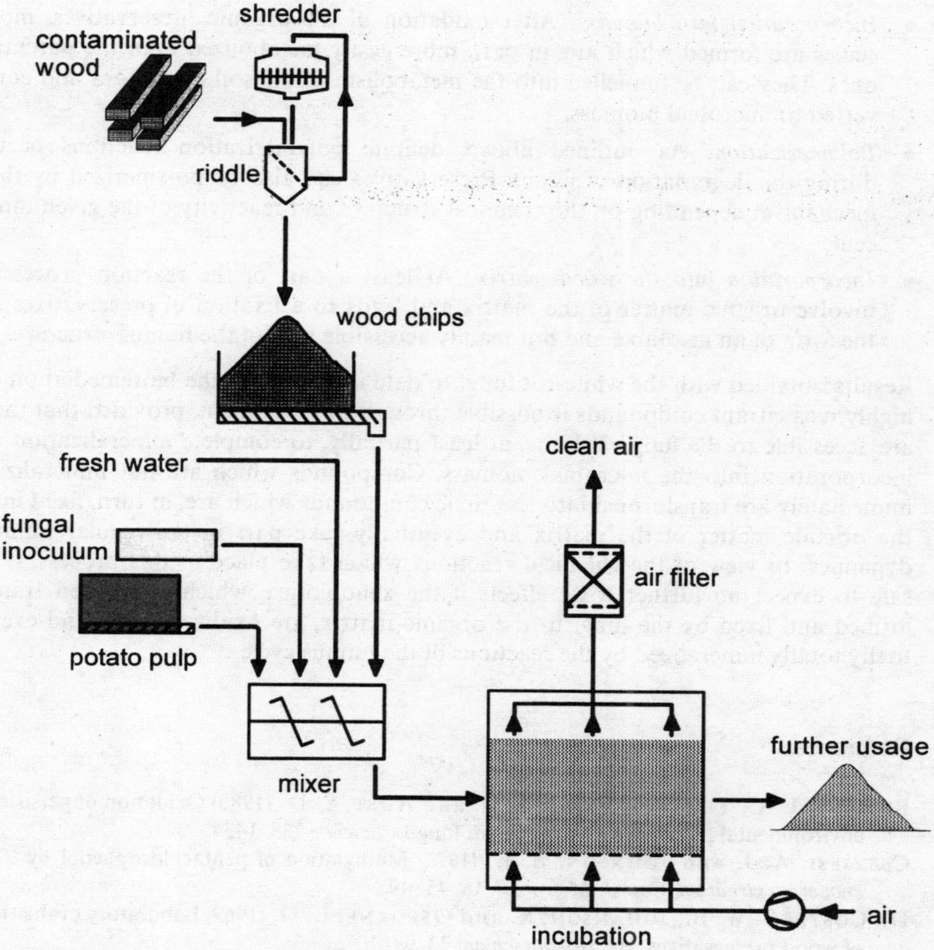

Figure 8.4 Process scheme for the remediation of contaminated timber with white-rot fungi

- After an incubation period of about 2–3 months, the material can be used for different purposes: the wood–mycelium mixture can be used as a basis for high quality compost or new wood composites may be produced from the contaminant-free material.

8.4.4 *The Fate of the Organic Compounds during the Fungal Treatment*

The ultimate fate of the compounds treated in this way depends on their chemical structure and, presumably, on the enzymes which are involved. Thus far several different possible chemical pathways have to be considered:

- *Complete mineralization*: This is the optimal fate when applying bioremediation. During the usual time frames in which bioremediation is applied this only happens to a small extent.

137

- *Incorporation into biomass:* After oxidation of the organic preservatives, molecules are formed which are, in part, more easily metabolized than the parental ones. They can be funnelled into the metabolism of the soil microflora and converted to microbial biomass.

- *Polymerization:* As outlined above, definite polymerization reactions occur during the degradation of lignin. Preservatives can also be polymerized by this mechanism depending on the chemical structure and reactivity of the given molecule.

- *Incorporation into the wood matrix:* At least a part of the reaction processes involve organic matter of the matrix and leads to a fixation of preservatives in the form of an insoluble and not readily accessible part of the humus structure.

Results obtained with the white-rot fungi to date indicate that the bioremediation of highly recalcitrant compounds is possible through this treatment, provided that they are accessible to the fungi. It leads, at least partially, to complete mineralization or incorporation into the microbial biomass. Compounds which are not mineralized immediately are transformed into less toxic compounds which are, in turn, fixed into the organic matter of the matrix and eventually take part in the regular humus dynamics. In view of the chemical reactions which take place in this process, it is safe to expect no further toxic effects if the xenobiotica, which have been transformed and fixed by the fungi to the organic matrix, are oxidized again and eventually totally mineralized by the reactions of the humus cycle.

References

BUMPUS, J. A., TIEN, M., WRIGHT, D. and AUST, S. D. (1985) Oxidation of persistent environmental pollutants by a white-rot fungus. *Science* **228**, 1434.

CSERJESI, A. J. and JOHNSON, E. L. (1972) Methylation of pentachlorophenol by *Trichoderma virgatum*. *Can. J. Microbiol.* **18**, 45–49.

DA COSTA, E. W. B., JOHANSON, R. and OSBORNE, L. D. (1969) Laboratory evaluation of wood preservatives, III. *Holzforschung* **23**, 99–107.

DRISKO, R. W. and O'NEILL, T. B. (1966) Microbiological metabolism of creosote. *Forest Prod. J.* **16**, 31–34.

DUNCAN, C. G. and DEVERALL, F. J. (1964) Degradation of wood preservatives by fungi. *Appl. Microbiol.* **12**, 57–62.

GRAHAM, R. D. (1973) History of wood preservation. In: Nicholas, D. D., ed., *Wood Deterioration and its Prevention by Preservative Treatments, Vol. 1, Degradation and Protection of Wood*, Syracuse, NY: Syracuse University Press, pp. 1–30.

GRAY, P. H. H. and THORNTON, H. G. (1928) Soil bacteria that decompose certain aromatic compounds. *Zbl. Bakteriol. Parasitenk.* Abt. II. **73**, 74–96.

HENNINGSON, B. and NILSSON, T. (1975) Some aspects on microflora and the decomposition of preservative-treated wood in ground contact. *Org. u. Holz*, Beihefte, **3**, 307–318.

HIGSON, F. K. (1991) Degradation of xenobiotics by white-rot fungi. In: Ware, G. W., ed., *Reviews of Environmental Contamination and Toxicology*, Vol. 122, New York, Berlin: Springer-Verlag.

HÖSLI, J. P. (1982) Wood preservation in the pre-industrial period. *Int. J. Wood Preserv.* **2**, 29–36.

HÜTTERMANN, A., MAJCHERCZYK, A. and GROTHEY, V. (1989) Verfahren zum Aufbereiten von Stroh als Substrat für die Anzucht von Pilzkulturen, German Patent PS 39 38 659.

KERNER-GANG, W. (1975) Einwirken von Mikroorganismen auf Steinkohlenteeröl. *Org. u. Holz*, Beihefte, **3**, 319–330.

KIRK, T. K. and FARRELL, R. L. (1987) Enzymatic 'combustion': The microbial degradation of lignin. *Annu. Rev. Microbiol.* **41**, 465–505.

LAMAR, R. T. and DIETRICH, D. M. (1992) Use of lignin-degrading fungi in the disposal of pentachlorophenol-treated wood. *J. Ind. Microbiol.* **9**, 181–191.

LEE, D.-H., TAKAHASHI, M. and TSUNODA, K. (1992) Fungal detoxification of organoiodine wood preservatives, I. *Holzforschung* **46**, 81–86.

LYR, H. (1962) Detoxification of heartwood toxins and chlorophenols by higher fungi. *Nature* **195**, 289–290.

LYR, H. (1963a) Enzymatische Detoxifikation chlorierte Phenole. *Phytopatol. Zeitschr.* **47**, 73–83.

LYR, H. (1963b) Die Aufnahme von radioaktivem Pentachlorphenol durch Schüttelmycel von *Trametes versicolor*. In: *Holzzerstörung duch Pilze*, International Symposium, 1962, Eberswalde, Berlin: Akademie Verlag, pp. 311–314.

MADHOSINGH, C. (1961) Tolerance of some fungi to a water-soluble preservative and its components. *Forest Prod. J.* **11**, 20–22.

MAJCHERCZYK, A. and HÜTTERMANN, A. (1993) Behandlung mit Weißfäulepilzen als Weg zur Rezyklisierung von Altholz. In: Hüttermann, A. and Kharazipour, A., eds, *Die Pflanzliche Zellwand als Vorbild für Holzwerkstoffe*, Frankfurt/M: Sauerländers Verlag, pp. 69–82.

MAJCHERCZYK, A., GROTHEY, V., HÜTTERMANN, A. and MAYER, F. (1991) Verfahren zur Anzucht von Pilzmycelien, German Patent PS 41 04 625.0.

MAJCHERCZYK, A., BEDAIWY, M., KÜHNE, A., KÖRNER, I., HADAR, Y. and HÜTTERMANN, A. (1996) The production of large amounts of fungal inoculum under unsterile conditions. In: Srebotnik, E. and Messner, K., eds, *Biotechnology in the Pulp and Paper Industry*, Wien: Facultas Universitätsverlag, pp. 199–204.

MARSDEN, D. H. (1951) Studies of *Hormodendrum resinae* Lindau, a common inhabitant of creosoted and coal-tar-treated wood. *Phytopathology* **41**, 658–659.

MARSDEN, D. H. (1954) Studies of the creosote fungus, *Hormodendrum resinae*. *Mycologia* **46**, 161–183.

MCBAIN, A., CUI, F., HERBERT, L. and RUDDICK, J. N. R. (1995) The microbial degradation of chlorophenolic preservatives in spent, pressure-treated timber. *Biodegradation* **6**, 47–55.

MOLL, F. (1920) Holzschutz, seine Entwicklung von der Urzeit bis zur Umwandlung des Handwerks in Fabrikbetriebe. *Beiträge zur Technik und Industrie* **10**, 66–92.

MORGAN, P., LEWIS, S. T. and WATKINSON, R. J. (1991) Comparison of abilities of white-rot fungi to mineralize selected xenobiotic compounds. *Appl. Microbiol. Biotechnol.* **34**, 693–696.

PASZCZYNSKI, A. and CRAWFORD, R. L. (1995) Potential for bioremediation of xenobiotic compounds by the white-rot fungus *Phanerochaete chrysosporium*. *Biotechnol. Progress* **11**, 368–379.

RACKHAM, H. (1945) *Pliny: Natural History*, Vol. 4, Cambridge, MA: Harvard University Press.

RICHARDSON, B. A. (1993) *Wood Preservation*, London: Chapman & Hall.

SHAH, M. M., BARR, D. P., CHUNG, N. and AUST, S. D. (1992) Use of white-rot fungi in the degradation of environmental chemicals. *Toxicol. Lett.* **64–65**, 493–501.

SINGH, A. P., HEDLEY, M. E., PAGE, D. R., HAN, C. S. and ATISONGKROH, K. (1992) Microbial degradation of CCA-treated cooling tower timbers. *IAWA Bull.* **13**, 215–231.

STRANKS, D. W. and HULME, M. A. (1975) The mechanism of biodegradation of wood preservatives. *Organ. u. Holz*, Beihefte, **3**, 345–353.

UNLIGIL, H. H. (1968) Depletion of pentachlorphenol by fungi. *Forest Prod. J.* **18**, 45–51.

Zabel, R. A. and Morrell, J. J. (1992) *Wood Microbiology, Decay and Its Prevention*, San Diego: Academic Press.
Zeddel, A. (1994) Abbau von polycyclischen aromatischen Kohlenwasserstoffen (PAKs) und polychlorierten Biphenylen (PCBs) durch Weißfäulepilze in Festphasensystemen, PhD Thesis, Göttingen, Germany.

Biotechnological Production of Wood Composites

ALIREZA KHARAZIPOUR AND ALOYS HÜTTERMANN

9.1 Biochemistry of Lignocellulose Synthesis

Lignocellulose, the woody plant cell wall, can be considered the best compound material available on earth. Its unparalleled strength is due to two different components being combined in its structural architecture: cellulose fibres and lignin, with hemicelluloses acting as a stabilizer between these two main components (Figure 9.1). The material has additional high tensile strength due to the fact that the lignocellulose fibres are twisted at a minimum of four morphological levels of construction. Pressure resistance is conveyed to the material by the incrustation of lignin into the space between the fibres at all levels of morphology and, as was reported more than 160 years ago by Theodor Hartig, lignin is also the glue which binds the woody cells together.

Cellulose and hemicelluloses are macromolecules which are composed of carbohydrate monomers. Cellulose is built only from glucose; hemicelluloses are more complex and composed of several different sugar monomers. In cellulose the building blocks are connected in the growing cell wall by stepwise enzymatic processes, with four reactions necessary to add one sugar unit to the growing polymer chain:

1   Activation of glucose to glucose-6-phosphate.
2   Isomerization of glucose-6-phosphate to glucose-1-phosphate.
3   Pyrophosphorylation with GTP to GDP-glucose.
4   Addition of the activated glucose molecule to the growing cellulose chain.

For the biosynthesis of lignin, the plant cell uses a completely different strategy (Higuchi, 1990; Dean and Eriksson, 1994). The phenolic monomers are polymerized in one step by an enzymatically catalyzed radical reaction, leading to the three-dimensional lignin macromolecule. The enzymes involved in this reaction are peroxidases and laccases.

By combination of the two different synthesis pathways – the orderly biosynthesis of the cellulose fibres and hemicelluloses, and the radical polymerization of

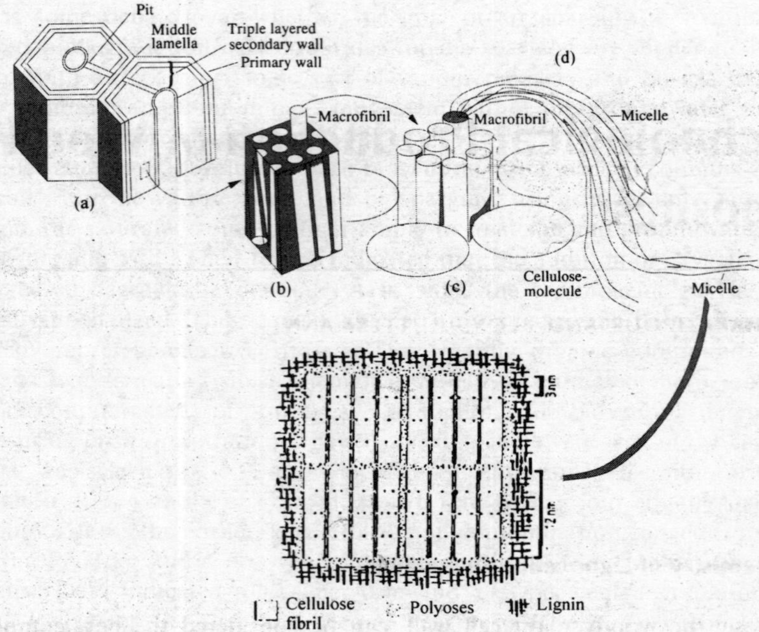

Figure 9.1 Structure of a woody cell wall

lignin – the composite material wood is formed, which is unparalleled in its technical properties by any synthetic material.

9.2 Wood Composites

The production of wood composites such as fibre or particle boards always follows the same basic process. First, solid wood is fragmented into smaller pieces as strands, chips or fibres. These are supplemented with a binder and pressed under heat to form a wood-like material. By this general process, the anisotropy of wood is reduced and wood of small dimensions or recycled wood can be converted to a product of much higher value. Only one component not present in the original wood is added: the binder, which currently may be urea–formaldehyde, phenol–formaldehyde, or isocyanate.

Two strategies have been developed which attempt to substitute these petrochemical adhesives with the natural binding material of the woody cell wall: lignin. Both try to follow the natural strategy of cell wall synthesis:

1 A two-component adhesive based on lignin as a binder and phenol-oxidase as a radical donor.

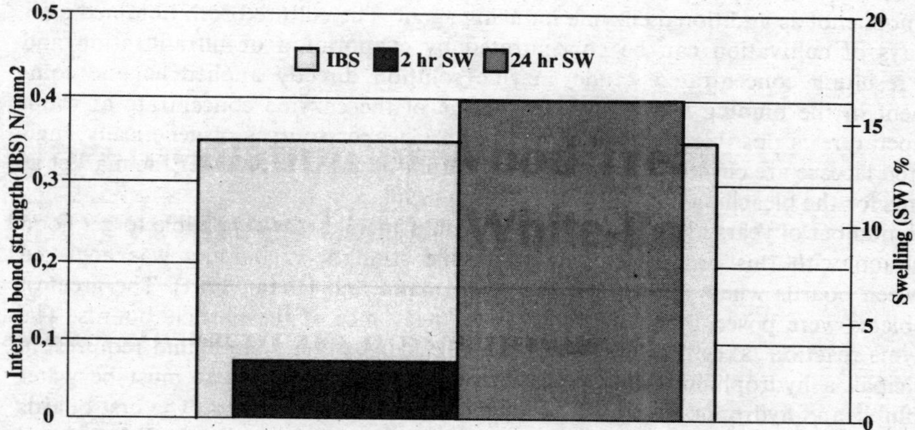

Figure 9.2   Technical properties of 19-mm particle boards, pressed for 3 minutes at 190°C with the lignin/laccase system as a binder, supplemented with 1 per cent isocyanate resin (phenylmethanediisocyanate, PMDI). The technical properties are comparable with standard urea–formaldehyde bonded boards and meet the German standards requirements

2   Enzymatic activation of the lignin at the surface of the wood fragments and subsequent use as an adhesive.

9.3   Two-component Adhesive Based on Lignin and Laccase

The use of enzymes for the bonding of wood particles was first suggested by Nimz *et al.* (1972) who used peroxidases for the cross-linking of lignin as a binding process. This invention, however, was not implemented at that time, probably because of the lack of suitable enzymes which were available in technical amounts. Application of a biological catalyst (enzyme) in a technical process, in this case particle board production, must meet the following criteria before it can be introduced:

1   The enzyme must be produced on cheap substrates in large quantities.

2   It must be able to be used in crude form without prior purification.

3   It must be reasonably stable and able to be stored at room temperature for at least 1 week.

4   Since particle boards are pressed at high temperatures, the enzyme must be heat tolerant and have a very high temperature optimum.

5   The design of the pressing process must follow the procedure which is commercially applied at present, with pressing times as short as possible.

6   The price of the final product must be competitive with the conventional petrochemical resins used today.

It took years to succeed in the development of a biotechnological system which meets all the requirements mentioned above. The 'heart' of the system is a Basidiomycete (*Trametes versicolor*), grown in a fermenter on waste lignin with a cheap aminophenol as additional enzyme inducing agent. The culture broth obtained after 4 days of cultivation can be concentrated by evaporation or ultrafiltration and the resulting concentrated, crude enzyme solution directly applied as one component in the binding system. Sterile storage of the enzyme concentrate at room temperature is possible for at least 1 month. Cheaper sources of genetically engineered laccase are currently being developed and these will soon enter the market as agents for the bleaching of pulp. (Fisher *et al.*, 1996).

A number of years ago it was shown that, in principle, it was possible to get good adhesion with this process, and good tensile strength of binding was achieved between boards when glued together (Hüttermann and Haars, 1981). The greatest problems were posed by the required water resistance of the particle boards. The enzyme reaction takes place in water; the enzyme is water soluble and requires, in principal, a hydrophilic substrate. The reaction product, however, must be water insoluble and hydrophobic in order to meet technical requirements. The first boards made with lignosulphonate had good tensile strengths but immediately disintegrated into pieces when they came into contact with water, because of the high hydrophilicity of the lignosulphonate. Current solutions to the problem involve the use of mainly water-insoluble lignins with the addition of small amounts of petrochemical resins. In this way, good quality particle boards can be achieved (Figure 9.2) with adhesives where more than 80 per cent of the petrochemicals are replaced by natural compounds and which are completely emission-free during both production and use (Kharazipour *et al.*, 1991).

Similar results were found by Yamaguchi *et al.* (1991, 1992) who applied dehydrogenative polymerization of vanillic acid with laccase for the binding of thermo-mechanical pulp and obtained an increase in plybond strength of the paperboard. The underlying mechanism for the increase in mechanical properties is the copolymerization of the new lignin with the residual lignin on the surface of the thermo-mechanical pulp (Yamaguchi *et al.*, 1994). Jin *et al.* (1990) used brown-rotted lignin as a binder for wood composites, together with either laccase or peroxidase.

The overall advantages of lignin based binders can be summarized as follows:

- Waste lignin, for which there is presently very little demand, could be used for a new technical process.

- The currently used petrochemical binders would be replaced by natural compounds.

- Lignin-based binders would have no emission problems either during production of the boards or in the homes of customers.

- Due to the high catalytic activity of the enzyme, the binding process could be performed under mild conditions without application of large amounts of highly reactive and potentially harmful chemicals.

9.4 Enzymatic Activation of Wood Fibres for the Production of Wood Composites

Several strategies have been developed to utilize the binding material which naturally glues wood fibres together for the binding of the wood composites. For

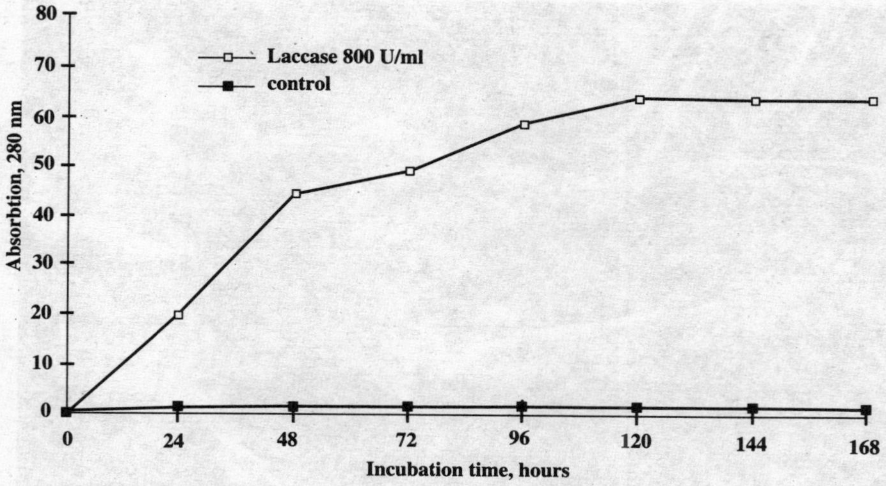

Figure 9.3 Kinetics of release of lignin from the fibre surface with respect to enzyme concentration and time of incubation

example, Klauditz and Stegmann (1955) demonstrated that water resistant bonds can be obtained by pressing woody material at high forces (100 kg cm^{-2}) and temperatures (200°C) for extended periods. During these conditions, binding is obtained by pyrolytic degradation of cell wall constituents. The authors calculated that small amounts of pyrolysis products (0.0572 g/100 g of wood fragments) would result in the formation of stable bonds and that the thickness of the glueline would be rather thin, 0.015 nm compared with 0.34 nm in resin-bonded particle boards. More recent reviews on this approach (Zavarin, 1984; Ellis and Paszner, 1994) have demonstrated that composites with reasonable mechanical properties can be achieved. Using modified methods, however, the problem which usually remains to be solved is the high swelling of these boards in the presence of water. Such composites have therefore not yet been produced commercially.

Indications that a pretreatment of board ingredients with fungal enzymes might not only lessen the mechanical strength of wood but may even convey intrinsic binding forces to the cell walls of fibres treated with fungi came from Kühne (1994). Wood chips incubated, via solid state fermentation, with either white- or brown-rot fungi have a lower energy demand when these wood chips are converted to fibres. There is also a decreased demand for petrochemical resins when the fibres are pressed to form boards (Körner, 1993).

During the processes for wood fibre production, the lignin of the middle lamella, which is the natural glue between the woody cells, is plastified at temperatures above its glass transition point in order to separate the cells. After cooling to room temperature, the lignin solidifies again and forms a glassy crust on the surface of the wood fibre. The crust forms a barrier which reduces the binding strength of any added resin (Wagenführ, 1988), resulting in the need for large amounts of binders for the production of fibre-boards.

To reactivate the fibre surface for bonding, the following requirements must be fulfilled:

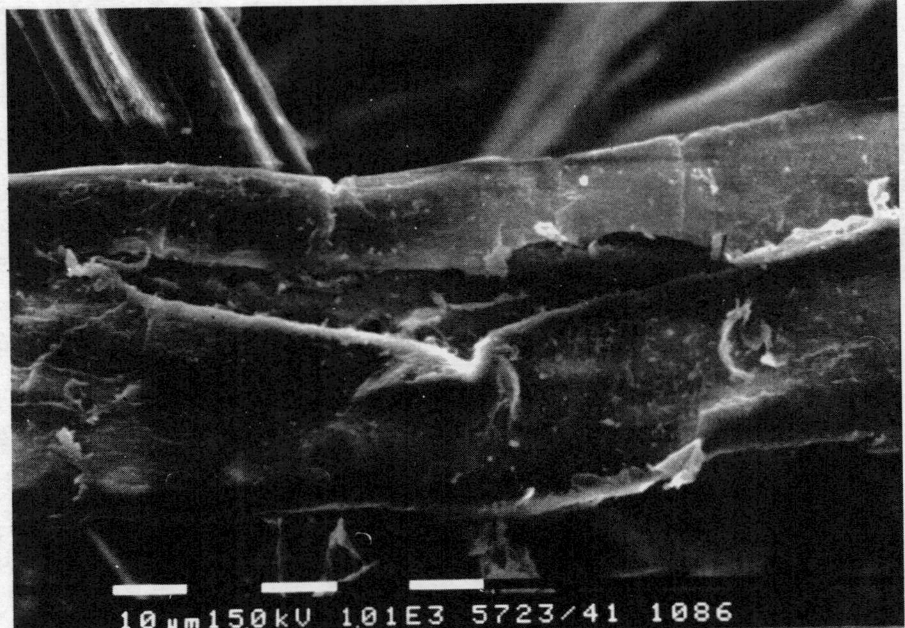

Figure 9.4a

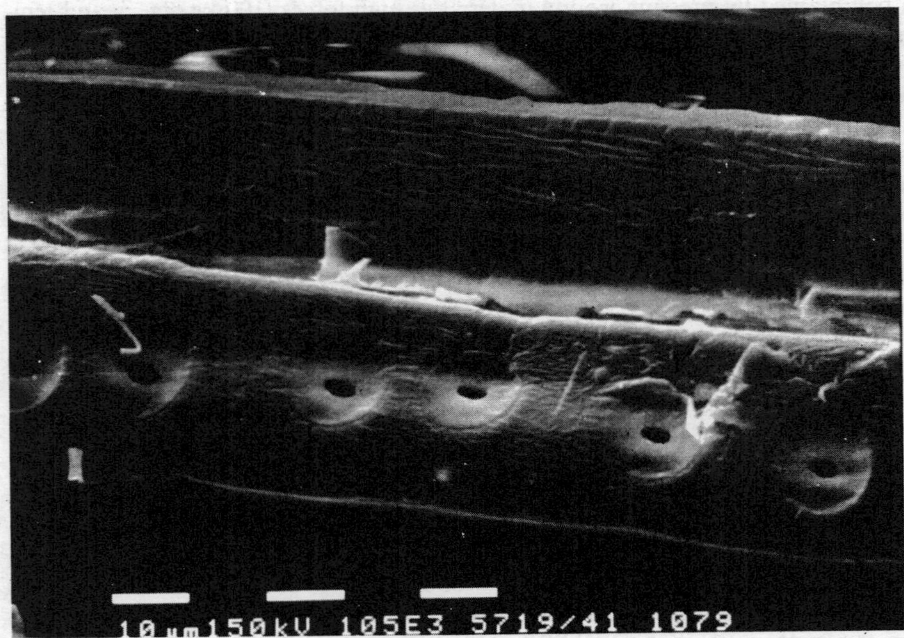

Figure 9.4b

Figure 9.4   Scanning electron microscope pictures of fibre surfaces. (a) Surface of a typical fibre obtained by the defibration process. The surface is covered by a thick crust of the former middle-lamella lignin. (b) Surface of a typical fibre after incubation with laccase solution for 72 hours. The crust of the former middle-lamella lignin is completely removed

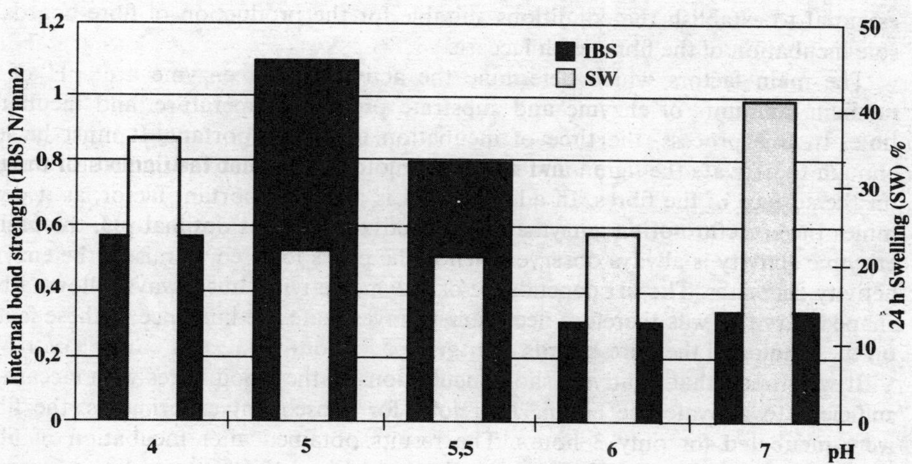

Figure 9.5 Technical properties of 5-mm MDF-boards, pressed for 3 minutes at 190°C; fibres were treated with laccase at different pH values during incubation

- the glassy middle lamella lignin must be transformed into an active state;
- the cellulose fibrils on the surface of the fibres must be loosened; and
- the activated lignin must be enzymatically polymerized to form stable bonds between the fibres.

Figure 9.3 shows that a treatment of woody fibres with laccase leads to a gradual release of lignin (measured by optical density at 280 nm) into the supernatant. The speed of the release of lignin from the fibre surface is dependent on the enzyme activity (i.e. concentration) of the solution. With high enzyme activity, maximal release is achieved after 2 days, with lower activities after 5 days. This reaction is also dependent on pH and temperature of the incubation (Kharazipour and Hüttermann, 1994). Inspection of the treated fibres by scanning electron microscopy revealed that the originally present crust of lignin (Figure 9.4) was removed completely by treatment with the enzyme.

These data indicate that the lignin from the fibre surface is susceptible to reaction with the enzyme. On the basis of our present knowledge of the reaction of laccase with lignin (Eriksson, 1990; Eriksson *et al.*, 1990; Youn *et al.*, 1995) the following interpretation of the above presented data can be offered: the substrate of the laccase reaction is the lignin of the fibre surface. During incubation, the following reactions take place: oxidation; partial polymerization with perhaps concomitant degradation and possible repolymerization; and cleavage of lignin–carbohydrate bonds.

The above data indicate that lignin from the fibre surface can be activated by the laccase reaction. Based on experience with use of the laccase–lignin system as a binder for particle-boards, further work was undertaken to test whether this reaction might serve as the basis for the formation of strong physico-chemical and perhaps even covalent bonds between fibres in wood composites. It was therefore essential to establish the conditions suitable for the production of fibre-boards by sole incubation of the fibres with laccase.

The main factors which determine the activity of an enzyme are: pH of the medium, amounts of enzyme and substrate present, temperature, and incubation time. In this process, the time of incubation is very important. It must be long enough to activate the lignin and not too long to ensure that the lignin still remains on the surface of the fibres. In addition, pH is a very important factor, as it determines the structure of the enzyme and its active centre. At optimal pH, the highest enzymic activity is always observed. When the pH is lowered or raised, the enzymic activity decreases. The pH dependence of enzyme activity thus always follows a bell-shaped curve. It was therefore necessary to investigate the influence of these factors on the binding of the fibre-boards.

It was found that relatively short incubations of the wood fibres with laccase are sufficient to activate the lignin. Therefore, for subsequent experiments, the fibres were incubated for only 3 hours. The results obtained after incubation of fibres with laccase at different pH values are shown in Figure 9.5. The mechanical properties of the boards follow exactly the pH curve of laccase: the maximal internal bond strength and the minimal swelling was obtained at pH 5, the pH optimum of this enzyme. These values are comparable with the ones obtained with urea–formaldehyde resins as binders for fibre-boards.

Peroxidase also appears able to activate the lignin at the fibre surface (Figure 9.6). A clear dependence of the mechanical properties on the pH value of the incubation medium was again found, but the best values were obtained at pH 6, which corresponds to the pH optimum of the Mn-peroxidase used as a catalyst. The

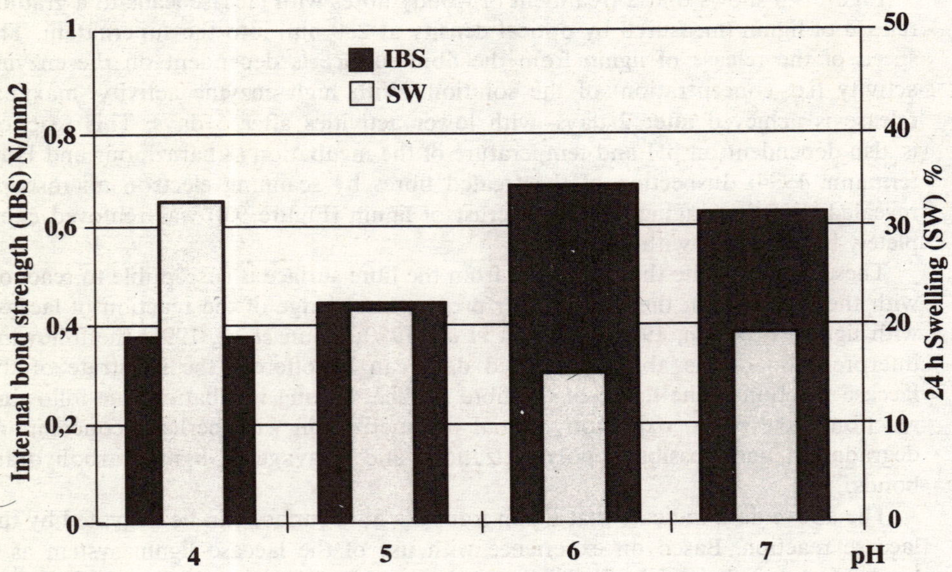

Figure 9.6 Technical properties of 5-mm MDF-boards, pressed for 3 minutes at 190°C; fibres were treated with Mn-peroxidase and H_2O_2 at different pH values during incubation

required German standards, DIN-Norm, for fibre-boards have already been fulfilled by the best laccase and peroxidase bound boards.

Studies on the biotechnological production of wood composites have shown that it is possible to obtain particle and MDF-boards which are bound mainly or even solely by lignin, the natural binder which glues the cells in the wood together when polymerized *in situ* by oxidative enzymes. Thus it may be possible to transfer the naturally occurring process of wood formation to the industrial process of wood composite production.

References

DEAN, J. F. D. and ERIKSSON, K.-E. (1994) Laccase and the deposition of lignin in vascular plants. *Holzforschung* **48**(S), 21–33.

ELLIS, S. and PASZNER, L. (1994) Activated self-bonding of wood and agricultural residues. *Holzforschung* **48**(S), 82–90.

ERIKSSON, K.-E. L. (1990) Biotechnology in the pulp and paper industry. *Wood Sci. Technol.* **24**, 70–102.

ERIKSSON, K.-E. L., BLANCHETTE, R. A. and ANDER, P. (1990) *Microbial and Enzymatic Degradation of Wood and Wood Components*, Berlin: Springer-Verlag, pp. 225–269.

FISCHER, K., FURUMOTO, H. and BROWN, I. D. (1996) Bleaching through enzymes via scaling of technology. In: Srebotnik, E. and Messner, K., eds, *Biotechnology in the Pulp and Paper Industry*, Wien: Facultas Universitätsverlag, pp. 39–42.

HIGUCHI, T. (1990) Lignin biochemistry: biosynthesis and biodegradation. *Wood Sci. Technol.* **24**, 23–63.

HÜTTERMANN, A. and HAARS, A. (1981) Mikrobielle Transformation von Ligninsulfonsäure. In: Hüttermann, A., ed., *Der Wald als Rohstoffquelle*, Frankfurt: Sauerländer, pp. 184–194.

JIN, L., SELLERS, T., SCHULTZ, T. P. and NICHOLAS, D. D. (1990) Utilization of lignin modified by brown-rot fungi. *Holzforschung* **44**, 207–210.

KHARAZIPOUR, A. and HÜTTERMAN, A. (1994) Enzymatische Behandlung von Holzfasern als Weg zu vollständig bindemittelfreien Holzwerkstoffen. In: Hüttermann, A. and Kharazipour, A., eds, *Die Pflanzliche Zellwand als Vorbild für Holzwerkstoffe*, Frankfurt: Sauerländer, pp. 83–98.

KHARAZIPOUR, A., HAARS, A., SHEKHOLESLAMI, M. and HÜTTERMANN, A. (1991) Enzymgebundene Holzwerkstoffe auf der Basis von Lignin und Phenoloxidasen. *Adhäsion* 30–36.

KLAUDITZ, W. and STEGMANN, G. (1955) Beiträge zur Kenntnis des Ablaufes und der Wirkung thermischer Reaktionen bei der Bildung von Holzwerkstoffen. *Holz Roh-Werkstoff* **13**, 434–440.

KÖRNER, I. (1993) Untersuchungen zur Inokulation und Fermentation von unsterilen Hackschlitzeln mit Braunfäulepilzen fürdie Herstellung von MDF-Platten, unpublished PhD thesis, Technische Universität Dresden.

KÜHNE, G. (1994) Modifizierung von Rohholz durch Pilze für die Holzwerkstoff-herstellung. In: Hüttermann, A. and Kharazipour, A., eds, *Die Pflanzliche Zellwand als Vorbild für Holzwerkstoffe*, Frankfurt: Sauerländer, pp. 55–68.

NIMZ, H., RAZVI, A., MARQUHARAB, I. and CLAD, W. (1972) Bindemittel bzw. Klebemittel zur Herstellung von Holzwerkstoffen sowie zur Verklebung von Werkstoffen verschiedener Art, German Patent DOS 222 1353.

WAGENFÜHR, A. (1988) Praxisrelevante Untersuchungen zur Nutzung biotechnologischer Wirkprinzipien bei der Holzwerkstoffherstellung, unpublished PhD thesis, Technische Universität Dresden.

YAMAGUCHI, H., NAGAMORI, N. and SAKATA, I. (1991) Applications of the dehydrogenative polymerization of vanillic acid to bonding of woody fibers. *Mokuzai Gakkaishi* **37**, 220–226.

YAMAGUCHI, H., MAEDA, Y. and SAKATA, I. (1992) Applications of the phenol dehydrogenative polymerization by laccase to bonding among woody fibres. *Mokuzai Gakkaishi* **38**, 931–937.

YAMAGUCHI, H., MAEDA, Y. and SAKATA, I. (1994) Bonding among woody fibers by use of enzymatic phenol dehydrogenative polymerization. Mechanism of generation of bonding strength. *Mokuzai Gakkaishi* **40**, 185–190.

YOUN, H.-D., HAH, Y. C. and KANG, S.-O. (1995) Role of laccase in lignin degradation by white-rot fungi. *FEMS Microbiol. Lett.* **132**, 183–188.

ZAVARIN, E. (1984) Activition of wood and nonconventional bonding. In: Rowell, R., ed., *The Chemistry of Solid Wood*. Advances in Chemistry Series 207, Washington DC: American Chemical Society, pp. 372–400.

Special (Secondary) Metabolites from Wood

JOHN R. OBST

10.1 Introduction

Flavonoids, lignans, terpenes, phenols, alkaloids, sterols, waxes, fats, tannins, sugars, gums, suberins, resin acids and carotenoids are among the many classes of compounds known as 'secondary metabolites'. This daunting array of substances, having different chemical, physical and biological properties, presents numerous challenges in the utilization of forest products. Do they also present opportunities?

Secondary metabolites are often defined on the basis of what they are not. To wit, primary metabolites are usually described as those substances that are the fundamental chemical units of living plant cells, such as nucleic acids, proteins and polysaccharides. Secondary metabolites may therefore be defined as being everything else that the organism produces. Intuitively, this definition is a little difficult to accept: why would a plant expend so much energy to produce materials that it does not need? Especially because some of these 'secondary' compounds are vital to its very existence.

In the realm of wood processing and utilization, there is a very pragmatic definition of secondary metabolites: they are everything that is not a structural polysaccharide or lignin. In this sense, secondary metabolites are often referred to as 'extraneous components' because they are mostly extraneous to the lignocellulosic cell wall and are concentrated in resin canals and cell lumina, especially those of ray parenchyma cells. These types of compounds are however actually found in all morphological regions and this definition cannot be strictly applied.

While such a definition emphasizing the structural components of wood is very functional, it can give the impression of demeaning the role of these 'extraneous components'. That is unfortunate, for if polysaccharides and lignins are the bones and flesh of woody tissue, it is the secondary metabolites that give woody plants their blood, soul and character. These components endow woods with their many colours and hues, scents and beauty. An excellent suggestion for recognizing their roles and impact is to refer to these materials as 'special metabolites' (Gottlieb, 1990).

Although much is known about the formation, composition and function of special metabolites, a question repeatedly heard is: 'Why are they formed?' Although not very satisfying, perhaps the best answer may be that the reason that many of these compounds are formed is a conundrum. It has been commonly taught that while animals have developed efficient systems to excrete unwanted by-products, plants have few such mechanisms and usually must alter and then store their wastes ('secondary metabolites') (Manitto, 1981). Others do not believe that the high energy cost of biosynthesizing most special metabolites is consistent with waste product removal. In any case, such processes have led to successful, serendipitous 'strategies'. For example, heartwood extractives retard wood decay, resin formation protects wounded tissues and toxic and antifeedant compounds in foliage and bark minimize insect and animal browsing through their poisonous, unpalatable or emetic properties. It is worth noting that these compounds may be toxic to the producing plant and the wide variety of phytochemicals produced may be part of a selection process to minimize plant toxicity while maximizing protection (Gottlieb, 1990).

Varying by species, woods may contain as little as 1 per cent or as much as one-third of their dry weights as special metabolites. Tropical and sub-tropical species typically contain greater amounts of extractives than do temperate zone woods. The concentration of special metabolites in trees is not uniform; generally, higher amounts occur in bark, heartwood, roots, branch bases and wound tissues. Variations also occur among species, from tree to tree, and from season to season. Most of the special products may be extracted with neutral solvents, and such extracts are sometimes the source of useful materials. It would be impossible to review adequately here the structures and properties of the thousands and thousands of compounds isolated and identified from trees. Instead, a brief overview is presented to sample the breadth and variety of special metabolites. Biosynthesis, isolation, purification, characterization, stereochemistry, reactions and derivatives generally will not be included. Fortunately, a number of treatises have been written on these subjects and the reader is referred to these sources, and the citations therein, for in-depth information. Some aspects of special metabolites, such as their production for use as medicinals, adhesives and preservatives, are discussed in other chapters.

10.2 Terpenes (Terpenoids)

Thousands of terpenes (terpenoids) have been isolated, purified and structurally elucidated, and this group represents the largest class of chemical compounds that occur as special metabolites. Terpenes are found throughout nature and occur in almost all plants. As a consequence, terpenes have been exploited since antiquity as perfumes, flavouring agents, waterproofing materials, insect repellents, fungicides and medicinals. Many continue in traditional uses while others provide the raw materials for the commercial synthesis of numerous high-value products.

Terpenes are derived from isoprene (isopentane) C_5 building blocks. By definition, a monoterpene is a compound of two isoprene-derived units totalling at least 10 carbon atoms. Hence, and perhaps a little confusing, the simplest terpenes, those

152

with a C_5 skeleton, are termed hemiterpenes. Hemiterpenes, such as isoprene, and prenyl and isoprene alcohols, generally do not accumulate in large amounts in woody plants and are not exploited as forest products.

Systematic chemical nomenclature is generally not used for terpenes due to the history of their isolation and their complexity; 'trivial' or common names are used instead. Monoterpenes may be either cyclic or acyclic (Figure 10.1). Examples of some of the acyclic monoterpenes (Dev, 1989) include citronellol (used in perfumes and isolated from *Chamaecyparis lawsoniana*), citronellal (the major component of citronella oil; used in soap perfumes and as an insect repellent) and linalool (which also may be synthesized from the more readily available β-pinene; used in the synthesis of fragrances and vitamins). Isomeric geraniol and nerol have sweet rose-like odours and are used as perfume bases.

An example of a single ring cyclic monoterpene is limonene, the major component of turpentines from some *Pinus* species (*P. pinea, pinceana* and *lumholtzii*). Other examples are α-terpinene, from *Pinus sylvestris*, terpinolene, a minor component in some heartwood oils and in some *Pinus* turpentines, and β-phellandrene, the major component of *Pinus contorta* turpentine (Figure 10.1). Bicyclic monoterpenes (Figure 10.1) are widely distributed; some occur in significant amounts and are extensively utilized. Abundant in turpentines from coniferous woods, α- and β-pinene, camphene and 3-carene are common examples of bicyclic monoterpenes (Dev, 1989). The composition of turpentines is species dependent (Fengel and Wegener, 1984). For example, the monoterpene fraction of *Pinus balfourniana* contains 81 per cent α-pinene and only 1.9 per cent β-pinene whereas that from *Pseudotsuga menziesii* contains 31 per cent α-pinene and 36 per cent β-pinene. Even greater variation is displayed by *Pinus heldreichii* which yields 82.8 per cent limonene and only 11.5 per cent of the pinenes. Other examples of monoterpenes include thujene, from *Boswellia serrata*; chaminic acid, present in the heartwood of *Chamaecyparis nootkatensis*; borneol, first isolated as crystals deposited by *Dryobalanops camphora*; and camphor, from *Cinnamomum camphora* (Dev, 1989) (Figure 10.1).

Considering all natural sources, over 2500 sesquiterpenes (three isoprene units, C_{15}) have been identified, forming the largest structural subclass of terpenes with at least 120 different skeletal types (Klyne and Buckingham, 1978; Dev, 1989). Cedrene, which is tricyclic, occurs in cedar wood oil (*Juniperus virginiana*). Eudesmol, bicyclic with two six-membered rings, has been isolated from the heartwood of *Callitropis araucriodes*. α-Santalol (Figure 10.1), which is tricyclic, is the major component of sandal oil (Srinivasan *et al.*, 1992) – an essential oil which is especially valuable in perfume formulations. Widdrol, bicyclic with six- and seven-membered rings, occurs in cedar woods (*Cedrus, Chamaecyparis, Juniperus*, etc.). Guaiol (Figure 10.1) is bicyclic with five- and seven-membered rings and comprises 30 per cent of the oil from guaiacum wood. Abscisic acid is a relatively simple monocyclic sesquiterpene but it has been found to be a natural plant growth regulatory hormone with widespread occurrence in the higher plants. A number of sesquiterpenes are primarily responsible for the odour of wood and have been tabulated (Imamura, 1989). Farnesene, which has an acyclic structure, is one of a suite of odiferous sesquiterpenes that are responsible for the characteristic aroma of *Juniperus* species.

Classes of special metabolites from trees have often been described on the basis of historic or common uses and isolations. 'Essential oil' is one such group. Generally obtained by steam distillation, these compounds are the source of the distinctive

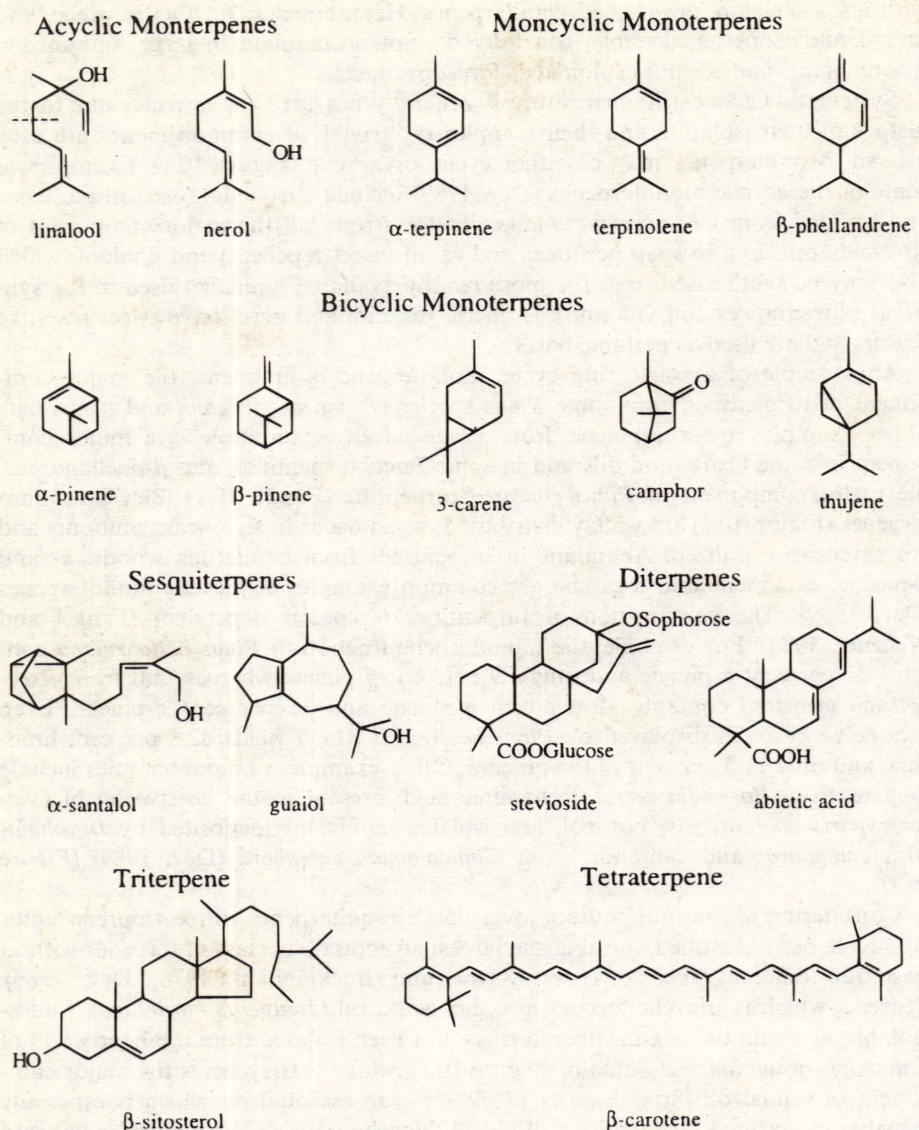

Figure 10.1 Terpenes are comprised of isoprene units which are generally linked in 'head to toe' fashion, as indicated by the dashed line on the structure of linalool. Although only a few examples of terpenes are presented, their wide range of composition and structural features are readily apparent

odour, or essence, of the tree. Essential oils that are particularly fragrant are valued as ingredients in perfumes. The variation in composition of these oils allows some wood identifications by odour (or analysis) and has seen limited utilization in taxonomic studies.

Given this relatively non-selective method of isolation and the awareness of the wide range of types of special metabolites, it might be a little surprising to learn that most essential oils are comprised mainly of terpenes. Although some essential oils are solids at room temperature, the use of the term 'oil' may be somewhat misleading as most of the oils are mixtures of low molecular weight, non-viscous, highly volatile compounds, typically mono- and sesquiterpenes, for example, 'oil of turpentine'. Monoterpenes, especially, are used in the synthesis of commercially important products.

Comprised of four isoprene units (C_{20}), diterpenes occur commonly in plants and woods. Phytol, an acyclic diterpene with a simple structure, is particularly ubiquitous as it comprises about one-third of the molecule chlorophyll and is found in the leaves of all green plants. Stevioside (Figure 10.1), isolated from the leaves of the shrub *Stevia rebaudiana* (Kohda *et al.*, 1976), is a terpene glucoside with the unique characteristic of being 300 times sweeter than sucrose (Noller, 1961). Stevioside-containing extracts have long been used as sweeteners and the pure compound is approved for use in food in several countries. Gibberellic acid, a tetracyclic lactone, belongs to a class of diterpenes (gibberellins), which occur widely in plants and function as phytohormones. Many of the gibberellins have one less carbon and are C_{19} compounds. Abietic, pimaric, communic and lambertianic acids are representative of a class of diterpenes known as resin acids (Figure 10.1). These acids commonly occur in rosins from gymnosperm woods, particularly the pines. Resin acids, which are obtained as by-products from the Kraft pulping of wood, are commonly used as sizing agents to adjust the absorption of water in paper products.

'Naval stores' is another of the historic categories of special metabolites (Zinkel and Russell, 1989). This term is derived from the use of pine tar and pitch to waterproof and caulk wooden boats and ships. Traditional sources of naval stores were from gums (pines were wounded and the exudate, called pine gum or oleoresin, was collected) and the destructive distillation or extraction of the heartwood from resinous pine stumps. These processes have been largely supplanted by the isolation of 'tall oil' from the Kraft (sulphate) pulping process which is used to produce paper-making fibres. The term 'tall oil' comes from the Swedish '*tallolja*' which means pine oil; care needs to be exercised not to confuse tall oil with terpineol which is commonly called pine oil. Naval stores chemicals are fractionated to yield turpentine and rosin. Turpentine composition from the same wood species varies depending on whether the isolation is from the exudate, wood or tall oil (Zinkel, 1981).

Sesterpenes (five isoprene units, C_{25}) occur uncommonly and initially were isolated from insect waxes and as fungal metabolites. They rarely occur in higher plants and are not commercially significant. In contrast, triterpenes (C_{30}) are widely distributed, especially among angiosperm plants (Mahato *et al.*, 1992). Examples of this class are the tetracyclic dammarenediol, isolated from *Shorea vulgaris* resins (Dev, 1989), and the pentacyclic betulin, a white pigment present in birch bark (*Betula alba*) and isolated from many other plants (Hayek *et al.*, 1989). Sterols, which are biosynthesized from squalene oxide, are found in a number of gymnosperm and angiosperm woods (*Larix*, *Abies*, *Picea*, *Pinus*, *Gmelina*, *Fagus*, *Quercus* and *Ulmus*). β-Sitosterol (Figure 10.1) is a commonly occurring sterol and is the major component of the sterol fraction from a number of conifers; campesterol, dihydrobrassicasterol, 24-methylstanol and 5-α-sitostanol often co-occur with β-sitosterol (Nes, 1989). Sterols may also occur as fatty acid esters (Dyas and Goad, 1993) or as glycosides. Some (termed saponins) contain several glycosides and produce a lather in water; others (functionally defined as cardiac glycosides) have particularly strong effects on the heart muscle and can function either as a medicine or as a poison. Although the concentration of sterols in heartwood is low (often less than 0.1 per cent) relatively large amounts of sterols may be isolated as by-products from Kraft pulping (tall oil).

Tetraterpenes (C_{40}) are comprised of the carotenoids, which include the red and yellow plant pigments (Robinson, 1991). The carotenoids, such as β-carotene (Figure 10.1), occur in leaves, flowers, seeds and fruits, but not in wood. Carotenoids are used as food colourings, in vitamin synthesis, and in cosmetics as suspensions, emulsions, lotions, lipsticks and powder bases.

Rubber and gutta-percha are polymeric isoprenes (containing 3000–6000 monomer units). The elastic properties of rubber are attributed to its all *cis* structure, whereas the all *trans* structure of gutta confers rigidity (Barlow, 1989; Robinson, 1991). The annual consumption of natural rubber is about 5.9 million tonnes (Reisch, 1996).

10.3   Phenolic Compounds

A number of simple phenols (Figure 10.2) have been isolated from various tissues of trees (Theander and Lundgren, 1989). Generally, members of this class have little or no value as individual, isolated, commercial products, but some, for example, safrole (oil of sassafras), eugenol (oil of cloves), methyl salicylate (oil of wintergreen) and cinnamic acid (oil of cinnamon), are relatively volatile, have characteristic aromas, and contribute significantly to the overall properties and qualities of essential oils. Some of these simple phenolics may also occur as glycosides, such as salicin, a bitter antipyretic which is found in relatively large amounts in many *Populus* and *Salix* species (Lee *et al.*, 1993). Coniferyl and sinapyl alcohols also occur as glucosides and it is this form that may be the immediate precursor to lignin.

Condensation reactions of simple phenolics are likely to be responsible for the formation of biphenyls. Ellagic acid is a dimer of gallic acid and is an important constituent of hydrolyzable tannins. Stilbenes, probably formed from cinnamic acid, are exemplified by pinosylvin (Figure 10.2) and resveratrol. Stilbenes are most commonly found in the heartwood of *Pinus* species and may occur with phenolic hydroxyls, methylated or as glycosides (Norin, 1989). Dimers, trimers and oligomers of stilbenes have been identified and have complex structures and stereochemistries. Many of the phenolics are implicated in various defence mechanisms.

Over 4000 different flavonoids have been isolated from plants and they commonly occur in foliage, bark, sapwood and heartwood in trees. Their function is thought to be that of providing resistance to attack by fungi and insects. This subclass name is derived from the flavan (2-phenylchroman) skeleton which gives rise to a number of structural variations. The flavonoid group includes chalcones, aurones, flavanones, flavones, isoflavones, flavanonols, flavonols, flavan-3,4-diols (leucoanthocyanidins), flavan-3-ols (catechins) and anthocyanidins.

Isoflavonoids have a unique structural variation; their basic skeleton is a 1,2-diphenylpropane in contrast to the 1,3-diphenylpropane skeleton of the flavonoids. Most common in legumes, isoflavonoids have been reported in the Pinaceae (Tahara and Ibrahim, 1995). Aurones and chalcones rarely occur in wood (Zavarin

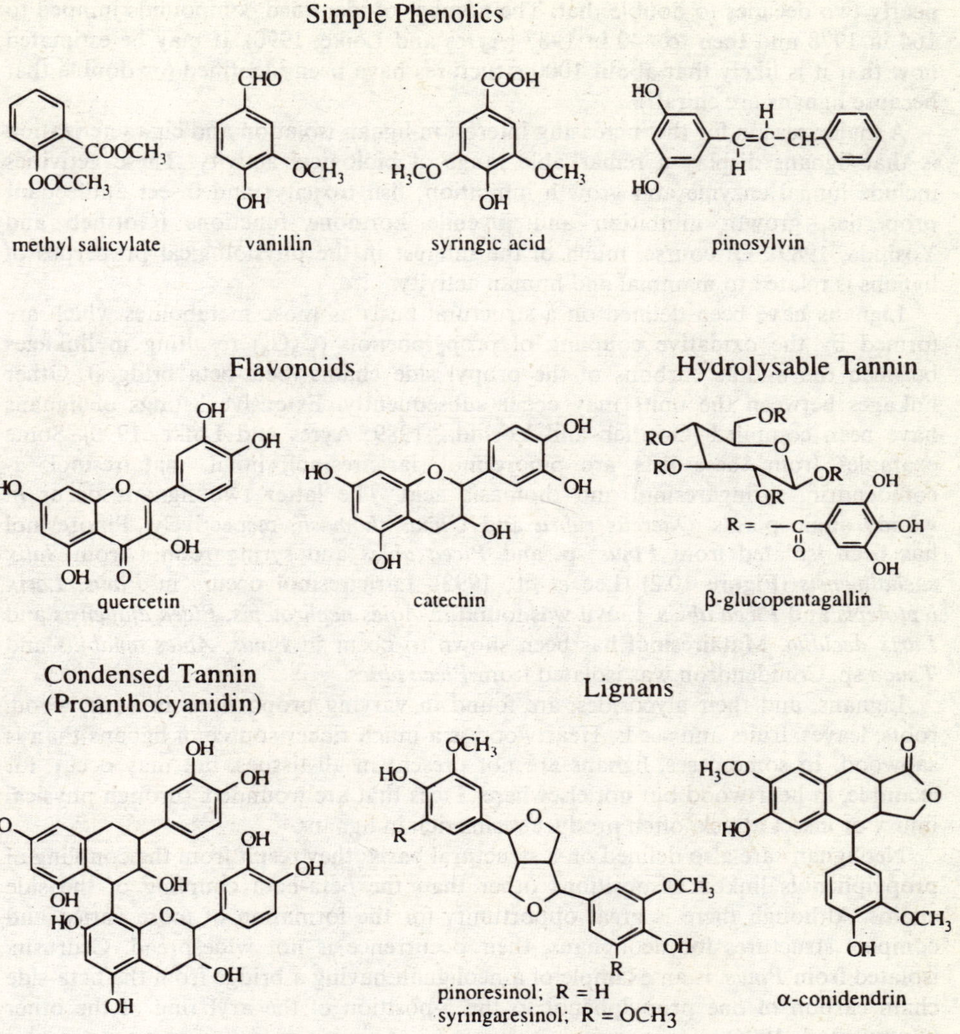

Simple Phenolics

methyl salicylate   vanillin   syringic acid   pinosylvin

Flavonoids

Hydrolysable Tannin

quercetin   catechin   β-glucopentagallin

Condensed Tannin (Proanthocyanidin)

Lignans

pinoresinol; R = H
syringaresinol; R = OCH$_3$

α-conidendrin

Figure 10.2 Many of the special metabolites of wood are phenolic in nature. Examples of a few of the various classifications are shown

and Cool, 1991). Red and blue pigments in flowers and fruits are due to anthocyanidins, but this group is absent in wood. Flavonoids reported in wood and bark have been summarized by Harborne (1989). Quercetin (Figure 10.2) is one of the most common flavonoids isolated from the bark of conifers. Some flavonoids, such as catechin (Figure 10.2), occur in both angiosperm and gymnosperm woods. Flavonoids may couple oxidatively to form dimers and oligomers (Hemingway, 1989).

10.4 Lignans (Neolignans and Norlignans)

The number of isolated and identified lignans has grown exponentially since the 1930s. In a review of lignans in 1936, only 14 compounds were described; it took nearly two decades to double that. The number of identified compounds jumped to 164 in 1978 and then to 440 in 1987 (Ayres and Loike, 1990). It may be estimated now that it is likely that about 1000 structures have been identified (or double that because lignans are chiral!).

A major reason for the increasing interest in lignan isolation and characterization is that lignans display a remarkable range of biological activity. These activities include fungal enzyme and growth inhibition; fish toxicity; and insect antifeedant properties, growth inhibition and juvenile hormone functions (Gottlieb and Yoshida, 1989). Of course, much of the interest in the physiological properties of lignans is related to mammal and human activity.

Lignans have been defined on a structural basis as those metabolites which are formed by the oxidative coupling of propylphenols (C_3C_6) resulting in linkages between the middle carbons of the propyl side chains (beta-beta bridges). Other linkages between the units may occur subsequently. Extensive listings of lignans have been compiled (Gottlieb and Yoshida, 1989; Ayres and Loike, 1990). Some examples from these lists are pinoresinol, lariciresinol, liovil, matairesinol, α-conidendrin, syringaresinol and thomasic acid. The latter two lignans occur in woody angiosperms, *Quercus rubra* and *Ulmus thomasii*, respectively. Pinoresinol has been isolated from *Pinus* sp. and *Picea abies* and syringaresinol from *Salix sachalinensis* (Figure 10.2) (Lee *et al.*, 1993). Lariciresinol occurs in *Pinus*, *Larix leptolepis* and *Picea abies*. Liovil was found in *Abies nephrolepis*, *Picea ajanensis* and *Larix decidua*. Matairesinol has been shown to occur in *Pinus*, *Abies amabilis* and *Tsuga* sp. Conidendron was isolated from *Picea abies*.

Lignans, and their glycosides, are found in varying proportions in bark, wood, roots, leaves, fruits and seeds. Heartwood is a much richer source of lignans than is sapwood. In some cases, lignans are not present in all tissues, but may occur, for example, in heartwood but not elsewhere. Trees that are wounded, through physical injury or insect attack, often produce resins rich in lignans.

Neolignans are also defined on a structural basis: they result from the coupling of propylphenols linked at positions other than the beta-beta coupling of the side chains. Although there is great opportunity for the formation of more varied and complex structures for neolignans, their occurrence is not widespread. Cedrusin, isolated from *Pinus*, is an example of a neolignan having a bridge from the beta-side chain carbon of one propylphenol to the 5-position of the aryl ring of the other (Castro *et al.*, 1996).

Norlignans (and norneolignans) are structurally similar to lignans but have one less carbon atom and are C_{17} structures. Norlignans are common in the Taxodiaceae (Castro *et al.*, 1996). Coupled through the beta-carbons and having one side chain with only two carbon atoms, yateresinol, found in *Libocedrus*, is a simple example of a norlignin. Norlignins are also sometimes termed conioids or sequirins (Zavarin and Cool, 1991).

10.5 Tannins

Tannins, which are oligomeric and polymeric phenolics, are widely distributed and are common both in gymnosperms and angiosperms. Species of *Acacia*, *Quercus*, *Betula*, *Salix* and *Pinus* are examples of tannin-containing trees. Considered to have antifeedant properties, tannins are found in bud and foliage tissues, seeds, bark, roots, sapwood and heartwood. Bark and heartwood are often the regions that contain the highest levels. In common with other plant phenolics, tannins are formed via the shikimic acid pathway.

The term 'tannin' is based on historic uses of plant materials to produce leather from animal skins. Because a wide variety of phenolics can act as tanning agents, more rigorous definitions have been employed. As a result, tannins have been separated into two classes based on their chemistry and origin. One class is the hydrolyzable tannins, which is further categorized as gallotannins (Figure 10.2) and ellagitannins; gallic and ellagic acids, respectively, are essential components. Such esters are easily hydrolyzed, hence the pertinent term 'hydrolyzable' tannins. Gallotannins consist of a saccharide core (usually glucose) with multiple esters of gallic acid, which in turn may be esterified by other gallic acids. Ellagitannins, which occur in dicotyledonous angiosperms, are monomeric and oligomeric, often containing gallate groups. Over 500 ellagitannins have been characterized (Quideau and Feldman, 1996). Hydrolyzable tannins have been described, classified, and their taxonomic significance discussed (Haslam, 1989; Okuda, *et al.*, 1993).

In contrast to hydrolyzable tannins are the 'condensed tannins'. These materials are not significantly hydrolyzed under the mild conditions used to degrade the hydrolyzable tannins. However, emphasizing their origin rather than their behaviour, this group is more correctly referred to as proanthocyanidins (or polyflavanoids) as they are derived from flavonoids (Porter, 1989). The proanthocyanidins are the most commonly found class of tannins and are composed of over a hundred different oligomeric and polymeric structures (Lewis and Yamamoto, 1989). Like many of the special metabolites, this class of compounds is proposed to be involved in plant defence mechanisms (Stafford, 1988). An example of a representative proanthocyanidin structure is presented in Figure 10.2 (Lewis and Yamamoto, 1989).

Discussions of the condensed tannins usually include mention of phlobaphenes. These are water-insoluble, condensed phenolics related to proanthocyanidins but have differing functionality, including methoxyl groups (Foo and Karchesy, 1989). As a class of natural products, phlobaphenes are poorly defined chemically. They are generally obtained as crude, heterogeneous products with complex structures and vary depending on the source. Phlobaphenes from Douglas fir (*Pseudotsuga menziesii*) bark may be reaction products of proanthocyanidins, carbohydrates, flavanoid methyl ethers, lignans and dihydroquercitin (Foo and Karchesy, 1989).

10.6 Carbohydrates

Although most of the carbohydrates in woody plants occur as structural polymers (cellulose and hemicelluloses), numerous other compounds such as sugars, oligosaccharides, alditols, cyclitols, and polysaccharides, are commonly found. Starch is ubiquitous in woody tissues, except in heartwood, as it functions as a food and chemical raw material reserve for trees. With the exception of the sago palm, trees have not been exploited as commercial sources of starch (BeMiller, 1989). However, non-structural polysaccharides in gums are utilized – as food additives and binding agents, and in adhesives, paints and varnishes. Such gums are not food reserves; they are formed in response to wounding and usually also contain terpenes and other compounds.

Gum arabic, a pale to orange-brown, slightly acidic polysaccharide solid, is used extensively in food products. It is frequently used as an emulsifier and to prevent the crystallization of confectionary sugar. It is also used to encapsulate flavours and as a glaze (Coppen, 1995). Gum arabic is obtained from *Acacia* species and its composition and properties vary with the sources. D-Galactose, L-arabinose, L-rhamnose, and D-glucuronic acid are formed on acid hydrolysis (Robinson, 1991). The Sudan is the major source of gum arabic, producing 33 227 tonnes in 1994. Other forest products gums include karaya (*Sterculia* species), carob (*Ceratonia siliqua*) and tara (*Caesalpinia spinosa*) (Coppen, 1995).

As intermediates in numerous biosyntheses, monosaccharides occur widely, but only in small amounts as they do not accumulate. Tree sap is a good source of oligosaccharides, particularly sucrose. Sugar maples (*Acer saccharum*) are tapped commercially and the sap is concentrated for use as a food (maple syrup) and flavour ingredient in various products. Interestingly, maple sap has almost no flavour. The unique flavour is developed after boiling the sap in air. While the sap contains mainly sucrose, many other components are present in small quantities, including carboxylic acids, phenols, phenylglycosides, oligosaccharides, vanillin, syringaldehyde, proteins and amino acids. This mixture apparently undergoes reactions, including oxidation, to form the flavour constituents. Some of these have been identified as methyl-, dimethyl-, trimethyl-, and ethyl-pyrazines (Kermasha *et al.*, 1995).

10.7 Cutin and Suberin

Cutin and suberin are often discussed together as they have similar waxy, water-repellent properties. Cutin impregnates the epidermal walls of leaves and stems and also forms a separate layer on the outer wall (cuticle) making the surface impermeable to water as well as forming a barrier to pathogens. Suberin is found in the cork tissue of bark.

Cutin is composed mainly of hydroxy and dihydroxy hexadecanoic and octadecanoic acids whereas suberin is composed of ω-hydroxy C_{16}–C_{24} acids, α,ω-dicarboxylic fatty acids, and C_{20}–C_{30} fatty acids and alcohols (Wallace and Fry, 1994). Phenolic compounds appear to be associated with both, but especially suberin. Suberized tissue has been described as being comprised of aliphatic domains, similar to cutin, linked to aromatic 'lignin-like' domains (Kolattukudy, 1984). However, suberin cannot be isolated intact and degradative analyses so far have not provided sufficient structural details. Thus, it has been stated that not only are the inter-unit bonding patterns unknown, but the monomer composition is also essentially unknown (Davin *et al.*, 1992). Although some efforts have been made in the past to isolate and utilize waxes from bark, cutin and suberin are generally unsuited for commercialization.

10.8 Other Metabolites (Aliphatic Compounds, Fatty Alcohols and Acids, and Nitrogenous Compounds)

Many of the simple chemicals do not accumulate in woody tissues. Compounds such as glycolic, glyceric, pyruvic and malonic acids are generally intermediary along pathways to form other products. However, other simple acids, such as oxalic, lactic and tartaric acids may be metabolic end-products. A product of photosynthesis and carbohydrate biosynthesis, shikimic acid is an important intermediate in the biosynthesis of aromatics (phenolics, lignin precursors, aryl amino acids, flavanoids, etc.)

Alkanes occur in woody tissues and sometimes accumulate. For example, *n*-heptane is a significant component in oleoresin turpentine from *Pinus sabiniana* (Zinkel, 1989). Long-chain hydrocarbons such as fatty alcohols and acids, fats (mono-, di- and triglycerides of fatty acids), and waxes are produced as special metabolites and accumulate. The role of fats and fatty acids is to provide food reserves, whereas waxes, comprised mainly of esters of saturated fatty acids and alcohols, and long-chain alkanes (C_{21}–C_{35}), primarily are present on foliage surfaces to reduce the evaporation of water. Common fatty acids are found in both angiosperm and gymnosperm woods, but linolenic and linolenic acids are found in higher concentrations in angiosperms. An isomeric form of linolenic acid, the *cis*-5,9, 12 acid, occurs in Pinaceae woods (Zinkel, 1989). A commercially significant source of fatty acids is tall oil.

Numerous nitrogen-containing compounds occur in wood and related tissues. That this type of compound is not present in large amounts is illustrated by nitrogen contents of about 0.1 per cent for wood. Some of this nitrogen is due to proteins and enzymes which are residuals left from cell growth and development. The occurrence of amino acids and proteins has been summarized by Durzan (1989). Although foliage and bark from some trees have been used as animal feeds, generally such applications are few and have limited potential.

There is however significant interest in an important class of compounds known as the alkaloids (Pelletier, 1983; Sakai *et al.*, 1989). Alkaloids, many of which have very complex structures, are relatively common. They occur in about 10 per cent of plant genera and have taxonomic relevance (Hegnauer, 1988). Alkaloids typically are found in higher concentrations in bark, seeds, roots and leaves than in wood. The traditional definition of alkaloids is broad and encompassing: they are alkali-like (basic) substances. However, this definition falls short because the functionality and chemistry of some alkaloids produce neutral and even acidic compounds. The alkaloids are comprised of a range of compounds with various functional groups including acids, amides, phenols, urethanes, pyrroles, piperidines, quinolizidines, acridones and indoles. Alkaloids have long been known both as medicines and poisons.

10.9 Application of Biotechnology

Changing the characteristics of plants to improve their cultivation, quality, growth rate, hardiness, and yield has been practised for thousands of years. Over the last century, an understanding of plant genetics and breeding transformed the simple practice of saving and sowing superior seeds into the sciences of agriculture, horticulture and silviculture. A quantum leap forward occurred just two decades ago as the ability to create transgenic plants was accomplished. Can this technique, or other methods of biotechnology, be applied to enhance the bounty and potential of special metabolites?

Cell and tissue cultures grow rapidly, compared with the many, many years of growth required for most trees. Cultures are renewable and could yield forest products without logging – which is sometimes controversial – and the resulting products could be labelled 'natural'. Thus, at first glance, it seems that plant cell cultures have significant potential for exploitation to produce useful quantities of valuable phytochemicals. While laboratory cell cultures can produce compounds such as terpenes, coumarins, anthraquinones, flavonoids, alkaloids and tannins (Tanaka *et al.*, 1995), there has been very limited commercial application in the production of special metabolites by this method (Charlwood and Rhodes, 1990).

There are a number of difficulties encountered in cultures from cambial and bark tissues. A major one is that plant cultures do not often produce the same compounds, or in the same abundance, as the mature plant. Most 'secondary' products are not formed in developing cells; few undifferentiated cell cultures are likely to form meaningful amounts of special metabolites. This problem can sometimes be addressed by introducing agents that act as inducers or elicitors. In trees, stress, such as wounding or infection, will usually trigger increased production of special metabolites. It might therefore be desirable to mimic this reaction in artificial systems. One interesting *in-vivo* application of this principle was to use bacteria to induce gall formation in *Taxus brevifolia* which resulted in the production of a higher level of taxanes (Stahlhut, 1994).

While many special metabolites have significant value, especially the industrially important terpenes and the unique essential oils, culture production methods are not likely to be economic in the near future. A significant barrier to commercialization is that it is estimated that the production of special metabolites becomes feasible only when selling costs are above $1000/kg (DiCosmo and Misawa, 1995). Some of the rare and structurally complicated metabolites that possess great medicinal value, such as the anticancer drug taxol, are however candidates for production via cell cultures.

Biotechnology in forestry and forest products is usually thought of in terms of forest tree improvement, micropropagation, plant breeding, herbicide tolerance, resistance to pathogens, improved pulp properties, bio-pulping and bleaching, and mill effluent treatment. There is great interest in employing genetic engineering to enhance plant resistance to insects since such an approach can be environmentally benign: it targets specific pests without the introduction of toxic materials into the environment. However, pest control by this method generally results in transgenic plants producing insecticidal proteins from gene transfers from bacteria (Boulter, 1993) and not from modifying amounts or types of special metabolites in the plant. Most of these efforts however have been with agricultural crops which are not particularly well-suited for emphasizing the potential role of special metabolites. Pest resistant forest trees are being sought, and although it is complicated to manipulate biosynthetic pathways to change special metabolite production, the development of pest resistance may still occur.

The utilization of most trees harvested for non-fuel purposes falls essentially into two categories: solid products (lumber, plywood, particle board, etc.) and paper products (writing and printing papers, corrugating and packaging papers, tissue, etc.). Because these are major uses, special metabolite applications are likely to be adjuncts to them, as, for example, the isolation of tall oil from pine Kraft pulping liquors. It would seem likely that if genetic transformations are to be applied to trees, the goal will be to improve their utilization as building and construction materials and as an improved raw material for the production of paper and paper products. That does not preclude increasing the concentration and/or distribution of special metabolites at the same time. Even when wood is used for construction and manufacturing, there is an opportunity since bark, which may be enriched in targeted special metabolites, can be obtained economically as a by-product of wood harvesting.

Wood pulping offers greater potential to isolate metabolites either through pre-extraction, especially if pulping with organic solvents becomes a reality, or post-separation or extraction. However, the value of pulp and paper is an overriding concern. Biotechnology in this industry is concerned almost exclusively with a few major research areas (Srebotnik and Messner, 1996). These are biopulping and bio-bleaching, paper deinking and recycling, effluent treatment, and fermentation of waste/by-product hemicelluloses; all employ microorganisms. Additionally, there are efforts being made through genetic engineering to modify lignin to create trees more amenable to pulping. Although research on gene identification to prevent disease and transformation of trees to impart disease and insect resistance is being conducted, biotechnology related to special metabolites is mainly restricted to fungal removal of pitch in paper making (Srebotnik and Messner, 1996). In this regard, it is important to recognize that all schemes to increase the production of special metabolites must not adversely impact the processing or use of wood in its major utilization for solid products and paper.

The production of natural rubber, latexes, gums, essential oils, and some other products from forest trees is generally not dependent on wood harvesting for other uses. However, these industries and applications are declining. These industries are often labour intensive, have limited markets, and usually are in direct competition with other materials or those synthesized from petrochemicals. It could however be argued that genetically improved stocks not only could ensure the supply of renew-able products of quality, but also maintain a way of life and improve the economic prospects of many workers. Alternatively, it could be argued that to offset the costs in the application of biotechnology, large, integrated and mechanized harvesting and processing systems would result, displacing a number of unskilled and semi-skilled labourers.

The attraction of applying biotechnology to special metabolites is that it would be a natural, renewable, 'green technology'. The disadvantages are the economics of new technologies which would be in direct competition with the existing processes, and with the chemical and petrochemical industries. The niche that forest products could provide is the utilization of special metabolites as by-products and/or as high-value, generally small volume, applications.

References

AYRES, D. C. and LOIKE, J. D. (1990) *Lignans*, Cambridge: Cambridge University Press.
BARLOW, F. W. (1989) Rubber, gutta and chicle. In: Rowe, J. W., ed., *Natural Products of Woody Plants II*, Berlin: Springer-Verlag, pp. 1028–1050.
BEMILLER, J. N. (1989) Carbohydrates. In: Rowe, J. W. ed., *Natural Products of Woody Plants I*, Berlin: Springer-Verlag, pp. 155–178.
BOULTER, D. (1993) Insect pest control by copying nature using genetically engineered crops. *Phytochemistry* **34**, 1453–1466.

CASTRO, M. A., GORDALIZA, M., MIGUEL DEL CORRAL, J. M. and SAN FELICIANO, A. (1996) The distribution of lignanoids in the order coniferae. *Phytochemistry* **41**, 995–1011.

CHARLWOOD, B. V. and RHODES, M. J. C. (1990) *Secondary Products from Plant Tissue Culture*, Oxford: Clarendon Press.

COPPEN, J. J. W. (1995) *Non-wood Forest Products: Gums, Resins and Latexes of Plant Origin*, Rome: Food and Agriculture Organization of the United Nations.

DAVIN, L. B., LEWIS, N. G. and UMEZAWA, T. (1992) Phenylpropanoid metabolism: biosynthesis of monolignols, lignans and neolignans, lignins and suberins. In: Stafford, H. A. and Ibrahim, R. K., eds, *Recent Advances in Phytochemistry*, Vol. 27, New York: Plenum Press, pp. 325–376.

DEV, S. (1989) Terpenoids. In: Rowe, J. W., ed., *Natural Products of Woody Plant II*, Berlin: Springer-Verlag, pp. 691–807.

DiCOSMO, F. and MISAWA, M. (1995) Plant cell and tissue culture: alternatives for metabolite production. *Biotechnol. Adv.* **13**, 425–453.

DURZAN, D. J. (1989) Nitrogenous extractives. In: Rowe, J. W., ed., *Natural Products of Woody Plants I*, Berlin: Springer-Verlag, pp. 179–200.

DYAS, L. and GOAD, L. J. (1993) Steryl fatty esters in plants. *Phytochemistry* **34**, 17–29.

FENGEL, D. and WEGENER, G. (1984) *Wood, Chemistry, Ultrastructure, Reactions*, Berlin: Walter de Gruyter.

FOO L. Y. and KARCHESY, J. J. (1989) Chemical nature of phlobaphenes. In: Hemingway, R. W. and Karchesy, J. J., eds, *Chemistry and Significance of Condensed Tannins*, New York: Plenum Press, pp. 109–118.

GOTTLIEB, O. R. (1990) Phytochemicals: differentiation and function. *Phytochemistry* **29**, 1715–1724.

GOTTLIEB, O. R. and YOSHIDA, M. (1989) Lignans. In: Rowe, J. W., ed., *Natural Products of Woody Plants I*, Berlin: Springer-Verlag, pp. 349–511.

HARBORNE, J. B. (1989) Flavonoids. In: Rowe, J. W., ed., *Natural Products of Woody Plants I*, Berlin: Springer-Verlag, pp. 533–570.

HASLAM, E. (1989) Gallic acid derivatives and hydrolyzable tannins. In: Rowe, J. W., ed., *Natural Products of Woody Plants I*, Berlin: Springer-Verlag, pp. 399–438.

HAYEK, E. W. H., JORDIS, U., MOCHE, W. and SAUTER, F. (1989) A bicentennial of betulin. *Phytochemistry* **28**, 2229–2242.

HEGNAUER, R. (1988) Biochemistry, distribution and taxonomic relevance of higher plant alkaloids. *Phytochemistry* **27**, 2423–2427.

HEMINGWAY, R. W. (1989) Biflavonoids and proanthocyanidins. In: Rowe, J. W., ed., *Natural Products of Woody Plants I*, Berlin: Springer-Verlag, pp. 571–651.

IMAMURA, H. (1989) Contribution of extractives to wood characteristics. In: Rowe, J. W., ed., *Natural Products of Woody Plants II*, Berlin: Springer-Verlag, pp. 843–860.

KERMASHA, S., GOETGHEBEUR, M. and DUMONT, J. (1995) Determination of phenolic compound profiles in maple products by high-performance liquid chromatography. *J. Agric. Food Chem.* **43**, 708–716.

KLYNE, W. and BUCKINGHAM, J. (1978) *Atlas of Stereochemistry, Absolute Configurations of Organic Molecules*, 2nd edition. London: Chapman and Hall.

KOHDA, H., KASAL, R., YAMASAKI, K., MURAKAMI, K. and TANAKA, O. (1976) New sweet diterpene glucosides form *Stevia rebaudiana*. *Phytochemistry* **15**, 981–983.

KOLATTUKUDY, P. E. (1984) Biochemistry and function of cutin and suberin. *Can. J. Bot.* **62**, 2918–2933.

LEE, H., WATANABE, N., SASAYA, T. and OZAWA, S. (1993) Extractives of short-rotation hardwood species. I. Phenolics of *Salix sachalinensis* Fr. Schm. *Mokuzai Gakkaishi* **39**, 1409–1414.

LEWIS, N. G. and YAMAMOTO, E. (1989) Tannins – their place in plant metabolism. In: Hemingway, R. W. and Karchesy, J. J., eds, *Chemistry and Significance of Condensed*

Tannins, New York: Plenum Press, pp. 23–46.

MAHATO, S. B., NANDY, A. K. and ROY, G. (1992) Triterpenoids. *Phytochemistry* **37**, 2199–2249.

MANITTO, P. (1981) *Biosynthesis of Natural Products*, Chichester: Ellis Horwood.

NES, W. R. (1989) Steroids. In: Rowe, J. W., ed., *Natural Products of Woody Plants II*, Berlin: Springer-Verlag, pp. 808–842.

NOLLER, C. R. (1961) *Chemistry of Organic Compounds*, Philadelphia: Saunders.

NORIN, T. (1989) Stilbenes, conioids and other polyaryl natural products. In: Rowe, J. W., ed., *Natural Products of Woody Plants I*, Berlin: Springer-Verlag, pp. 512–533.

OKUDA, T., YOSHIDA, T. and HATANO, T. (1993) Classification of oligomeric hydrolyzable tannins and specificity of their occurrence in plants. *Phytochemistry* **32**, 507–521.

PELLETIER, S. W. (ed.) (1983) *Alkaloids, Chemical and Biological Perspectives*, Vol. 1–6, New York: John Wiley & Sons.

PORTER, L. J. (1989) Condensed tannins. In: Rowe, J. W., ed., *Natural Products of Woody Plants I*, Berlin: Springer-Verlag, pp. 651–690.

QUIDEAU, S. and FELDMAN, K. S. (1996) Ellagitannin chemistry. *Chem. Rev.* **96**, 475–503.

REISCH, M. S. (1996) Thermoplastic elastomers target rubber and plastics markets. *Chem. Eng. News* **74**, 10–14.

ROBINSON, T. (1991) *The Organic Constituents of Higher Plants*, North Amherst: Cordus Press.

SAKAI, S.-I., AIMI, N., YAMANAKA, E. and YAMAGUCHI, K. (1989) The alkaloids. In: Rowe, J. W., ed., *Natural Products of Woody Plants I*, Berlin: Springer-Verlag, pp. 200–258.

SREBOTNIK, E. and MESSNER, K. (1996) (eds) *Biotechnology in the Pulp and Paper Industry, Recent Advances in Applied and Fundamental Research*, Vienna: Facultas-Universitatsverlag.

SRINIVASAN, V. V., SIVARAMAKRISHNAN, V. R., RANGASWAMY, C. R., ANANTHA-PADMANABHA, H. S. and SHANKARANARAYANA, K. H. (1992) *Sandal (Santalum album l.)*, Dehra Dun: Indian Council of Forestry Research and Education.

STAFFORD, H. A. (1988) Proanthocyanidins and the lignin connection. *Phytochemistry* **27**, 1–6.

STAHLHUT, R. W. (1994) *In vivo* production of taxanes, US Patent 5279953.

TAHARA, S. and IBRAHIM, R. K. (1995) Prenylated isoflavanoids – an update. *Phytochemistry* **38**, 1073–1094.

TANAKA, N., SHIMOMURA, K. and ISHIMARU, K. (1995) Tannin production in callus cultures of *Quercus acutissima*. *Phytochemistry* **40**, 1151–1154.

THEANDER, O. and LUNDGREN, L. N. (1989) Monoaryl natural products. In: Rowe, J. W. ed., *Natural Products of Woody Plants I*, Berlin: Springer-Verlag, pp. 369–399.

WALLACE, G. and FRY, S. C. (1994) Phenolic components of the plant cell wall. *Int. Rev. Cytol.* **151**, 229–267.

ZAVARIN, E. and COOL, L. (1991) Extraneous materials from wood. In: Lewin, M. and Goldstein, I. S., eds, *Wood Structure and Composition*, New York: Marcel Dekker, pp. 321–407.

ZINKEL, D. F. (1981) Turpentine, rosin and fatty acids from conifers. In: Goldstein, I. S., ed., *Organic Chemicals from Biomass*, Boca Raton: CRC Press, pp. 163–187.

ZINKEL, D. F. (1989) Fats and fatty acids. In: Rowe, J. W., ed., *Natural Products of Woody Plants I*, Berlin: Springer-Verlag, pp. 369–399.

ZINKEL, D. F. and RUSSELL, J. (1989) *Naval Stores: Production, Chemistry, Utilization*, New York: Pulp Chemicals Association.

Wood/Bark Extracts as Adhesives and Preservatives

ANTONIO PIZZI

11.1 Introduction

The loose term 'renewable resources adhesives' has been used to identify polymeric compounds of natural, vegetable origin that have been modified and/or adapted to the same use as some classes of purely synthetic adhesives (Pizzi, 1991). Two main classes of these adhesives currently exist: one already extensively commercialized in the southern hemisphere and the other on the slow way to commercialization. These two types of resin are tannin based adhesives (Pizzi, 1980b, 1983) and lignin adhesives (Nimz, 1983; Glasser and Sarkanen, 1989; Lewis and Lantzy, 1989; Tahir and Sellers, 1990; Pizzi, 1994c,d). Both types are aimed primarily as substitutes for synthetic phenolic resins. In some aspects, such as performance, they closely mimic, or are even superior to, synthetic phenolic adhesives, while in others they behave vastly differently from their synthetic counterparts. In this chapter we focus on tannin-based adhesives because: tannins can properly be termed extractives; and such adhesives have already been in extensive industrial use in the southern hemisphere, in certain fields of application, for the last 25 years. Tannins are generally extracted from the bark of certain tree species as for pine tannin and mimosa or wattle tannin; or from their wood, as for quebracho tannin from South America. They are generally extracted by washing them out of the lignocellulosic material by infusion of the latter and subequent washing with hot water or with hot diluted solutions of sodium sulphite or metabisulphite. Tannin adhesives are of particular interest not only for their excellent performance in various applications but also for their mostly environment-friendly composition. Lignin adhesives are not dealt with here, since lignin is not an extractive but rather a main wood constituent. The reader is referred to specialized literature reviews of the field (Nimz, 1983; Glasser and Sarkanen, 1989; Pizzi, 1994d). Applications of wood/bark extracts in the preservation field are more rare and they are only briefly included in this chapter.

11.2 Tannin-based Adhesives – Theoretical Considerations

The word tannin has been used loosely to define two different classes of chemical compounds of mainly phenolic nature: hydrolyzable tannins and condensed tannins.

The former, including chestnut, myrabolans (*Terminalia* and *Phyllantus* tree species), and dividi (*Caesalpina coraria*) extracts, are mixtures of simple phenols such as pyrogallol and ellagic acid and of esters of a sugar, mainly glucose, with gallic and digallic acids (Pizzi, 1983). They have been used successfully as partial substitutes (up to 50 per cent) of phenol in the manufacture of phenol–formaldehyde resins (Kulvik, 1975, 1976). Their chemical behaviour towards formaldehyde is analogous to that of simple phenols of low reactivity, and their moderate use as phenol substitutes in the above-mentioned resins does not present difficulties. Their lack of macromolecular structure in their natural state, the low level of phenol substitution they allow, and their low nucleophilicity, limited worldwide production, and higher price somewhat decrease their chemical and economic interest. Condensed tannins, on the other hand, constituting more than 90 per cent of the total world production of commercial tannins (approximately 200 000 tons per year), are both chemically and economically more interesting for the preparation of adhesives and resins. Condensed tannins and their flavonoid precursors are known for their wide distribution in nature and particularly for their substantial concentration in the wood and bark of various trees. These include various *Acacia* (wattle or mimosa bark extract), *Schinopsis* (quebracho wood extract), and *Tsuga* (hemlock bark extract) species, from which commercial tannin extracts are manufactured, and various pine bark extract species. Where bark and wood of trees were found to be particularly rich sources of condensed tannins, commercial development ensued through large-scale afforestation and/or industrial extraction, mainly for use in leather tanning. The main producers of commercial condensed tannin extracts are, in relative order of importance: Argentina, South Africa, Brazil, Paraguay, Zimbabwe, Indonesia, Kenya and Chile. The production of tannins for leather manufacture reached its peak immediately after World War II and has since progressively declined. This decline of their traditional market, coupled with the increased price and decreased availability of synthetic phenolic materials due to the advent of the energy crisis of the early 1970s, stimulated fundamental and applied research on the use of such tannins as a source of condensed phenolics.

11.2.1 Condensed Tannins

The structure of the flavonoid constituting the main monomer of condensed tannins may be represented as shown in Figure 11.1 (Roux, 1965; Roux and Paulus, 1961). Tannins in which the A-rings of the structural unit present only one -OH group are said to present resorcinol-like A-rings and reactivity, this being mainly the case with

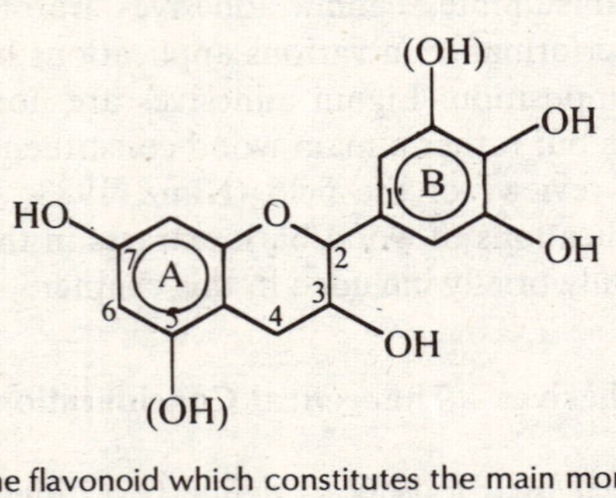

Figure 11.1 Structure of the flavonoid which constitutes the main monomer of condensed tannins

mimosa and quebracho tannins which are represented by the formula shown in Figure 11.1 when the A-ring -OH group in parentheses is not present on the structure. Tannins, however, which present two -OH groups on the A-rings of their structural unit are said to present phloroglucinol-like A-rings and reactivity. These names are derived from resorcinol (1,3-dihydroxybenzene) and phloroglucinol (1,3,5-trihydroxybenzene), two particularly reactive phenols. B-rings are called catecholic or pyrogallic when they present, respectively, two or three -OH groups. These names are derived from catechol (1,2-dihydroxybenzene) and pyrogallol (1,2,3-trihydroxybenzene). The flavonoid unit shown in Figure 11.1 is repeated 2 to 11 times in mimosa tannin, with an average degree of polymerization of 4 to 5, and up to 30 times for pine tannins, with an average degree of polymerization of 6 to 7 (Thompson and Pizzi, 1995).

The nucleophilic centres on the A-ring of a flavonoid unit tend to be more reactive than those found on the B-ring. This is due to the vicinal hydroxyl substituents, which cause general activation in the B-ring without any localized effects such as those found in the A-ring. Formaldehyde reacts with tannins to produce polymerization through methylene bridge linkages at reactive positions on the flavonoid molecules, mainly the A-rings. The reactive positions of the A-rings are either sites 6 or 8 (according to the type of tannin) of internal flavonoid units and both positions 6 and 8 of the upper terminal flavonoid units. The A-rings of mimosa and quebracho tannins show reactivity towards formaldehyde comparable with that of resorcinol (Pizzi, 1983).

Assuming the reactivity of phenol to be 1, that of resorcinol to be 10, and that of phloroglucinol to be 100, the resorcinol-like A-rings have a reactivity of 8–9, while the phloroglucinol-like A-rings present reactivities of approximately 40. However, because of their size and shape, the tannin molecules become immobile at a low level of condensation with formaldehyde, with the result that the available reactive sites are too far apart for further methylene bridge formation. This may lead to incomplete polymerization and therefore weakness. Bridging agents with longer molecules should allow the distances that are too long for methylene bridges to be spanned. Alternatively, other techniques can be used to solve this problem.

In condensed tannins from mimosa bark the main polyphenolic pattern is represented by flavonoid analogues based on resorcinol A-rings and pyrogallol B-rings. These constitute about 70 per cent of mimosa tannins. The secondary but parallel pattern is based on resorcinol A-rings and catechol B-rings (Roux, 1965; Pizzi, 1983). These tannins represent about 25 per cent of the total mimosa bark tannin fraction. The remaining parts of the condensed tannin extract are the 'non-tannins' (Roux, 1965). They may be subdivided into carbohydrates, hydrocolloid gums, and small amino and imino acid fractions (Roux, 1965; Pizzi, 1983). The hydrocolloid gums vary in concentration from 3–6 per cent and contribute significantly to the viscosity of the extract despite their low concentration. Similar flavonoid A- and B-ring patterns also exist in quebracho wood extract (*Schinopsis, Balansae* and *Lorentzii*) (King and White, 1957; King *et al.*, 1961; Roux and Paulus, 1961), but with probably a much lower proportion of the phloroglucinol A-ring pattern (Clark-Lewis and Roux, 1959; King *et al.*, 1961; Roux *et al.*, 1975; Abe *et al.*, 1987). Similar patterns to wattle (mimosa) and quebracho are found in hemlock and Douglas fir bark extracts but completely different patterns and relationships exist in pine tannins (Porter, 1974; Hemingway and McGraw, 1976; Rossouw *et al.*, 1980). Pine tannins have only two main patterns: one represented by flavonoid analogues based on phloroglucinol A-rings and catechol B-rings, the other present in much lower proportion, represented by phloroglucinol A-rings and phenol B-rings. The A-rings of pine tannins therefore possess only the phloroglucinol type of structure, which is much more reactive towards formaldehyde than a resorcinol-type structure, with important consequences for the use of these tannins as adhesives (see Figure 11.2).

In condensed polyflavonoid tannin molecules the A-rings of the constituent flavonoid units retain only one highly reactive nucleophilic centre, the remainder accommodating the interflavonoid bonds. Only the A-rings show reactivity towards formaldehyde sufficient to ensure reaction and cross-linking in the application of tannins as wood adhesives. Pyrogallol or catechol B-rings are by comparison unreactive and may be activated by anion formation only at relatively high pH (Hillis

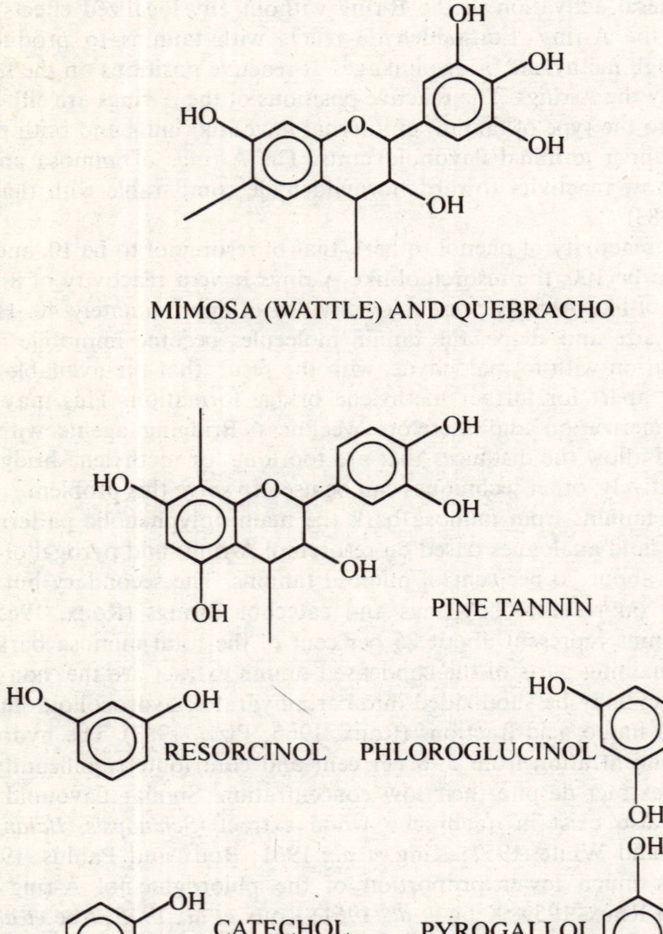

MIMOSA (WATTLE) AND QUEBRACHO

PINE TANNIN

RESORCINOL   PHLOROGLUCINOL

CATECHOL   PYROGALLOL

Figure 11.2 Structure of some tannins and related substances

and Urbach, 1959a,b; Roux *et al.*, 1975). Hence the B-rings do not participate in the reaction except at high pH values (pH 10). However at high pH the reactivity towards formaldehyde of the A-rings is so extreme that the tannin–formaldehyde adhesives prepared have unacceptably short pot lives (Pizzi and Scharfetter, 1978). As a result in general tannin adhesives practice, only the A-rings are used to cross-link the network. With regard to the pH dependence of the reaction with formalde-hyde, it is generally accepted that the reaction rate of wattle tannins with formaldehyde is slowest in the pH range 4.0–4.5 (Plomley, 1966); for pine tannins, the range is between 3.3 and 3.9 (Pizzi, 1994c,d). Due to the vast difference in their reactivities, mimosa and quebracho tannins are generally used as adhesives in the pH range 6.5–8, while pine tannin is used in the pH range 4.5–6.

Formaldehyde is generally the aldehyde used in the preparation, setting, and curing of tannin adhesives. It is normally added to the tannin extract solution at the required pH, preferably in its polymeric form of paraformaldehyde, which is capable of fairly rapid depolymerization under alkaline conditions or as urea–formalin con-centrates. Hexamethylenetetramine (hexamine) may also be added to resins due to its formaldehyde-releasing action under heat. Hexamine is, however, unstable in acid medium (Saayman, 1967) but becomes more stable with increased pH values. Hence under alkaline conditions the liberation of formaldehyde might not be as rapid and efficient as required. There is also some debate as to whether the bonds formed with hexamine as hardener are as boil resistant as those formed by parafor-maldehyde (McLean and Gardner, 1952; Herrick and Bock, 1958; Pizzi, 1983; Pizzi *et al.*, 1994). The reaction of formaldehyde with tannins may be controlled by the addition of alcohols to the system. Under these circumstances some of the formalde-hyde is stabilized by the formation of hemiacetals (e.g. $CH_2(OH)(OCH_3)$) if methanol is used) (Scharfetter *et al.*, 1977). When the adhesive is cured at an elevated tem-perature, the alcohol is driven off at a fairly constant rate and formaldehyde is progressively released from the hemiacetal. This ensures that less formaldehyde is volatilized when the reactants reach curing temperature and that the pot life of the adhesive is extended. Other aldehydes have also been substituted for formaldehyde (Plomley, 1966; Pizzi and Scharfetter, 1978; Pizzi, 1983), for example, acetaldehyde, furfural and glyoxal.

During the reaction of polyflavonoid tannins with formaldehyde, two reactions are present:

1　The reaction of formaldehyde with tannin and with low-molecular-weight tannin–aldehyde condensates, which are responsible for the formaldehyde con-sumption.

2　The liberation of formaldehyde, available again for reaction. The latter reaction is probably due to the breakdown of unstable $-CH_2-O-CH_2-$ ether bridges ini-tially formed to $-CH_2-$ linked compounds.

Ether-bridged compounds have been isolated for the phenol–formaldehyde reaction (Megson, 1958). The presence of ether-bridged compounds is detected in the tannin–formaldehyde reaction by a surge in the concentration of formaldehyde observed in kinetic curves due to methylene ether bridges decomposition (Rossouw *et al.*, 1980).

When heated in the presence of strong mineral acids, condensed tannins are subject to two competitive reactions. One is degradative leading to lower-molecular-weight products; the second is condensative (a result of hydrolysis of heterocyclic rings–hydroxybenzyl ether links) (Roux *et al.*, 1975). In the latter reaction the *p*-

hydroxybenzyl carbonium ions created condense randomly with nucleophilic centres on other tannin units to form insoluble condensation products called phlobaphenes or 'tanner's red' which precipitates out (Freudenberg and de Lama, 1958; Brown and Cummings, 1959; Brown *et al.*, 1961; Roux *et al.*, 1975). Other modes of condensation (e.g. free-radical coupling of B-ring catechol units) cannot be excluded in the presence of atmospheric oxygen (Kennedy *et al.*, 1984; Masson *et al.*, 1996, 1997; Merlin and Pizzi, 1996). In predominantly aqueous conditions, phlobaphene formation or formation of insoluble condensates predominates. These reactions, characteristic of tannins and not of synthetic phenolic resins, must be taken into account when formulating tannin adhesives.

Sulphitation of tannin is one of the oldest and most useful reactions in flavonoid chemistry and slightly sulphited water is sometimes used to increase tannin extraction from the bark containing it. In certain types of adhesives sulphitation can offer the important advantages of allowing higher concentration of tannin phenolics (due to enhanced solubility and decreased viscosity) and of higher moisture retention (allowing slower adhesive film dry-out and hence longer assembly times) (Pizzi, 1983). However there are distinct disadvantages in that sulphonate groups promote sensitivity to moisture, resulting in adhesive deterioration, and bad water resistance of the cured glue line even when cross-linking is quite adequate (Dalton, 1950, 1953; Parrish, 1958; Pizzi, 1979c).

11.3 Industrial Thermosetting Tannin Adhesives for Wood Panels

11.3.1 *Tannins*

From the information above it is clear that tannins are a phenolic polymeric material and thus, like phenol, will react with formaldehyde: the similarity however ends there. Their high reactivity renders impossible the maintenance of a methylol group on a flavonoid tannin for more than an hour at ambient temperature and prevents the application to tannins of the recognized technology developed for synthetic phenol–formaldehyde adhesives as used in board manufacture. In order to become a useful industrial raw material, novel technologies are needed to cope with both the reactivity and the natural variability of the tannins.

Tannin Purity

The purity of vegetable tannin extracts varies considerably. Commercial wattle bark extracts normally contain 70–80 per cent active phenolic ingredients. The non-tannins fraction, consisting mainly of simple sugars and high-molecular-weight hydrocolloid gums, does not participate in the resin formation with formaldehyde. Sugars reduce the strength and water resistance by dilution of resin solids and the reduction is in direct proportion to the amount added. By contrast, the hydrocolloid gums have a much more marked effect on both strength and water resistance of the adhesive (Scharfetter *et al.*, 1977; Pizzi, 1978b, 1983; Pizzi and Stephanou, 1994a,b). If it is assumed that the non-tannins in tannin extracts have a similar influence on adhesive properties, it can be expected that unfortified tannin–formaldehyde networks will achieve about 80 per cent of the performance shown by synthetic adhesives.

In many glued wood products, the demands on the glue line are so high that unmodified tannin adhesives are unsuitable. The possibility of refining extracts has proved fruitless largely because the intimate association between the various constituents makes industrial fractionation difficult and expensive. Fortification is in many cases the most practical approach to reducing the effect of impurities. Fortification generally consists of copolymerization of the tannin with phenolic or aminoplastic resins, this being exclusively done during their hardening even in cases where the fortifier is added to the tannin as a separate component long before hardening (Pizzi, 1977, 1983; Pizzi and Scharfetter, 1977, 1978; Scharfetter *et al.*, 1977). Addition of such fortifying synthetic resins can be carried out during manufacture of the adhesive resin, during glue-mix assembly, just before use, or during adhesive use.

The presence of non-tannins in tannin extracts, mainly monomeric and polymeric sugars, can drastically decrease the performance of a tannin adhesive, especially if the amount of these sugars is 20 per cent or more. Addition of a synthetic resin to the tannin then helps in a simple manner to decrease the percentage of interfering non-tannins below 20 per cent of the total mix. If added in sufficient quantity, various synthetic resins have been found effective in reducing the non-tannin fraction to below 20 per cent of the total mix and in overcoming other structural problems (Pizzi, 1983). The main resins used are phenol–formaldehyde and urea–formaldehyde resols with a medium to high content of methylol groups. Even the oldest formulation in which the tannin is fortified by just mixing it with 3–5 per cent resorcinol has long been discontinued in its country of origin to introduce better performance and lower cost formulations; it has however been taken up by some tannin producers and is still used quite extensively in other parts of the world. The synthetic resins added as fortifiers can fulfil several functions such as those of hardeners, fortifiers or both. Generally, they are used as fortifiers between 10 and 20 per cent of total adhesive solids, and paraformaldehyde is used as the hardener; the addition of synthetic resin fortifiers is most common for marine-grade plywood adhesives. These fortifiers are particularly suitable for resorcinolic types of condensed tannins, such as mimosa. They can be copolymerized with the tannins during resin manufacture, during use, or both. Copolymerization and curing are based on the condensation of the tannin with the methylol groups carried by the synthetic resin. Since tannin molecules are generally large, the rate of molecular growth in relation to the rate of linkage is high, so that tannin adhesives generally tend to have fast gelling and curing times and shorter pot lives than those of synthetic phenolic adhesives. From the point of view of reactivity, phloroglucinol tannins such as pine tannins are much faster than mainly resorcinol tannins such as mimosa. The usual ways of reducing reactivity and to lengthen adhesive pot life are:

1 To add alcohols to the adhesive mix to form hemiacetals with formaldehyde which will retard the tannin–formaldehyde reaction.

2 To adjust the adhesive's pH to produce the required pot life and rate of curing.

3 To use hexamine as hardener, which can give a very long pot life at ambient temperature but retain a fast curing time at higher temperatures.

Tannin Viscosity

The viscosity of bark extracts is strongly dependent on concentration. The viscosity increases very rapidly above a concentration of 50 per cent. Compared with synthetic resins, the tannin extracts are more viscous at the concentrations normally required in adhesives. High viscosity of aqueous solutions of condensed tannins is due to the following causes, in order of importance:

1 Presence of high-molecular-weight hydrocolloid gums in the extract (Pizzi, 1978a,b, 1983). The viscosity is directly proportional to the amount of gums present in the extract (Pizzi, 1983).

2 Tannin–tannin, tannin–gum and gum–gum hydrogen bonding. Aqueous tannin extract solutions are not true solutions, but rather colloidal suspensions in which water access to all parts of the molecules present is very slow. Hence, it is difficult to eliminate intermolecular hydrogen bonds and colloidal interactions by dilution only (Pizzi, 1978a,b, 1983; Pizzi and Stephanou, 1994b; Kim *et al.*, 1995a,b; Kim and Mainwaring, 1996; Masson *et al.*, 1996, 1997).

3 Presence of high-molecular-weight tannins in the extract (Pizzi, 1983).

In plywood adhesives, which are generally applied to the wood veneers by means of roller spreaders, the high viscosity of the tannin extract is generally not critical and can be manipulated by dilution as the roller spreader can handle even fairly high adhesive viscosities. In the case of particleboard adhesives, decrease of viscosity is an important prerequisite as too high a viscosity will not allow pumping and spraying of the adhesive on the wood particles. For particleboard application, decrease of the viscosity of the tannin extracts by dilution with water is only possible to a very limited, and insufficient, extent. Too high an amount of water in the resinated particleboard will cause: steam blisters in the hot press, degrading the final, finished panel; or the use of considerably longer pressing times, in order to avoid steam blistering, rendering board production with tannin adhesives uneconomical. Small amounts of viscosity reducers which decrease or destroy the colloidal interactions of concentrated tannin solutions, for example, urea (Pizzi, 1978b; Kim and Mainwaring, 1996), phenol and naphthalene (Pizzi, 1978b), and glycol ethers (Dombo, 1994), are sometimes used to good effect especially when a non-modified tannin formulation is used.

Tannin Modification

When reacted with formaldehyde, unmodified condensed tannins give adhesives having characteristics that do not suit particleboard manufacture: high viscosity, low strength, and poor water resistance. The most commonly used process to eliminate these disadvantages consists of a series of subsequent acid and alkaline treatments of the tannin extract, causing hydrolysis of the gums to simple sugars and some tannin structural changes. Viscosity, strength, and water resistance of the unfortified tannin–formaldehyde adhesive are all improved (Pizzi, 1983). Such treatments may also cause partial rearrangement of the flavonoid molecules liberating resorcinol *in situ*, increasing its number of reactive sites and rendering it more cross-linked with formaldehyde, and ultimately yielding an adhesive which, without addition of any fortifier resins, gives truly excellent performance for exterior-grade particleboard (Pizzi, 1978a,b, 1983, 1994d).

This modification can only be introduced to a limited extent, to avoid precipitation of the tannin from solution by the formation of phlobaphenes. Typical particleboard results obtained are shown in Table 11.1. Particular glueing and pressing

Table 11.1 Typical industrial, exterior-grade particleboard results obtained with: unfortified mimosa and quebracho tannin–formaldehyde adhesives obtained by acid–alkali treatment; pine tannin–formaldehyde adhesives fortified with isocyanates; and synthetic resin adhesives as controls

	Panel density (g/cm^3)	Original internal bond (IB) tensile perpendicular (MPa)	IB after a 2-h boil + drying (MPa)	Cyclic test swelling after five cycles (%)
TF unfortified	0.690	0.80	0.45	3.0
TF + isocyanate	0.690	0.84	0.43	3.0
Synthetic PF	0.690	0.65	0.38	5.0
Synthetic MF	0.690	0.75	0.05	15.0

PF, phenol–formaldehyde adhesive; MF, melamine–formaldehyde adhesive

techniques have been developed for tannin particleboard adhesives (Pizzi, 1978a, 1979a, 1994d) to achieve pressing times much faster than those obtained with synthetic phenol–formaldehyde adhesives. Pressing times of 7 s/mm of panel thickness have been achieved and press times of 9 s/mm at 190–200°C press temperature are in daily operation: these pressing times are comparable with those obtainable with urea–formaldehyde or melamine–formaldehyde resins. The success of these simple types of particleboard adhesive relies heavily on industrial application technology rather than just on the preparation technology of the adhesive itself (Pizzi, 1978a, 1979a, 1994d). A much higher moisture content of the resinated chips is tolerable with these adhesives than with any of the synthetic phenol–formaldehyde (PF) and aminoresin adhesives.

The preferred hardeners to obtain exterior grade properties are mainly paraformaldehyde, or formurea (a urea-stabilized formaldehyde solution); formalin solution is never used as it badly affects the weather and water durability of the panels.

The best adhesive formulation for phloroglucinolic tannins such as pine tannin extracts is a comparatively new formulation that is also capable of giving excellent results when using resorcinolic tannins (e.g. wattle extract) (Pizzi, 1981, 1982, 1994d, Pizzi *et al.*, 1993). The adhesive glue mix consists of an unmodified tannin extract 40–50 per cent solution to which has been added paraformaldehyde and polymeric non-emulsifiable 4,4′-diphenylmethane diisocyanate (commercial MDI) (Pizzi, 1982, 1994d; Pizzi *et al.*, 1993). The proportion of tannin extract solids to MDI is in the range 70:30 to 90:10 based on mass (Pizzi *et al.*, 1993). A competitive reaction between the isocyanate·groups of MDI and the reactive tannin phenolic rings for the unstable methylol group formed initially on the tannin by reaction with formaldehyde occurs: tannin–tannin methylene bridges and tannin–MDI urethane bridges produced as a consequence of this competitive series of reactions yield the hardened adhesive network. The properties of the particleboard manufactured with this system using pine tannin adhesives are listed in Table 11.1.

The results obtainable with this system are hence quite good and not too different from the results obtainable with some of the other tannin adhesives already described (Table 11.1). In the case of a phloroglucinolic tannin extract being used, no pH adjustment of the solution is needed. One point that was given close consideration is the deactivating effect of water on the isocyanate group of MDI. It has been found that the amount of deactivation by water of this group when in a concentrated solution (50 per cent or over) of a phenol is much lower than previously thought (Pizzi, 1980a,b, 1981, 1982; Pizzi and Walton, 1992; Pizzi *et al.*, 1993). This is the reason why aqueous tannin extract solutions and MDI can be reacted without substantial MDI deactivation by the water present.

Today factories with experience in this type of pine tannin adhesive are starting to use the pine tannin by itself, without MDI, with addition of only paraformaldehyde as a hardener. Alternatively paraformaldehyde in suspension in a solution of urea can be used. Very low formaldehyde emission panels can be obtained just by manipulation of the pH and other application parameters (Pizzi and Stephanou, 1994a). It must be stressed that the technology of tannin adhesives is so different from that of synthetic wood adhesives that factories often need to undergo an 'education' period before being capable of handling them.

There are other, newer systems of tannin adhesives which are still at the development stage. Among these must be mentioned the adhesive system based on the autocondensation of the tannin itself, without any addition of formaldehyde or any other aldehyde (Meikleham *et al.*, 1994; Pizzi and Meikleham, 1995; Pizzi *et al.*, 1995a,b; Masson *et al.*, 1996, 1997; Merlin and Pizzi, 1996). This system is based on the autocondensation and hardening reactions tannins undergo when in very alkaline conditions. Such conditions are sufficient for the more reactive tannins, such as pine tannin, but weak Lewis acids catalysts, such as finely divided silica or silicic acid in solution, are needed for the slower tannins such as mimosa and quebracho to obtain acceptable internal bond strength from the particleboard prepared in this manner. This system can only produce internal grade panel products. The interest in such an adhesive system is due to the fact that it utilizes a totally natural, non-toxic, environmentally-friendly adhesive which does not emit formaldehyde because it does not contain any.

A second system which holds considerable promise, and is at the conditions development stage, is that based on a tannin solution coupled with hexamine as hardener (Pizzi, 1994a,b,d; Pizzi *et al.*, 1994; Heinrich *et al.*, 1996). This system became of particular interest after the discovery that when hexamine is in the presence of very reactive phenolic species it does not, in the main, have the time to decompose to formaldehyde but reacts earlier at the level of Hultzsch methylene bases $^{+}CH_2\text{-}NH\text{-}CH_2^{+}$ (Hultzsch, 1950; Megson, 1958; Pizzi and Tekely, 1995, 1996). As a consequence there is only minimal formaldehyde emission from the panels bonded with it (Pizzi *et al.*, 1994). Under correct conditions exterior grade panels are produced, and very little noxious emission occurs at any stage of the manufacture of the boards (Pizzi *et al.*, 1994). What has proven difficult with this system is to find sets of conditions under which to minimize the variability of results and exterior durability obtained. The system works very well at much lower press temperature than traditional synthetic or tannin adhesives; pressing times are however comparable (Heinrich *et al.*, 1996). Undue increases in press temperature yield worse strength results and much worse emission results as hexamine decomposition to formaldehyde is accelerated and maximized.

A new technology which is in the early stages of development is based on liquefaction of tannin, or the bark/wood rich in tannin, directly into phenol (Santana *et al.*, 1995, 1996). This approach is attractive as standard phenol–formaldehyde resols preparation technology can be used, but it is reported to be able to substitute only up to 33 per cent of the phenol in a phenol–formaldehyde resin with bark or with tannins, a relatively low level of substitution. There is no doubt that in applications other than tannin adhesives this method may well be of interest, as has been clearly shown by the very useful pioneering work on wood 'phenolation' by Shiraishi's group in Japan (Alma *et al.*, 1996), and that latecomers to this business might find it of interest for tannin adhesives as they can apply a better known technology. The advantage of such an approach is that waste bark, rather than extracted tannin, can be used. However, such an approach is expensive as an additional operation, phenolation needs to be carried out, and only a low level of phenol substitution with tannin is achieved. Even more important, this system denies two of the main advantages of real tannin-based (mostly tannin) adhesives: their environment-friendly, non-toxic natural materials label (namely, they do not leach phenol); and their very rapid curing time which, by yielding shorter press times than synthetic adhesives, make tannin-based adhesives an attractive economic proposition. The excuses often offered for using tannin dissolved in phenol to produce phenol–tannin–formaldehyde resins, namely that tannin-based adhesives have very short pot lives and negligible shelf lives (Santana *et al.*, 1996), result from misunderstanding that such problems have already been solved at industrial level, as otherwise the commercialization of tannin-based adhesives could not have ensued.

Finally, it is necessary to discuss the oldest system of utilization of tannins for wood adhesives, namely their addition (4–8 per cent) to synthetic phenol–formaldehyde adhesives for marine plywood as an accelerator of hardening. The tannin is directly added to the glue-mix, generally included in the powder filler, which makes it easy to use. This market has traditionally been dominated for the last 25 years by quebracho tannin from South America although all the other condensed tannins work equally well for the purpose.

11.3.2 Corrugated Cardboard Adhesives

The adhesives developed for the manufacture of damp-ply-resistant corrugated cardboard are based on the addition of 4–8 per cent spray-dried mimosa or quebracho extract, 0.5–1.0 per cent urea–formaldehyde resin (UF), and 0.5–1.0 per cent formaldehyde to the starch-based adhesive used for bonding corrugated cardboard (McKenzie and Yuritta, 1974; Custers *et al.*, 1979). The tannin–urea–formaldehyde copolymer is formed *in situ*, and any free formaldehyde left in the glue line is absorbed by the tannin extract. The tannin extract powder should be added at a level of 4–5 per cent of the total starch content of the mix (i.e., carrier plus slurry). Successful results can be achieved in the range of 2–12 per cent of the total starch content, but 4 per cent is the recommended starting level. The final level is determined by the degree of water hardness and desired bond quality. This tannin extract–UF fortifier system is highly flexible and can be adopted to damp-proof a multitude of basic starch formulations.

11.3.3 Cold-Setting Laminating and Fingerjointing Adhesives for Wood

A series of different resins are prepared by copolymerization of resorcinol with resorcinolic A-rings of polyflavonoids, such as condensed tannins (Pizzi and Roux, 1978a,b; Pizzi 1979b, 1983, 1994d) (see Figure 11.3). The copolymers formed have

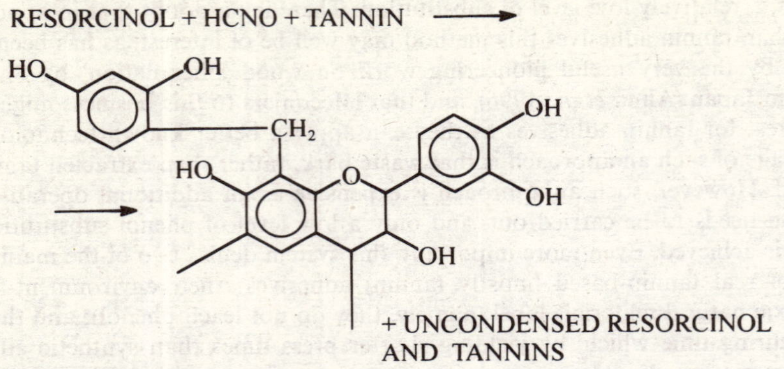

RESORCINOL + HCNO + TANNIN $\longrightarrow$

+ UNCONDENSED RESORCINOL
AND TANNINS

Figure 11.3 Production of cold-setting wood adhesives

been used as cold-setting exterior-grade structural wood adhesives complying with
the relevant international specifications. Several formulations are used. The system
most commonly used commercially relies on the simultaneous copolymerization of
resorcinol and of the resorcinolic A-rings of the tannin, due to their comparable
reactivities towards formaldehyde.

The final mixture of the products of this system is an adhesive that can be set and
cured at ambient temperature by the addition of paraformaldehyde. Addition of
vegetable flours such as wood flour and coconut shell flour as fillers is also necessary. Other cold-set systems exist and are described in the more specialized literature (Pizzi and Roux, 1978a,b; Pizzi, 1979b, 1983). The typical results obtainable
with these adhesives are indicated in Table 11.2. A particularly interesting system
now used extensively in several southern hemisphere countries is the so-called 'honeymoon' fast-setting, separate-application system (Pizzi *et al.*, 1980; Pizzi and
Cameron, 1984, 1989). In this system one of the surfaces to be mated in the joint is
spread with a standard synthetic phenol–resorcinol–formaldehyde adhesive plus
paraformaldehyde hardener and fillers. The second surface is spread with a 50 per
cent tannin solution at pH 12. When the two surfaces are fingerjointed, sufficient
strength develops for components to be installed within 30 minutes. In addition
laminated beams (glulam) need to be clamped for only 2.5 to 3 hours instead of the
traditional 16 to 24 hours, with a consequent considerable increase in factory productivity. It is also the only cold-setting system which can give unabated performance at temperatures as low as 5°C (compared with the 20°C for traditional
cold-sets) and with an amount of resorcinol which is half that of any other adhesive
(hence it is also cheaper). This adhesive system also provides full weather- and boil-

Table 11.2 Results of tannin–resorcinol–formaldhyde cold-setting adhesives used on beech strips

	Dry	After a 24-h cold-water soak	After a 6-h boil
Tensile strength (N)	3200–3800	2300–2900	2200–2800
Wood failure (%)	90–100	75–100	80–100

proof capabilities and has been used industrially in a few countries for more than 15 years.

Another separate application adhesive method similar to the more classical 'honeymoon' adhesives described above is the application of the 'greenweld' system (New Zealand Forest Research Institute, 1992), originally developed for synthetic phenol–resorcinol–formaldehyde (PRF) cold-set adhesives, to classical tannin–resorcinol–formaldehyde cold-set adhesives, or even to PRF/tannin extract honeymoon adhesives. In the 'greenweld' system ammonia is the chemical which accelerates hardening by both increasing the pH of the system and functioning as a cross-linking accelerator by reaction with methylol groups of the resin. The same accelerating effect of ammonia is well-known in both synthetic phenolic resol and novolak resins (Pizzi, 1983). This system works well but suffers from the drawback that ammonia is not only volatile but toxic: this will limit its application in sophisticated, environment-conscious industrial markets.

11.3.4 Tyre Cord Adhesives

Another application of condensed tannin extracts that has proved technically successful is as tyre cord adhesives. Both thermosetting tannin formulations (Chung and Hamed, 1989) and tannin–resorcinol–formaldehyde formulations (Saayman, 1975) have shown good results experimentally, although this technology is not used at present as it offers no economic or other advantages of significance over existing binders.

11.3.5 Foundry Core Binders

The latest application of tannin-based resins has been for foundry core sand binders. This application has been traditionally dominated by synthetic phenolic resins and by furanic and furanic/phenolic resins. A formulation based on furfuryl alcohol, tannin extract and small amounts of an aldehyde other than formaldehyde has now been sold in the North American market with some success (McKillip, 1993). The tannin extract used is one produced free of carbohydrates and is thus more expensive. Carbohydrates need to be eliminated as their caramelization and vaporization at the high foundry temperatures induce pinholing on the surface of the metal piece produced. The formulation has excellent performance, low cost and is now used at the rate of several hundred tons of tannin per year with consumption still growing.

11.3.6 Wood Preservatives Based on Tannins

Wood preservatives based on flavonoid and hydrolyzable tannins have also been developed (Laks *et al.*, 1988; Dirol, 1994; Pizzi and Baecker, 1996). They are based on the complexing capacity for a metal with the vicinal hydroxy groups on the B-ring of flavonoids and the ring of hydrolyzable tannins. The metal used as a biocide is copper, (see Figure 11.4).

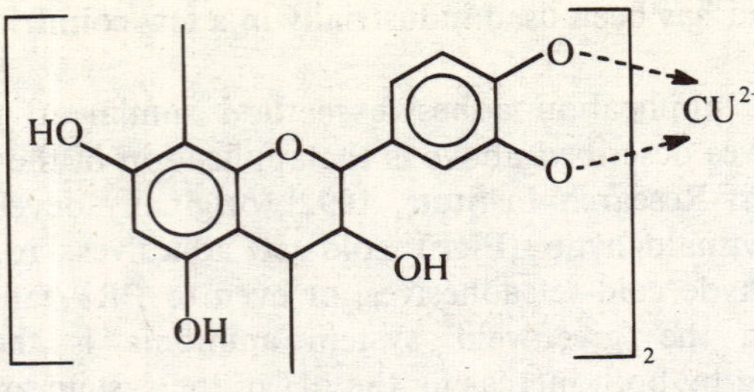

Figure 11.4 Structure of a tannin-based wood preservative

The results reported are acceptable but much poorer than long-term ground contact wood preservatives such as copper–chromium arsenates (CCA). Such preservatives, however, are acceptable for aerial, non-ground-contact applications. A copper tannate preservative using an hydrolyzable tannin, chestnut tannin, is in the early stages of introduction to the industrial market in France (Dirol, 1994). Complexes of the non-toxic preservative boron with flavonoid tannins are at the experimental stage. Such complexes have proved unsuitable as preservatives because they are still too easily washed out of the wood (Dirol, 1994). However, a new reaction between boron compounds and tannins, which was discovered for tannin adhesives applications, can be used to improve the fix and retard leaching of boron from the treated wood (Meikleham *et al.*, 1994). The results obtained indicate a non-ground contact but effective wood preservative with the advantage that such material is non-toxic to humans (Pizzi and Baecker, 1996).

There are many applications of tannin resins and of tannins themselves other than adhesives and preservatives. Throughout the whole field of industrial technologies, such as coatings, antioxidants, rigid and floral foams, rural roads stabilizers, ore flotation agents, ion-exchange resins, special mineral sequestrating agents, industrial water purification flocculants, textile dyes, food additives and pharmaceuticals, there are opportunities and future developments for these materials are likely.

References

ABE, I., FUNAOKA, M. and KODAMA, M. (1987) *Mokuzai Gakkaishi* **33**, 582–592.

ALMA, M. H., YOSHOKA, M., YAO, Y. and SHIRAISHI, N. (1996) *Holzforschung* **50**, 85–90.

BROWN, R. and CUMMINGS, W. (1959) *J. Chem. Soc.* 4302–4305.

BROWN, R., CUMMINGS, W. and NEWBOULD, J. (1961) *J. Chem. Soc.* 3677–3680.

CHUNG, K. H. and HAMED, G. R. (1989) In: Hemingway, R. W. and Karchesy, J. J., eds., *Chemistry and Significance of Condensed Tannins*, New York: Plenum Press.

CLARK-LEWIS, J. W. and ROUX, D. G. (1959) *J. Chem. Soc.* 1402–1407.

CUSTERS, P. A. J. L., RUSHBROOK, R., PIZZI, A. and KNAUFF, C. J. (1979) *Holzforschung Holzverwertung* **31**, 131–134.

DIROL, D. (1994) *CTBA Yearly Technical Committee Report*, Paris, France.

DALTON, L. K. (1950) *Aust. J. Appl. Sci.* **1**, 54–56.

DALTON, L. K. (1953) *Aust. J. Appl. Sci.* **4**, 54–57.

DOMBO, B. (1994) Private communication.

FREUDENBERG, K. and DE LAMA, J. M. A. (1958) *Annalen* **612**, 78–80.

GLASSER, W. G. and SARKANEN, S. (1989) *Lignin, Properties and Materials*, ACS Symposium Series 397, Washington DC: American Chemical Society.

HEINRICH, H., PICHELIN, F. and PIZZI, A. (1996) *Holz Roh Werkstoff*, Kurz Originalia **54**, 262.

HEMINGWAY, R. W. and McGRAW, G. W. (1976) *Appl. Polymer Symp.* 28.

HERRICK, F. W. and BOCK, L. H. (1958) *Forest Prod. J.* **8**, 269–273.

HILLIS, W. E. and URBACH, G. (1959a) *J. Appl. Chem.*, 474.

HILLIS, W. E. and URBACH, G. (1959b) *J. Appl. Chem.*, 665.

HULTZSCH, K. (1950) *Chemie der Phenolharze*, Berlin: Springer-Verlag.

KENNEDY, J. A., MUNRO, M. H. G., POWELL, H. K. J., PORTER, L. J. and FOO, L. Y. (1984) *Aust. J. Chem.* 885–890.

KIM, S.-R. and MAINWARING, D. A. (1996) *Holzforschung* **50**, 42–48.

KIM, S.-R., SARATCHANDRA, K. and MAINWARING, D. A. (1995a) *J. Appl. Polymer Sci.* **56**, 909–914.

KIM, S.-R., SARATCHANDRA, K. and MAINWARING, D. A. (1995b) *J. Appl. Polymer Sci.* **56**, 915–921.

KING, H. G. C. and WHITE, T. (1957) *J. Soc. Leather Traders Chem.* **41**, 368-376.

KING, H. G. C., WHITE, T. and HUGES, R. B. (1961) *J. Chem. Soc.*, 3234–3240.

KULVIK, E. (1975) *Adhesives Age* **18**, 3–4.

KULVIK, E. (1976) *Adhesives Age* **19**, 3–5.

LAKS, P. E., McKAIG, P. A. and HEMINGWAY, R. W. (1988) *Holzforschung* **42**, 299–304.

LEWIS, N. G. and LANTZY, T. R. (1989) Chapter 2. In: Hemingway, R. W., Conner, A. H. and Branham, S. J., eds, *Adhesives from Renewable Resources*, ACS Symposium Series 385, Washington, DC: American Chemical Society.

MASSON, E., MERLIN, A. and PIZZI, A. (1996) *J. Appl. Polymer Sci.* **60**, 263–269

MASSON, E., PIZZI, A. and MERLIN, A. (1997) *J. Appl. Polymer Sci.* **64**, 243–265.

McKENZIE, A. E. and YURITTA, Y. P. (1974) *Appita*, 26–28.

McKILLIP, W. (1993) Private communications.

McLEAN, H. and GARDNER, J. A. F. (1952) *Pulp Paper Mag. Can.* August.

MEGSON, N. J. L. (1958) *Phenolic Resins Chemistry*, Sevenoaks, UK: Butterworth.

MEIKLEHAM, N., PIZZI, A. and STEPHANOU, A. (1994) *J. Appl. Polymer Sci.* **54**, 1827–1845.

MERLIN, A. and PIZZI, A. (1996) *J. Appl. Polymer Sci.* **59**, 945–952.

NEW ZEALAND FOREST RESEARCH INSTITUTE (1992) Rotorua, New Zealand.

NIMZ, H. H. (1983) Lignin-based wood adhesives. In: Pizzi, A., ed., *Wood Adhesives Chemistry and Technology*, Vol. 1, New York: Marcel Dekker.

PARRISH, J. R. (1958) *J. S. African Forest Assoc.* **32**, 26–34.

PIZZI, A. (1977) *Adhesives Age* **20**, 27–30.

PIZZI, A. (1978a) *Adhesives Age* **21**, 32–35.

PIZZI, A. (1978b) *Forest Prod. J.* **28**, 42–47.

PIZZI, A. (1979a) *Holzforschung Holzverwertung* **31**, 85–86.

PIZZI, A. (1979b) *J. Appl. Polymer Sci.* **23**, 2777–2792.

PIZZI, A. (1979c) *Colloid. Polymer Sci.* **257**, 37–41.

PIZZI, A. (1980a) *J. Appl. Polymer Sci.* **25**, 2123–2126.

PIZZI, A. (1980b) *J. Macromol. Sci. Rev.* **C18**, 247–312.

PIZZI, A. (1981) *J. Macromol. Sci. Chem. Ed.* **A16**, 1243–1247.

PIZZI, A. (1982) *Holz Roh Werkstoff* **40**, 293–300.

PIZZI, A. (1983) Phenolic resin wood adhesives, and tannin-based wood adhesives, Chapters 3 and 4. In: Pizzi, A., ed., *Wood Adhesives Chemistry and Technology*, Vol. 1, New York: Marcel Dekker.

PIZZI, A. (1991) *Holzforschung Holzverwertung* **43**, 83–87.

PIZZI, A. (1994a) *Holz Roh Werkstoff* Kurz Originalia, **52**, 286.

PIZZI, A. (1994b) *Holz Roh Werkstoff* Kurz Originalia, **52**, 229.

PIZZI, A. (1994c) Natural phenolic adhesives 1, Tannin. In: Pizzi, A. and Mittal, K. L., eds, *Handbook of Adhesives Technology*, New York: Marcel Dekker, pp. 347–358.

PIZZI, A. (1994d) *Advanced Wood Adhesives Technology*, New York: Marcel Dekker.

PIZZI, A. and BAECKER, A. (1996) *Holzforschung* **50**, 507–510.

PIZZI, A. and CAMERON, F. A. (1984) *Forest Prod. J* **34**, 61–66.

PIZZI, A. and CAMERON, F. A. (1989) Fast-setting adhesives for fingerjointing and glulam. In: Pizzi, A., ed., *Wood Adhesives Chemistry and Technology*, Vol. 2, New York: Marcel Dekker, pp. 229–305.

PIZZI, A. and MEIKLEHAM, N. (1995) *J. Appl. Polymer Sci.* **55**, 1265–1269.

PIZZI, A. and ROUX, D. G. (1978a) *J. Appl. Polymer Sci.* **22**, 1945–1954.

PIZZI, A. and ROUX, D. G. (1978b) *J. Appl. Polymer Sci.* **22**, 2717–2718.

PIZZI, A. and SCHARFETTER, H. O. (1977) *CSIR Special Report*, HOUT I38, Pretoria, South Africa.

PIZZI, A. and SCHARFETTER, H. O. (1978) *J. Appl. Polymer Sci.* **22**, 1745–1761.

PIZZI, A. and STEPHANOU, A. (1994a) *Holz Roh Werkstoff* **52**, 218–222.

PIZZI, A. and STEPHANOU, A. (1994b) *J. Appl. Polymer Sci.* **51**, 2125–2130.

PIZZI, A. and TEKELY, P. (1995) *J. Appl. Polymer Sci.* **56**, 1645–1650.

PIZZI, A. and TEKELY, P. (1996) *Holzforschung* **50**, 277–281.

PIZZI, A. and WALTON, T. (1992) *Holzforschung* **46**, 541–547.

PIZZI, A., ROSSOUW, D. DU T., KNUFFEL, W. and SINGMIN, M. (1980) *Holzforschung Holzverwertung* **32**, 140–151.

PIZZI, A., VON LEYSER, E. P., VALENZUELA, J. and CLARK, J. G. (1993) *Holzforschung* **47**, 164–172.

PIZZI, A., VALENZUELA, J. and WESTERMAYER, C. (1994) *Holz Roh Werkstoff* **52**, 311–315.

PIZZI, A., MEIKLEHAM, N. and STEPHANOU, A. (1995a) *J. Appl. Polymer Sci.* **55**, 929–933.

PIZZI, A., MEIKLEHAM, N., DOMBO, B. and ROLL, W. (1995b) *Holz Roh Werkstoff* **53**, 201–204.

PLOMLEY, K. F. (1966) Paper No. 39, Division of Australian Forest Products Technology.

PORTER, L. J. (1974) *NZ J. Sci.* **17**, 213–217.

ROSSOUW, D. DU T., PIZZI, A. and McGILLIVRAY, G. (1980) *J. Polymer Sci. Chem. Ed.* **18**, 3323–3335.

ROUX, D. G. (1965) *Modern Applications of Mimosa Extract*, Grahamstown, South Africa: Leather Industries Research Institute, pp. 34–41.

ROUX, D. G. and PAULUS, E. (1961) *Biochem. J.* **78**, 785–790.

ROUX, D. G., FERREIRA, D., HUNDT, H. K. L. and MALAN, E. (1975) *Appl. Polymer Symp.* **28**, 335–355.

SAAYMAN, H. M. (1967) *LIRI Research Bulletin No. 466*, Grahamstown, South Africa: Leather Industries Research Institute.

SAAYMAN, H. M. (1975) Unpublished results.

SCHARFETTER, H. O., PIZZI, A. and ROSSOUW, D. DU T. (1977) *IUFRO Conference on Wood Gluing*, Merida, Venezuela, October.

SANTANA, M. A. E., BAUMANN, M. and CONNER, A. H. (1995) *Holzforschung* **49**, 146–150.

SANTANA, M. A. E., BAUMANN, M. and CONNER, A. H. (1996) *J. Wood Chem. Tech.* **16**, 1–19.

TAHIR, P. M. and SELLERS, JR, T. (1990) *Proceedings, Division 5, 19th IUFRO World Congress*, Montreal, Quebec, Canada, pp. 207–214.

THOMPSON, D. and PIZZI, A. (1995) *J. Appl. Polymer Sci.* **55**, 107–112.

Ethanol Production from Forest Product Wastes

JACK N. SADDLER AND DAVID J. GREGG

12.1 Introduction

Over the last 25 years there has been considerable interest in the potential of producing fuel ethanol from biomass. The OPEC oil embargo of the 1970s resulted in a marked increase in the price of oil and influenced countries such as Brazil, Canada, Finland, Japan, Sweden and the USA to plan for better liquid fuel self-sufficiency (Klyosov, 1986; Vallander and Eriksson, 1990). More recently research into fuels from renewable resources has been driven by environmental concerns, particularly the role of fossil fuel contribution to poor air quality and global warming (von Sivers and Zacchi, 1993). As a result there has been considerable research and discussion about the environmental advantages of using ethanol as a transportation fuel and as a gasoline supplement.

The benefits of using ethanol and the ether form of ethanol, ethyl *tert*-butyl ether (ETBE), as an alternative to gasoline, have been discussed in detail (Wyman and Hinman, 1990). Ethanol is a clean-burning, high-octane fuel that can be readily substituted for gasoline and its combustion results in significant reductions of toxic emissions such as formaldehyde, benzene and 1,3-butadiene (Chang *et al.*, 1991).

Blends of ethanol or ETBE with gasoline increase the octane of the mixture and can improve performance. Ethanol blends cause internal-combustion gasoline engines to run with leaner fuel mixtures and they reduce carbon monoxide emissions by 10–30 per cent. ETBE in gasoline also reduces carbon monoxide emissions and it further lowers the Reid vapour pressure of gasoline, thereby decreasing the release of smog-forming compounds including ozone. Ozone is recognized as being one of the most pervasive and persistant urban air-quality problems. Consequently, urban areas in the USA with heavy air pollution, such as areas of California and Colorado, have led the way towards implementation of ethanol (and other oxygenated fuel) vehicle regulations.

When ethanol is produced from renewable sources such as biomass it can both decrease urban air pollution and reduce the accumulation of carbon dioxide, one of the greenhouse gases. Thus replacement of gasoline with ethanol, derived from renewable biomass feedstocks that sequester CO_2 during growth, is expected to reduce CO_2 emissions by 90–100 per cent (Chang *et al.*, 1991; DeLuchi, 1993). It has also been estimated that enough neat ethanol could be made from the cellulosic biomass residues that are currently available within the USA today to potentially replace twice the amount of gasoline consumed within the USA in 1994 (Wyman, 1995).

12.2   Feedstocks

12.2.1   *Sugar and Starch*

Ethanol, when used as a transportation fuel, can be generated from a number of feedstocks which are usually categorized into sugar, starch and lignocellulosic based materials. Sugar crops including sugar cane, sugar beets and sweet sorghum produce monomeric sugars (glucose, fructose and sucrose) that can be directly fermented to ethanol. Starch crops, including grains (corn, wheat, barley, grain sorghum) and tubers (potatoes, sweet potatoes) require an extra processing step, called hydrolysis, prior to fermentation. Hydrolysis converts the complex sugars or starches in the grains and tubers to monomeric sugars suitable for fermentation. Fuel ethanol is currently produced from both sugar and starch based feedstocks with approximately 3 billion US gallons of ethanol produced from sugar cane in Brazil (Wyman, 1993), 1 billion US gallons from corn in the USA, and 5.5 million US gallons from both corn and wheat in Canada. However, the use of these feedstocks for fuel production competes with other higher value usages such as food production. Current production of fuel ethanol is often based on excess agricultural production and it is generally recognized that this volume is too small in comparison with the anticipated levels of production required for total conversion of transportation fuel markets from gasoline to ethanol. It is also apparent that there is the potential for competition with food production for both the sugar and starch feedstocks and that prime agricultural lands normally required to produce the foodstuffs should not be diverted for fuel production.

12.2.2   *Lignocellulosics*

Lignocellulosic biomass is typically composed of a complex mixture of three polymers – cellulose, hemicellulose and lignin – and a small amount of other compounds that are loosely termed extractives (Tshiteya, 1992). The fermentation of sugars derived from lignocellulosic feedstocks has proven to be more of a process design and operating challenge than traditional sugar or starch based processes. For example, there is a considerably wider variation in the type and nature of the processes and equipment needed to convert lignocellulosic feedstocks to ethanol. Although sugar, starch and lignocellulosic substrates can have compositional variability due to variations in the species of feedstock used, growing site, climate, age and the part of the plant used, lignocellulosic feedstocks have the following additional problems: proportional variability within the mixture of the three major components; differences in the types and amounts of extractives; and natural variability in the monomeric sugars that make up the hemicellulose component.

12.3 Acid vs. Enzymatic Hydrolysis of Lignocellulosics

12.3.1 *Acid Hydrolysis of Lignocellulosics*

Acid hydrolysis of biomass feedstocks has been studied and practised commercially for many years (Klyosov, 1986). Although several types of acid, including sulphurous, sulphuric, hydrochloric, hydrofluoric, phosphoric, nitric and formic have been used for hydrolysis, there are essentially two types of acid hydrolysis process, termed dilute and concentrated, with a number of generic processes associated with each of these two options (Klyosov, 1986; Wayman and Parekh, 1990; Biohol Developments *et al.*, 1991; Tshiteya, 1992). Although the conversion of lignocellulosics to glucose via acid hydrolysis has been proven technically on a large commercial scale (Tshiteya, 1992) it is still considered that enzymatic hydrolysis has the potential to surpass greatly the efficiency of acid hydrolysis (B. H. Levelton & Associates *et al.*, 1989). Major technical problems that have yet to be resolved using acid hydrolysis are: the corrosion of the reaction vessels; degradation of product sugars resulting in low yields; the need for neutralization before subsequent bioconversion; formation of numerous environmentally noxious by-products; high capital and operating cost; and solvent losses (Frings and Coombs, 1992; Tshiteya, 1992; Wright and Feinberg, 1992). Although a considerable amount of research is still directed towards these problem areas, the long-term outlook for acid based biomass to ethanol processes is still not overly optimistic as various economic projections indicate that several major technical problems have yet to be resolved before the economics of an acid based process are significantly more attractive (Wayman and Parekh, 1990; Tshiteya, 1992; Wright and Feinberg, 1992; von Sivers and Zacchi, 1995).

12.3.2 *Enzymatic Hydrolysis of Lignocellulosics*

Enzymatic hydrolysis processes are relatively new (research started in the 1970s) and integrated processes have yet to be proven both technically and economically. Enzyme based hydrolysis of lignocellulosics tends to show better promise than acid based hydrolysis, primarily because of the potential for higher sugar yields and the production of less toxic effluent streams. However, any proposed enzyme based bioconverison process is generally more complex than an acid hydrolysis process as the enzymes tend to show more substrate specificity and require more carefully controlled reaction conditions. To date, a truly 'generic' enzymatically-based biomass-to-ethanol process has been difficult to identify because of the heavy influence that the type of feedstock, type of by-products, number of unproven processes and equipment currently available will have on the design of such a process. However, it is generally acknowledged that a generic enzymatic-based process would include the following steps: pretreatment, fractionation, enzyme production, enzyme hydrolysis, fermentation, ethanol and other by-product recovery, and waste treatment (Gregg and Saddler, 1995b). The pioneering work that has been done at the pilot or demonstration scale has shown that the subprocess steps are all strongly interdependent (Pourquie *et al.*, 1988; Matsui, 1991; Hayn *et al.*, 1993). Thus it has been extremely difficult to identify the relative technical or economic merits of each of the subprocess variations and their subsequent influence on the final production cost of ethanol.

Some of the process steps, such as fermentation and ethanol recovery, have been commercialized and often used as component steps in other industries such as brewery and distillery plants. These component process steps therefore have an established technoeconomic baseline and can be considered to be mature technologies. The less mature process steps – pretreatment, fractionation, hydrolysis and pentose fermentation – have generally been compared on a relative technical or economic basis using laboratory, pilot-plant equipment or techno-economic modelling to simulate the entire, integrated process. There have been a number of pilot plants, two fully-integrated pilot plants (NEDO, 1986–91, Japan; and NREL, 1994 to present, USA) and several smaller non-integrated pilot plants (Institut Français du Petrole's Souston Plant, France; Voest-Alpine's VABIO plant, 1982–88, Austria; and IOGEN's plant, 1989, Canada) built and operated over the last 5 years. They have been primarily used to test the technical and economic feasibility of various aspects of the enzymatic conversion process (Pourquie *et al.*, 1988; Matsui, 1991; Hayn *et al.*, 1993). There have also been a number of attempts to model techno-economically the various process scenarios using laboratory/pilot-plant equipment (Douglas, 1989; Nguyen and Saddler, 1991; Hinman *et al.*, 1992; von Sivers, 1995; Gregg, 1996; Saddler and Gregg, 1996).

12.4 The Major Component Steps in an Enzyme Based Biomass-to-Ethanol Process

12.4.1 Pretreatment

Pretreatment is the process step required to make the relatively recalcitrant lignocellulosic material more easily digestible to the hydrolytic enzymes while preserving the yield of the original carbohydrates for fermentation. This can be accomplished by various mechanisms such as the removal of the lignin sheath, reduction of cellulose crystallinity, or by increasing the surface area that is accessible to the enzymes. However, inhibitory breakdown products can be formed if the pretreatment conditions used are too harsh. The nature of the lignocellulosic substrate used has a major impact on this process step as a certain pretreatment, which may be effective with one lignocellulosic substrate, may prove to be ineffective on another. A number of reviews have covered pretreatment in detail and these have generally separated the different types of pretreatment into physical, chemical, biological and combinations of these methods (Millet *et al.*, 1976; Chang *et al.*, 1980; Horton *et al.*, 1980; Puls and Dietrich, 1980; Su *et al.*, 1980; Klyosov, 1986; Grethlein and Converse, 1991; Schell *et al.*, 1991). There appear to be four main pretreatment methods currently being researched and commercialized: organosolv (Paszner and Cho, 1988; Aziz and Sarkanen, 1989), steam explosion (Gregg and Saddler, 1995a), dilute-acid prehydrolysis (Torget *et al.*, 1991) and ammonia fibre explosion (AFEX) (Holtzapple *et al.*, 1991). Of these various options, only the steam-explosion process has resulted in the substantial commercialization and sale of reactors by companies such as Stake Technology of Canada.

12.4.2 Fractionation

Fractionation is the subprocess step that separates the lignocellulosic slurry obtained after the pretreatment into the three main fractions of cellulose, hemicellulose and lignin. Subsequent fractionation after pretreatment is generally recognized as an efficient way of providing for separate processing of the individual components while recovering most of the material available in the original feedstock. There are generally two unit operations within the process which provide separation of both the hemicellulose and lignin components. The major product derived after these two fractionation steps is a cellulose rich residue which is subsequently hydrolyzed enzymatically. Although pretreatment and fractionation can be carried out simultaneously in processes such as organosolv, in processes such as steam explosion or acid hydrolysis it is usually carried out as two separate sub-process steps. Due to the inter-related nature of the enzymatic based biomass-to-ethanol process, the type of feedstock and pretreatment method used will probably determine the fractionation procedure that is adopted (Wong *et al.*, 1988; Schwald *et al.*, 1989a,b; Wayman and Parekh, 1990; Ramos *et al.*, 1992). However, the greater the number of washing steps required, the higher the capital and operating costs will become, and the more likely that the net return on the investment will be negative.

Hemicellulose, once it has been solubilized through pretreatment by SO_2-steaming or dilute acid (Wayman and Parekh, 1990), can be processed in a number of different ways. Currently, the xylose derived from agricultural or hardwood hemicelluloses cannot be readily fermented to ethanol. This will be discussed more fully in the subsequent fermentation section. Although there has been a considerable amount of work carried out on other hemicellulose derived products such as furfural, xylitol and single-cell protein this has not lead to any significant commercial products, partly because the few high-value products that were identified have too small a market volume.

After pretreatment, lignin can generally be extracted from most lignocellulosic residues by a sodium hydroxide wash. Lignin has a high heat content and has been used traditionally in the pulp and paper industry as a boiler fuel for process heat or cogeneration of steam and electricity. The potential for by-product utilization is immense because approximately 1 kg of lignin is produced per litre of ethanol. Although there have been a large number of high-value lignin-derived products identified, the operation of a few commercial-scale plants would probably saturate the world market for most of these applications.

12.4.3 *Hydrolysis*

The enzymatic hydrolysis of the cellulosic component of lignocellulosics requires the use of a complete cellulase enzyme complex containing various endoglucanases, exo-glucanases and cellobiases (Coughlan, 1989). Various groups, due primarily to technical and economical reasons, have advocated the on-site production of cellulases rather than using the commercial cellulases sold by companies such as Novo–Nordisk, Genencor, Primalco and at least a further seven companies worldwide. These companies currently produce and market different types of cellulases for applications in areas such as textiles, detergents, animal feed and pulp and paper processing. Cellulases are synthesized and excreted by various microorganisms with certain fungi, primarily the *Trichoderma* species (notably *T. reesei*), among the most efficient producers (Phillippidis, 1994). Generally, the amount of cellulase produced by wild-type strains of fungi or bacteria is too low to support an economical industrial process (Pourquie and Warzywoda, 1993). For this reason strain improvement programmes were initiated in the mid-to-late 1970s and still continue to this day. Strains originally isolated using agar plating and isolation techniques have been improved by various methods, such as increasing the production of the whole cellulase complexes, increasing the resistance to glucose repression and enhancing pH and temperature tolerance. Certain strains, such as *T. reesei* (CL 847) have remained remarkably stable, in a genetic sense, over the whole transition from laboratory-scale to precommercial production scale (30 000-litre fermenter). However, continued genetic improvement through both molecular biology and further use of traditional mutagenesis and screening protocols has not significantly improved the multifactorial characters such as specific activity, productivity or yields (Baker *et al.*, 1995).

Cellulase production is now possible at a cost well below expectations based on earlier economic studies (Phillippidis, 1994). However, the cost of supplying enzymes for a biomass-to-ethanol process is still too high to allow the economic production of chemicals or fuels from lignocellulosic derived sugars. Although on-site production of hemicellulose and glucose hydrolysates could provide a cheap source of substrates, the cost and effectiveness of on-site enzyme production has yet to be proven (Pourquie and Warzywoda, 1993). Consequently, a significant amount of work on the production, mechanism and effectiveness of cellulase is continuing in this area. It can therefore be expected that, with increasing volumes of enzyme sales and the increased number of new applications in areas such as pulp and paper, the commercial cost of enzymes will continue to drop.

As mentioned earlier, enzymatic hydrolysis is accomplished through the synergistic action of several cellulase components (Coughlan, 1989). Synergistic action means that the combined activity of the enzymes is greater than the sum of each of the components. The current mechanistic model (Wright *et al.*, 1990) primarily involves three main groups of enzymes that are required before cellulose can be hydrolyzed effectively to glucose. The first component includes the endoglucanases, which attack the amorphous cellulose in a random action, producing more free ends. This enhances the action of the second group of enzymes, the exoglucanases, which remove cellobiose units from both the reducing and non-reducing ends of cellulose chains. The third component includes the β-glucosidases which split the cellobiose units into monomeric glucose units. Each of the enzyme components is influenced by end-product inhibition and consequently the build-up of any of the products from any of the enzymatic reactions results in inhibition of the overall cellulose hydrolysis reaction.

The cellulase enzyme complex has been isolated from a wide variety of organisms including anaerobic protozoa, aerobic fungi, and aerobic and anaerobic bacteria (Isaacs, 1984). Enzymes from *Trichoderma reesei*, which is an aerobic, mesophilic fungus, are the most extensively studied cellulases, essentially because all the necessary enzyme components for cellulose hydrolysis are produced extracellularly in high concentrations (Tshiteya, 1992) and the organism can be grown in submerged aerobic culture. Furthermore, the cellulase complex produced is resistant to chemical inhibitors and remains stable for up to 48 hours at 50°C (Wayman and Parekh, 1990). Although maximum cellulase activity for most fungal derived cellulases occurs at 50 ± 5°C and a pH of 4.0–5.0, cellulase complexes are known to differ substantially in their pH and temperature tolerance and in the ratio and amount of the different cellulase components. For example, *T. reesei* wild-type preparations are deficient in cellobiase, a β-glucosidase type enzyme, and result in an accumulation of cellobiose unless supplemented with the enzyme. However, there are induced mutant strains of *T. reesei* that display high β-glucosidase activity (Tshiteya, 1992).

The high specificity of the cellulase enzymatic reaction should theoretically result in efficient hydrolysis of cellulose to glucose. In practice, the yields of glucose are influenced by many factors that can impact on both the rate and extent of hydrolysis (Klyosov, 1986). For example, it is known that the hydrolysis rate progressively declines with time due to a range of substrate and enzyme related factors. It has been shown that the surface area available for enzyme–substrate interaction is influenced by cellulose pore size and the shielding effects of hemicellulose and lignin (Isaacs, 1984). The crystalline structure of cellulose excludes water molecules as well as any larger molecules, including the cellulase enzymes, and thus reduces the available surface area. Although it has been suggested that the crystalline regions of cellulose will be hydrolyzed at a much slower rate than the amorphous cellulose, due to the greater stability resulting from the interchain hydrogen bonding, this has not proven to be the case. Analysis of residual substrates has shown that the crystallinity remains unchanged as hydrolysis proceeds (Ramos *et al.*, 1992).

As well as the surface area of the substrate limiting hydrolysis, both the rate and extent of the reaction are influenced by enzyme factors such as end-product inhibition, the irreversible and/or non-specific adsorption of cellulases onto the substrate, and the inactivation of key components of the cellulase complex (Ramos *et al.*, 1992). There are various ways in which the effectiveness of the cellulases could be enhanced. For example, the low specific activity of commercial cellulase preparations has led to the requirement for high cellulase enzyme loadings. As a result, a considerable amount of research is now focused on learning more about the enzymatic and inhibitory mechanisms associated with the cellulase complex, screening for strains with enhanced enzymatic features (higher productivity, yield, resistance to end-product inhibition and higher specific activity) and the designing and testing of alternative operational and flow characteristics (e.g., fed-batch reactors, high enzyme concentrations, enzyme recycling) to enhance the activity and reuse of the cellulases.

12.4 Fermentation

12.4.1 Hexose Fermentation

The fermentation of glucose to ethanol was one of the first complex biological and chemical processes mastered by man. Alcohols became an important fuel and chemical feedstock in the mid-nineteenth century, predating the petrochemical industries of today. For example, most solvents and chemicals such as acetone, butanol and ethanol were originally produced from a number of carbohydrate feedstocks using a variety of conversion processes. However, with the rapid growth of the petroleum and petrochemical industry following World War I, fermentation research and development has been restricted primarily to the brewing and distilling industries and the last two decades of fermentation research have tried to extend the limits of traditional technologies rather than develop radically different processes. By optimizing the typical batch cycle, a batch fermentation which took 7 days in 1978 was reduced to 3 days by 1986. At the same time companies such as

Melle-Boinot developed a system which operated in a mode between a batch and continuous fermentation and reduced the fermentation time to 16 hours. This was accomplished by recycling the yeast and residual fermentation substrate from earlier batches, thereby decreasing the volume of the fermenter and consequently increasing yields. Simultaneous ethanol distillation and advances in antibiotic research have also aided in the development of continuous fermentation processes. The ability to draw off the ethanol continuously as it is produced enables the system to keep a high level of productivity by removing the product which can slow down the metabolism of the yeast and eventually kill it. Advances in antibiotics were also required as continuous fermentation systems are vulnerable to outbreaks of infection and subsequent costly shutdowns. The Biostil process incorporates many of these developments (Wayman and Parekh, 1990) while other methods and technologies continue to be researched to increase the efficiency of ethanol production. Strategies include the development of flocculating yeasts, extractive fermentation, yeast immobilization, and continued research into modifying the various organisms through various classical and genetic engineering techniques.

12.4.2 *Pentose Fermentation*

The importance of being able to ferment the hemicellulose pentose derived sugars to the economics of the biomass-to-ethanol process has been emphasized by a number of authors (Douglas, 1989; Tshiteya, 1992; Beck, 1993; Hahn-Hagerdal *et al.*, 1993). There are several reasons why the fermentation of pentoses to ethanol continues to be researched aggressively. These include the relative ease by which pentose can be recovered from most lignocellulosic substrates by methods such as dilute acid hydrolysis and acid catalyzed steam explosion; the rapidly expanding knowledge base on the biochemistry of ethanol production from xylose continues to offer potential ways of enhancing fermentation efficiency. Finally, as mentioned above, a major driving force is the economic incentive that the extra ethanol from the xylose fraction of many biomass sources has on the overall ethanol process (Beck, 1993). The identification of a xylose-fermenting yeast, *Pachysolen tannophilus*, in 1981 first indicated that ethanol could be derived from xylose and allowed projections to be made of the economic viability of converting both cellulose and hemicellulose derived sugars to ethanol. Other pentose fermenting organisms such as *Pichia stipitis* and *Candidae shihatae* have also been identified and characterized over the last 5 years (Hahn-Hagerdal *et al.*, 1993). The focus of most of this research has been to understand the limitations of the xylose fermentation and to use modern genetic engineering techniques to improve the ethanol yields and productivity of the yeasts. Some recent work has looked at the application of xylose-fermenting organisms in processing schemes using actual lignocellulosic sugar substrates. This has indicated that various problems such as inhibitory products and fermentation of mixed sugar streams have yet to be fully resolved.

Several groups have been trying to manipulate certain bacteria and yeasts genetically so that they could incorporate the genes necessary for xylose fermentation to ethanol. Metabolic engineering has been applied to *Erwina chrysanthemi* (Tolan and Finn, 1987a), *Escherichia coli* (Neale *et al.*, 1988; Alterthum and Ingram, 1989; Ohta *et al.*, 1990) *Klebsiella oxytoca* (Ohta *et al.*, 1991), *Klebsiella planticola* (Tolan and Finn, 1987b; Feldmann *et al.*, 1989), and *Zymomonas mobilis* (Zhang *et al.*, 1995) and to *Saccharomyces cerevisiae* (Ho *et al.*, 1997). So far the organisms which are most capable of fermenting both hexoses and pentoses have been *E. coli*, *K. oxytoca*, *Z. mobilis* and *S. cerevisiae*. Enteric bacteria such as *E. coli* do not grow well on xylose, even though they possess the necessary genes for xylose uptake and utilization. Normally, anaerobic fermentation of sugars with *E. coli* results in a range of products including lactate, acetate, succinate and formate with ethanol being only a minor product (Gottschalk, 1979). Although engineered organisms can directly ferment xylose and other five-carbon sugars they tend to prefer a neutral pH. However, a neutral pH creates a greater demand for base and a higher potential for contamination by other organisms.

12.4.3 Ethanol Recovery

The beer resulting from a typical glucose fermentation usually contains about 8–12 per cent ethanol by volume. This dilute concentration is primarily a consequence of ethanol end-product inhibition. It is not possible to produce anhydrous ethanol by simple distillation as an azeotrope is formed at 95% ethanol; the azeotrope has a lower vapour pressure than either ethanol or water and is therefore preferentially distilled. A second step using a third component (e.g. benzene) to form another azeotrope with one or both of the original components has traditionally been employed (Perry and Chilton, 1973). A considerable amount of work continues to be carried out to try to reduce the effects of end-product inhibition and the costs of conventional distillation (vapour recompression, cascade pressurization, super-critical fluid carbon dioxide) and dehydration (carbon dioxide extraction, solvent extraction, extractive fermentation, pervaporation, molecular sieves, adsorption).

12.5 Current and Future Status of Enzymatic Hydrolysis

As is apparent in this brief review, the bioconversion of lignocellulosics to ethanol is a complicated and strongly interdependent series of process steps. Currently there are no true examples of commercial or totally integrated demonstration-sized plants which can convert lignocellulosic materials to ethanol. Although some of the current and past pilot plants have been able to demonstrate the successful operation of entire sets of subprocess steps, these plants have not been able to operate continuously over a prolonged period of time. Hopefully, plants such as those at NREL in Colorado, USA, and others that will be potentially constructed in Sweden and Canada will provide the technical and operational experience that will allow the establishment of true commercial plants.

As a less expensive option, many researchers have attempted to assess the current techno-economic status and future potential of the various bioconversion processes by building techno-economic models based on information obtained from both laboratory and pilot studies and from experience gained in the operation of similar processes in other industries, for example, ethanol from grain/corn. At this point in time, the comparison of these models appears to be the most direct way of providing a relative subprocess cost estimation. Although it is often difficult to compare model results directly as the basic tenets on which the models are based can differ substantially, there is general agreement that an enzyme based biomass-to-ethanol process could produce ethanol for about US$0.50 per litre based on the recovery of most of the cellulose and hemicellulose derived sugars and an energy credit for burned lignin (Gregg and Saddler, 1996; von Sivers and Zacchi, 1996). More optimistic studies incorporating advances in genetic engineering, ethanol recovery and by-product credits have estimated that US$0.13–0.18 per litre of ethanol can be achieved (Hinman *et al.*, 1992; Lynd *et al.*, 1996). Although continued research will undoubtedly decrease the cost of producing biomass-to-ethanol it is probable that the eventual increase, both environmentally and economically, in the cost of petroleum derived fuels will provide the necessary incentive to establish biomass-to-ethanol processes.

References

ALTERTHUM, F. and INGRAM, L. O. (1989) Efficient ethanol production from glucose, lactose and xylose by recombinant *Escherichia coli*. *Appl. Microbiol. Biotechnol.* **55**, 1943–1948.

AZIZ, S. and SARKANEN, K. (1989) Organosolv pulping – a review. *Tappi* **39**, 169–175.

BAKER, J. O., ADNEY, W. S., THOMAS, S. R., NIEVES, R. A., CHOU, Y.-C., VINZANT, T. B., TUCKER, C. S., LAYMON, R. A. and HIMMEL, M. E. (1995) Synergism between purified bacterial and fungal cellulases. In: Saddler, J. N. and Penner, M. H., eds, *Enzymatic Degradation of Insoluble Carbohydrates*, Washington, DC: ACS, pp. 113–141.

BECK, M. J. (1993) Fermentation of pentoses from wood hydrolysates. In: Saddler, J. N., ed, *Bioconversion of Forest and Agricultural Plant Residues*, Wallingford, UK: CAB International, pp. 211–229.

BIOHOL DEVELOPMENTS, O'BOYLE, A., GOOD, D., POTTS, D., FEIN, J., GRIFFITH, R., BECK, M.-J., DAHLGREN, D. and WALLIN, T. (1991) A process for the efficient conversion of waste pine to ethanol, Final Report No. Contract File 23216-9-9018/01-SZ, 12 September, Energy, Mines and Resources Canada.

CHANG, M. M., CHOU, T. Y. C. and TSAO, G. T. (1980) Structure, pretreatment and hydrolysis of cellulose. *Adv. Biochem. Eng.* **20**, 15–42.

CHANG, T. Y., HAMMERLE, R. H., JAPAR, S. M. and SAMEEN, I. T. (1991) Alternative transportation fuels and air quality. *Environ. Sci. Technol.* **25**, 1190–1197.

COUGHLAN, M. P. (1989) *Enzyme Systems for Lignocellulosic Degradation*, London: Elsevier Applied Science.

DeLUCHI, M. A. (1993) Greenhouse gas emissions from the use of new fuels for transportation and electricity. *Transport. Res.* **27**, 187–194.

DOUGLAS, L. J. (1989) A technical and economic evaluation of wood conversion processes, Final Report No. DSS Contract File 23283-8-6091, Energy, Mines and Resources Canada.

FELDMANN, S., SPRENGER, G. A. and SAHM, H. (1989) Ethanol production from xylose with a pyruvate-formate-lyase mutant of *Klebsiella planticola* carrying a pyruvate-decarboxylase gene from *Zymomonas mobilis*. *Appl. Microbiol. Biotechnol.* **31**, 152–157.

FRINGS, R. M. and COOMBS, J. (1992) Wastewaters from the bioconversion of biomass – utilisation and treatment – literature review and problem analysis, International Energy Agency, IEA/BE T7/A7 Report Number 4, 1992/04.

GOTTSCHALK, G. (1979) *Bacterial Metabolism*, New York: Springer-Verlag.

GREGG, D. J. (1996) The development of a techno-economic model to assess the effect of various process options on a wood-to-ethanol process, M.ASc. Thesis, Department of Wood Science, University of British Columbia.

GREGG, D. and SADDLER, J. N. (1995a) A techno-economic assessment of the pretreatment and fractionation steps of a biomass-to-ethanol process. *Appl. Biochem. Biotechnol.* **57/58**, 711–727.

GREGG, D. J. and SADDLER, J. N. (1995b) Bioconversion of lignocellulosic residue to ethanol: process flowsheet development. *Biomass and Bioenergy* **9**, 287–302.

GREGG, D. J. and SADDLER, J. N. (1996) The IEA Network; Biotechnology for the conversion of lignocellulosics, Biomass-to-Ethanol Process Development, European Motor Biofuels Forum, Graz, Austria.

GRETHLEIN, H. E. and CONVERSE, A. O. (1991) Common aspects of acid prehydrolysis and steam explosion for pretreating wood. *Biores. Technol.* **36**, 77–82.

HAHN-HAGERDAL, B., HALLBONN, J., JEPPSON, H., OLSSON, L., SKOOG, K. and WALFRIDSSON, M. (1993) Pentose fermentation to alcohol. In: Saddler, J. N., ed., *Bioconversion of Forest and Agricultural Plant Residues*, Wallingford, UK: CAB International, pp. 231–290.

HAYN, M., STEINER, W., KLINGER, R., STEINMULLER, H., SINNER, M. and ESTERBAUER, H. (1993) Basic research and pilot studies on the enzymatic conversion of lignocellulosics. In: Saddler, J. N., ed., *Bioconversion of Forest and Agricultural Residues*, Wallingford: CAB International, pp. 33–72.

HINMAN, N. D., SCHELL, D. J., RILEY, C. J., BERGERON, P. W. and WALTER, P. J. (1992) Preliminary estimate of the cost of ethanol production for SSF technology. *Appl. Biochem. Biotech.* **34/35**, 639–649.

HO, N. W. Y., TOON, S., CHEN, Z. D., BRAINARD, A., LUMPKIN, R. E., RILEY, C. J. and PHILIPPIDIS, G. (1997) Further improvements of recombinant *Sacchoromyces* yeast for xylose fermentation. *Appl. Biochem. Biotechnol.* (in press).

HOLTZAPPLE, M. T., JUN, J., ASHOK, G., PATIBANDLA, S. L. and DALE, B. D. (1991) The ammonia freeze explosion (AFEX) process: a practical lignocellulosic pretreatment. *Appl. Biochem. Biotechnol.* **28/29**, 59–74.

HORTON, G. L., RIVERS, D. B. and EMERT, G. H. (1980) Preparation of cellulosics for enzymatic conversion. *Ind. Eng. Chem. Prod. Res. Devel.* **19**, 422–429.

ISAACS, S. H. (1984) Ethanol production by enzymatic hydrolysis – parametric analysis of a base-case process, SERI-Chem Systems Ltd.

KLYOSOV, A. A. (1986) Enzymatic conversion of cellulosic materials to sugar and alcohol – the technology and its implications. *Appl. Biochem. Biotechnol.* **12**, 260–270.

LEVELTON, B. H. & ASSOCIATES LTD, EDWARDS, W. C. and QUAN, R. G. (1989) A study of the merits and difficulties of coupling a softwood sawmill and an ethanol plant, Final Report, 30 September, Energy, Mines and Resources Canada.

LYND, L. R., ELANDER, R. T. and WYMAN, C. E. (1996) Likely features and costs of mature biomass ethanol technology. *Appl. Biochem. Biotechnol.* **57/58**, 741–761.

MATSUI, S. (1991) Development of fuel alcohol technologies: research and development of a total system using woody materials, *Eleventh Annual Conference on Alcohol and Biomass Energy Technologies*, Tokyo, Japan, New Energy and Industrial Technology Development Organization (NEDO).

MILLET, M. A., BAKER, A. J. and SATTER, L. D. (1976) Physical and chemical pretreatments for enhancing cellulose saccharification. *Biotechnol. Bioeng.* **6**, 125–153.

NEALE, A. D., SCOPES, R. K. and KELLY, J. M. (1988) Alcohol production from glucose and xylose using *Escherichia coli* containing *Zymomonas mobilis* genes. *Appl. Microbiol. Biotechnol.* **29**, 162–167.

NGUYEN, Q. A. and SADDLER, J. N. (1991) An integrated model for the technical and economic evaluation of an enzymatic biomass conversion process. *Biores. Technol.* **85**, 275–282.

OHTA, K., BEALL, D. S., MEIJA, J. P., SHANMUGAN, K. T. and INGRAM, L. O. (1990) Effects of environmental conditions on xylose fermentation by recombinant *Escherichia coli*. *Appl. Environ. Microbiol.* **56**, 463–465.

OHTA, K., BEALL, D. S., MEIJA, J. P., SHANMUGAN, K. T. and INGRAM, L. O. (1991) Metabolic engineering of *Klebsiella oxytoca* M5A1 for ethanol production from xylose and glucose. *Appl. Environ. Microbiol.* **57**, 2810–2815.

PASZNER, L. and CHO, H. J. (1988) High efficiency conversion of lignocellulosics to sugars for liquid fuel production by the ACOS process. *Energy Exploit. Explor.* **6**, 39–60.

PERRY, R. H. and CHILTON, C. H. (1973) *Chemical Engineers' Handbook*, New York: McGraw-Hill.

PHILLIPPIDIS, G. P. (1994) Cellulase production technology: evaluation of current status. In: Himmel, M. E., Baker, J. O. and Overend R. P., eds, *Enzymatic Conversion of Biomass for Fuels Production*, Washington, DC: American Chemical Society, p. 499.

POURQUIE, J. and WARZYWODA, M. (1993) Cellulase production by *Trichoderma reesei*. In: Saddler, J. N., ed., *Bioconversion of Forest and Agricultural Plant Residues*, Wallingford: CAB International, pp. 107–116.

POURQUIE, J., WARZYWODA, M., CHEVRON, F., THERY, M., LONCHAMP, D. and VANDECASTEELE, J. P. (1988) Scale up of cellulase production and utilization. In: Aubert, J. P., Beguin, P. and Millet, J., eds, *Biochemistry and Genetics of Cellulose Degradation*, London: Academic Press, pp. 71–86.

PULS, J. and DIETRICH, H. H. (1980) Separation of lignocellulosics into highly accessible fibre materials and hemicellulose fraction by the steaming-extraction process. In: *Energy from Biomass*, Brighton, UK, pp. 348–353.

RAMOS, L. P., BREUIL, C. and SADDLER, J. N. (1992) Comparison of steam pretreatment of eucalyptus, aspen and spruce wood chips and their enzymatic hydrolysis. *Appl. Biochem. Biotechnol.* **34/35**, 37–47.

SADDLER, J. N. and GREGG, D. J. (1996) Techno-economic assessment of the pretreatment and hydrolysis of wood for ethanol production, *9th European Bioenergy Conference*, Copenhagen, Denmark.

SCHELL, D. J., TORGET, R., POWER, A., WALTER, P. J., GROHMANN, K. and HINMAN, N. D. (1991) A technical and economic analysis of acid catalysed steam explosion and dilute sulphuric acid pretreatments using wheat straw or aspen wood chips. *Appl. Biochem. Biotechnol.* **28/29**, 87–97.

SCHWALD, W., BREUIL, C., BROWNELL, H. H., CHAN, M. and SADDLER, J. N. (1989a) Assessment of pretreatment conditions to obtain fast complete hydrolysis on high concentrations. *Appl. Biochem. Biotechnol.* **20/21**, 29–44.

SCHWALD, W., SMARIDGE, T., CHAN, M., BREUIL, C. and SADDLER, J. N. (1989b) The influence of SO_2 impregnation and fractionation on product recovery and enzymic hydrolysis of steam-treated sprucewood. In: Coughlan, M. P., ed, *Enzyme Systems for Lignocellulosic Degradation*, New York: Elsevier, pp. 231–242.

SU, T. M., LAMED, R. J., LOBOS, J., BRENNON, M., SMITH, J. F., TABOR, D. and BROOKS, R. (1980) Final Report Period 1 December 1979–31 December 1980, US Department of Energy, Final Report Number SERI/TR-8271-1-77.

TOLAN, J. S. and FINN, R. K. (1987a) Fermentation of D-xylose and L-arabinose to ethanol by *Erwinia chrysanthemi*, *Appl. Environ. Microbiol.* **53**, 2033–2038.

TOLAN, J. S. and FINN, R. K. (1987b) Fermentation of D-xylose to ethanol by genetically modified *Klebsiella planticola*. *Appl. Environ. Microbiol.* **53**, 2039–2044.

TORGET, R., WALTER, P. J., HIMMEL, M. and GROHMANN, K. (1991) Dilute-acid pretreatment of corn residues and short rotation woody crops. *Appl. Biochem. Biotechnol.* **28/29**, 75–86.

TSHITEYA, R. M. (1992) Conversion technologies – biomass to ethanol – alcohol fuels, National Renewable Energy Lab, Reference Report No. 3, September.

VALLANDER, L. and ERIKSSON, K.-E. (1990) Production of ethanol from lignocellulosic materials: state of the art. In: Fiechter, A., ed., *Advances in Biochemical Engineering and Biotechnology*, Berlin: Springer-Verlag, pp. 63–95.

VON SIVERS, M. (1995) Ethanol from wood – a technical and economic evaluation of ethanol production processes, Licentiate Dissertation, Department of Chemical Engineering I, Lund Institute of Technology.

VON SIVERS, M. and ZACCHI, G. (1993) A techno-economical comparison of three processes for the production of ethanol from wood, Lund Institute of Technology – Department of Chemical Engineering I, Project No. LUTKDH/(TKKA-7006)/1-27/(1993).

VON SIVERS, M. and ZACCHI, G. (1995) A techno-economic comparison of three processess for the production of ethanol from pine. *Biores. Technol.* **51**, 43–52.

VON SIVERS, M. and ZACCHI, G. (1996) Ethanol from lignocellulosics: a review of the economy. *Biores. Technol.* **56**, 131–140.

WAYMAN, M. and PAREKH, S. R. (1990) *Biotechnology of Biomass Conversion*, Milton Keynes: Open University Press.

WONG, K. K. Y., DEVERELL, K. F., MACKIE, K. L., CLARK, T. A. and DONALDSON, L. A. (1988) The relationship between fiber porosity and cellulose digestibility in steamed-exploded *Pinus radiata. Biotechnol. Bioeng.* **31**, 447–456.

WRIGHT, J. D. and FEINBERG, D. A. (1992) A comparison of the production of methanol and ethanol from biomass, Final Report Contract No: 23218-1-9201/01-SQ, March, Energy, Mines and Resources Canada.

WRIGHT, J. D., WYMAN, C. E. and GROHMANN, K. (1990) Simultaneous saccharification and fermentation of lignocellulose: process evaluation. *Appl. Biochem. Biotechnol.* **19**, 75–90.

WYMAN, C. E. (1993) An overview of ethanol production for transportation fuels. *First Biomass Conference of the Americas: Energy, Environment, Agriculture and Industry*, Burlingtion, VT.

WYMAN, C. E. (1995) Biomass derived oxygenates for transportation fuels. *Second Biomass Conference of the Americas: Energy, Environment, Agriculture, and Industry*, Portland, OR.

WYMAN, C. E. and HINMAN, N. D. (1990) Ethanol: fundamentals of production from renewable feedstocks and use as a transportation fuel. *Appl. Biochem. Biotechnol.* **24/25**, 735–758.

ZHANG, M., EDDY, C., DEANDA, K., FINKELSTEIN, M. and PICATAGGIO, S. (1995) Metabolic engineering of a pentose metabolism pathway in ethanologenic *Zymomonas mobilis. Science* **267**, 240–243.

Production of Mushrooms from Wood Waste Substrates

FREDERICK C. MILLER

13.1 Introduction

Many fungi in nature have the ability to grow on wood as a substrate. Among the fungi that grow on wood are a smaller number of fungi that produce fruiting bodies desirable for human consumption. Mushroom fruiting bodies are the fleshy sexual reproductive structures of fungi. In Asia there has been a long tradition of harvesting wild wood decaying mushrooms which are much prized as culinary delicacies and promoters of good health. Cultivation of wood decay fungi in the Orient has been practised for a thousand years (Chang, 1993), but in more recent years cultivation has both intensified and spread globally. Cultivation of various mushrooms on wood substrates has become common in Europe and North America since the 1980s.

Fungi currently grown on substrates based on wood waste include shiitake (*Lentinula edodes*), enoke or enokitake (*Flammulina veluipes*), nameko (*Pholiota nameko*), shimeji (*Hypsizygus marmoreus*), jelly fungus or wood ear (*Auricularia auricula* and other species), and white jelly fungus or silver ear (*Tremella aurantialba*). Other common names often encountered for these mushrooms include oak mushroom and black forest mushroom for shiitake, winter mushroom for enoki, viscid mushroom for nameko, and Jew's ear for jelly fungus. World production for 1991 was 526 000 tonnes for shiitake, 187 000 tonnes for enoki, and 40 000 tonnes for nameko (Chang and Miles, 1991), and by the late 1980s 172 000 tonnes for jelly fungi (Luo, 1993). Production of these mushrooms is rising rapidly, however, and in the USA alone production in the 1993–94 season for non-*Agaricus* (white button mushroom) increased 75 per cent over the previous season (USDA, 1994).

Maitake or hen-of-the-woods (*Grifola frondosa*) is a mushroom grown on wood based substrates. Cultivation of maitake was first reported in 1983 (Zhao and Yang, 1983). Maitake is of commercial interest not only as an edible fungus, but also for medical applications in lowering of blood pressure and cholesterol levels (Lee, 1996). Production of maitake is currently very small. Bunashimeji (*Hypsizigus marmoreus*) is another wood decay mushroom that has only recently been cultivated (Tsuneda, 1994). In Japan this mushroom was not cultivated in the 1970s, yet in 1991 the production in Japan was 36 623 tonnes. In China the veiled lady mushroom

(*Dictyophora indusiata*) is grown on a wood waste substrate (Yang and Jong, 1987) but production figures do not appear to be available.

Cookeina sulcipes is a prized edible mushroom of the tropics that grows on wood and especially the wood of cocoa trees. While not yet cultivated, research is ongoing to bring this mushroom into commercial production using waste wood from cocoa plantations (Vazquez *et al.*, 1995a).

Pleurotus or oyster mushroom (*Pleurotus sajor-caju* and other species) is a major commercial mushroom with a world production of 800 000 tonnes in 1990 (Chang and Miles, 1991). *Pleurotus* is a wood decay fungus that is grown to some extent on wood wastes (Royse, 1992), but in commercial practice straw is used overwhelmingly in European and North American practice (LaBorde, 1989). In Japan and other parts of Asia, however, wood waste is the substrate used for *Pleurotus* cultivation (Tsuneda, 1994). *Pleurotus* can be grown on softwoods such as mulberry or poplar but not so well on more dense hardwoods such as oak (Khan and Khatoon, 1989).

There are also non-culinary mushrooms grown on wood wastes for the production of medically active compounds. Reshi or Lin-Chi (*Ganoderma lucidum*) has been used in the Orient for centuries as a folk remedy for various illness and is now cultivated for medical use (Su *et al.*, 1993; Royse, 1995). Other species in the genus *Ganoderma* are also of medical interest because of their production of physiological active triterperoids (Su *et al.*, 1993). Production of *Ganoderma* sp. and other wood decay mushrooms for medical purposes will certainly increase (Chang, 1993).

Classifying edible mushrooms by ecological habitat, Chang (1993) reported that those fungi which grow above ground on wood are the easiest of all fungi to induce to produce fruiting bodies in cultivation. As cultivation methods and techniques improve, it is likely that additional types of wood decaying mushrooms will be brought to market successfully. Market availability at affordable prices has been a considerable factor in the tremendously increased consumption of mushrooms in the past 50 years.

Mushrooms are prized for their culinary appeal, but the recent greater perception of mushrooms as a healthy food has further increased consumer demand. In recent years annual production of mushrooms has increased globally in the range of 24 per cent a year (Chang and Miles, 1991). For shiitake alone, production has steadily increased from 10 000 tonnes in 1950 (Royse *et al.*, 1985) to 526 000 tonnes by 1991 (Chang and Miles, 1991). All these trends indicate that production of wood decay mushrooms will continue to increase, and create a large demand for wood waste materials. Projecting from Chang and Miles' (1991) figures, current world production of wood decay mushrooms is somewhere around 3 million tonnes. Assuming that an average 2 kg of dry substrate is required to produce 1 kg of mushrooms, the annual demand for wood wastes for mushroom growing could be greater than 5 million tonnes.

13.2 Production Methods

13.2.1 *Natural Log Production*

While production of many mushrooms on wood waste is now carried out, shiitake cultivation has always led the way in methods development. The oldest cultivation method for the production of mushrooms from wood decay fungi is based on the use of natural logs. Natural log cultivation of shiitake started many centuries ago in China and later spread to Japan. In the earliest techniques, logs were cut and stacked outdoors and growers depended on wind-borne spores for passive inoculation. Later a practice of notching the logs with an axe to break the bark surface to improve chances of inoculation was used. In the 1920s Kitayima invented a direct inoculation method where a pure spawn culture was placed (spawning) into holes drilled into the logs (Royse, 1995). This method, still used by many growers around the world, was a major improvement that greatly increased the reliability of natural log cultivation methods.

Figure 13.1 illustrates the growing of shiitake on natural logs. Natural log growing methods are quite slow, requiring at least 8 to 12 months from inoculation to the first flush of mushrooms, and cropping of logs will occur over about 5 years (Nutalaya and Pataragetvit, 1981; Tsuneda, 1994). Oak logs are usually spawned in early spring, and the inoculated logs are placed in an environment that encourages mycelial growth, a temperature range of 22–26°C being preferred. After a grow-out period the logs are transferred to a growing yard where conditions will favour the development of fruiting bodies. A good growing yard is a mixed forest of deciduous and evergreens where enough sunlight will reach the logs to promote primordia formation, yet there is enough shade to maintain high moisture contents and autumn and spring temperatures of 5–15°C. Most cropping of natural logs occurs in the autumn and spring, and through the winter in warmer climates.

Figure 13.1   Shiitake being grown on natural oak logs showing the manner in which the logs are stacked in the growing yard (photograph by D. J. Royse)

13.2.2 *Production on Sawdust Based Substrates*

Culture Methods for Shiitake

In what is considered the first report of cultivation of *Lentinus* sp. on wood waste, Etter (1929) in the 1920s grew mushrooms on a synthetic medium of cottonwood dust supplemented with cornmeal, cornstarch and liquid malt. Serious cultivation of shiitake on sawdust based media, however, did not get underway until the late 1970s with technology developments in Taiwan, Japan and mainland China (Miller and Jong, 1987). Ho (1989) stated that growing shiitake on sawdust media in Taiwan commenced in 1978. In early work, Han *et al.* (1981) carried out extensive studies on the physiology and ecology of growing shiitake on sawdust based media and predicted (quite correctly) that the use of such substrates would be developed and widely used in the future. The big breakthrough for sawdust substrates was the use of plastic bags, in which the substrate could be compressed, sterilized, inoculated, and grown out. This method of mixing together loose substrates in a plastic bag and then allowing mycelial growth to knit the materials into a solid block became known as 'artificial log' or 'bag' culture. Once artificial log methods became commercially successful for shiitake, similar methods became quickly adapted for the cultivation of many other wood decay fungi (Tsuneda, 1994).

Typical bag culture of shiitake (Han *et al.*, 1981; Miller and Jong, 1987; Royse, 1995) entails mixing the sawdust substrate and other amendments together and adding water until a moisture content of about 60 per cent is achieved. Bags filled with substrate are compressed, placed on racks, and autoclaved for about 2 hours to sterilize the substrate. After cooling down, inoculation is made with a shiitake spawn. Spawn is allowed to run (grow out) for 20 to 25 days, during which the mycelia knit the substrate together and a thick white mat forms along the outer surface of the artificial log. At this time the plastic bag is removed, and the outer surface ages to become brown and leathery, and fruiting body primordia form.

During the 4-week browning period logs are sprayed with water once or twice a day, and a light–dark cycle is imposed to stimulate primordia formation. Shiitake has a specific requirement for light to fruit, requiring periods of light of 370–420 μm range and intensity of 400–500 lux (Han *et al.*, 1981; Miller and Jong, 1987). At the end of the browning period logs are soaked in cold water for about 3–4 hours to stimulate mushroom formation, and placed in temperature controlled growing rooms where mushrooms will be ready for harvest in about 9–11 days. Figure 13.2 shows typical production of shiitake on a sawdust based artificial log. After a harvest, logs are soaked for about 12 hours to replace lost water, and cropped again for a series of cycles. Production can continue for perhaps 2 months depending on disease pressure and farm management requirements.

Kalberer (1989) was the first to report good results by pasteurizing a sawdust based shiitake substrate at 95°C instead of the usual sterilization before inoculation. Kalberer's main motive was to avoid the high costs associated with autoclaving, as these costs are prohibitive for smaller farms. This practice of pasteurizing sawdust based shiitake media apparently works very well in commercial practice, although reports of this practice are lacking in the literature. Based on some commercial experience, pasteurization at 60°C works even better than the higher temperature Kalberer recommended, because the lower temperature greatly reduces problems with other disease and weed fungi. Naturally occurring populations can survive

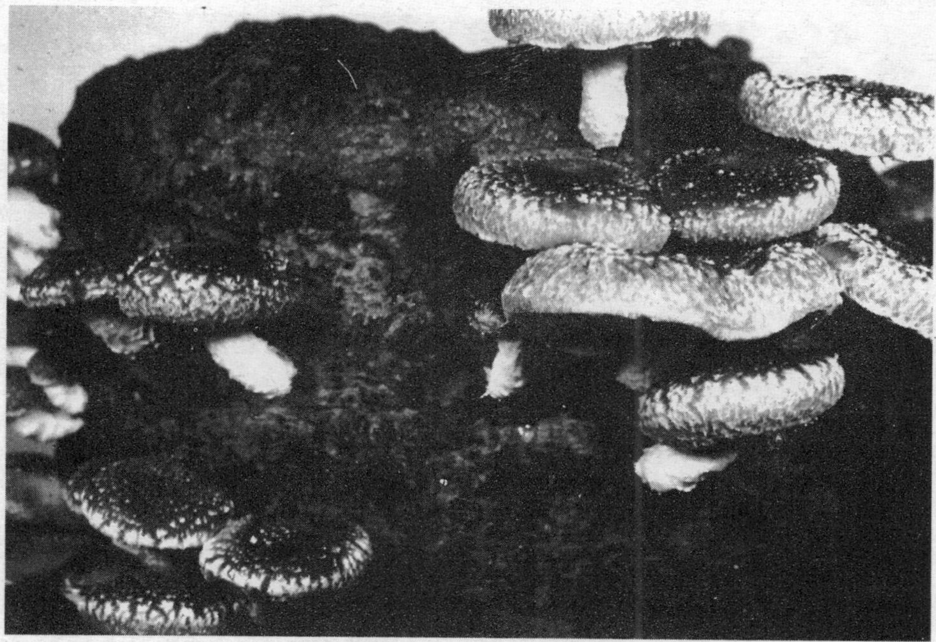

Figure 13.2   Shiitake being grown on an artificial log made of oak sawdust and supplements. The shiitake spawn was inoculated into the sawdust media and grown out in a plastic bag, with the bag being completely removed when the artificial log is sufficiently mature (photograph by D. J. Royse)

pasteurization and thereby make it more difficult for other disease populations to become established. This pattern of successional exclusion by established populations has been observed in other mushroom growing situations (Miller, 1992; Miller and Spear, 1995). Sterilization is still preferred for media with high nutritional supplementation or for slow growing shiitake strains (Lee, 1996).

Another shiitake growing variation popular in Japan is the use of wide-mouthed plastic bottles to contain the sawdust based substrate (Tsuneda, 1994). The use of plastic bottles allows more highly mechanical methods of substrate handling and inoculation. Crop cycles are normally somewhat longer in bottle culture than in bags.

Although the use of sawdust based media for bag and bottle culture has become a major form of production, natural log cultivation methods are still quite common. The main disadvantage of the sawdust based media cultivation methods is that, in Asia, the quality of mushrooms grown in this manner is considered inferior (Chang, 1993).

Methods for Other Wood Decay Fungi

The use of sawdust based media in bag or bottle growing culture has developed greatly in the past two decades (Tsuneda, 1994; Royse, 1995). Sawdust based media are now commonly used for the cultivation of all the wood decay fungi mentioned in the Introduction. While general cultivation methods for other wood decay fungi follow a pattern similar to that given for shiitake growing, production cultures of

Figure 13.3 Maitake fruiting on an oak sawdust based substrate. The bag culture method shown here is typical for growing most wood decay mushrooms (photograph by D. J. Royse)

these other mushrooms are normally left in bottles or bags and mushrooms are harvested only from the top surface of the substrate culture. Figure 13.3 shows maitake in a typical bag culture. Cultivation methods also vary for each specific fungus as each has its own growing requirements. Variations occur in growing time periods, moisture requirements, light (some fungi have no light requirement for fructification), temperature, and periodic changes of environmental conditions to induce fruiting.

13.3 Formulations for Growing Based on Wood Waste

Formulations for growing wood decay mushrooms commercially vary a good deal. Some variation is related to specific substrate requirements of a particular mushroom species or strain, while other variations are based on what materials are cheap and in good local supply. For some mushrooms sawdust must be from specific kinds of trees, while other mushrooms grow equally well on all sawdusts. For most mushrooms sawdust is a prime media component, but various other materials are added for nutritional and structural reasons, and also because empirical observation has shown that the addition of certain materials improves commercial success.

Shiitake media based on red or white oak sawdust have been reported to be good (Miller and Jong, 1987; Pettipher, 1988; Lee, 1996), although pararubber tree (*Hevea brasiliensis*) (Triratana and Tantikanjana, 1989), hornbeam (Ho, 1989), alder and birch (Itavaara, 1989), beech (Kawai *et al.*, 1995) or spruce (Kalberer, 1989) work well. Maple and birch sawdust also give results similar to oak (Miller and

Jong, 1987). Among 18 broad leafed trees common to Taiwan, Han *et al.* (1981) found most worked equally well except for *Acacia confusa*, *Zelkora serrata* and *Chamaecyparis formosensis* which gave poor results.

Sawdust media for shiitake growing is usually supplemented with a readily available carbohydrate source and a nitrogen and other nutrient source (Miller and Jong, 1987). These supplements include millet, rye, corn, glucose, oatmeal, rice bran and wheat bran (Royse and Bahler, 1986; Miller and Jong, 1987; Royse, 1995). These supplements can be added to the sawdust from a 10–40 per cent ratio, although 20 per cent appears to be about average. Other materials added in smaller quantities are calcium carbonate and gypsum to adjust pH into a favourable range of 4.5 to 6.0.

Without much variation, sawdust media as developed for shiitake appear to work well for other wood decay mushrooms such as jelly fungi (Luo, 1993), white jelly fungi (Xiang, 1991) and enokitake, nameko, shimeji (Royse, 1995) and maitake (Lee, 1996). Indeed, most scientific and technical literature is focused on shiitake substrates, with little or nothing published on the production of substrates for other wood decay fungi. This lack of literature indicates that sawdust media developed for shiitake work well enough for other wood decay fungi so that additional substrate development research is not necessary.

One wood decay mushroom that stands apart in terms of substrates for cultivation is *Pleurotus* spp. (oyster mushroom), in that it appears to be cultivatable on almost any lignocellulosic material. While *Pleurotus* is commonly grown on chopped wheaten straw in Europe and North America, in some parts of the United States cotton seed hulls are the primary substrate (Royse, 1993; 1995). In Japan, *Pleurotus* is grown in bottle or plastic bag culture using a redwood sawdust based media (Tsuneda, 1994). *Pleurotus* can also be grown on sugar cane bagasse (Alum and Khan, 1989; Bononi *et al.*, 1991), corn stover (Bassous *et al.*, 1989), textile wastes (Khan and Siddiqui, 1989), coffee pulp (Martinez-Carrera, 1989), and castor and mulberry wood (Madan *et al.*, 1987). It has been reported that *Pleurotus* can be grown on spent shiitake oak sawdust substrate with the addition of small amounts of wheat bran and millet (Royse, 1992). This ability of *Pleurotus* to grow on such a wide variety of materials, and even on materials used previously to grow other wood decay mushrooms, gives an indication of the very aggressive degradative abilities of *Pleurotus*.

The jelly fungi (*Auricularia* spp.) contain a number of tropical species that can be grown on the same sawdust substrates as shiitake, but can also be grown on wood from a very wide variety of trees (Vazquez *et al.*, 1995b). Materials that have worked well for jelly fungi production include mixtures of coffee tree sawdust, coffee pulp, corn cobs, and a powder made from *Leucaena* leaves. This ease of growth on various substrates coupled with the good eating qualities and shelf life characteristics of the jelly fungi accounts for the great increase in production in the past few years.

13.4 Biochemistry and Ecophysiology

Wood decay fungi can be divided into white-rot and brown-rot fungi based on the biochemistry of mechanisms of decay (Highley *et al.*, 1994). White-rot fungi utilize cellulose and non-cellulosic polysaccharides at about the same rate while lignin is used more rapidly on a relative basis. (Non-cellulosic polysaccharides are referred to as hemicellulose by some authors, but this usage is imprecise (Ilyama *et al.*, 1994).) Hemicellulose traditionally refers to an analytical fraction of unknown composition and not to the specific structures we now know for the non-cellulosic polysaccharides. Brown-rot fungi utilize only the cellulosic and non-cellulosic polysaccharides in wood and leave the lignin undigested, but to some extent modified. White-rot fungi depolymerize wood substances at a rate matching metabolic uptake. Brown-rot fungi rapidly depolymerize holocellulose in wood, and degradation products are formed much faster than metabolic utilization. Being a white-rot or brown-rot fungus *per se* does not provide a specific ecological advantage as either type can out-compete and replace the other dependent on the specific species involved (Owens *et al.*, 1994). From ecological principles (Miller, 1992) a prediction can be made that commercial culture of brown-rot fungi will be more troubled by diseases and competitor organisms because of the available substrate liberated during degradation not taken up by the cultured brown-rot fungus.

The distinction between white-rot and brown-rot fungi is not as clear cut for all wood decay mushrooms as it might appear, however, because lignin degradation ability for different fungi, and even strains within a single species, can vary appreciably (Savoie *et al.*, 1995). In the common button mushroom (*Agaricus bisporus*) there has been conflicting evidence for lignin degradation for decades that is only now being addressed by improved analytical procedures (Ilyama *et al.*, 1994). Shiitake can produce lignin degrading enzymes induced by the presence of lignin, but sensitivity to induction varies greatly among various shiitake strains (Savoie *et al.*, 1995). *Lentinus lepideus* however, while in the same genus as shiitake, is a brown-rot fungus (Collett, 1992) and cannot produce lignin degrading enzymes.

Enokitake (*Flammulina veluipes*) is a brown-rot fungus (Zadrazil, 1993), but most other edible fungi grown on sawdust substrates are white-rot fungi. Many wood decay fungi grown on sawdust substrates have not been characterized as white- or brown-rot fungi because such issues are not significant *per se* to mushroom farmers. Almost all biochemical research on white-rot fungi has been carried out because of their ability to convert lignocellulosic materials into animal feeds (Hadar *et al.*, 1993; Zadrazil, 1993), biodegrade complex organic materials in the environment (Lamar and Evans, 1993; Barr and Aust, 1994; Reddy, 1995), and behave as biopulping and biobleaching agents (Katagiri *et al.*, 1995). Mushrooms grown on wood waste substrates known to be white-rot fungi include shiitake (Tsuneda, 1994), *Pleurotus* (Buswell *et al.*, 1993), nameko (Yang *et al.*, 1993) and jelly fungi (Luo, 1993).

Major enzyme systems in wood decay mushrooms (Buswell *et al.*, 1993) can be classed as:

1 *Cellulases*: endoglucanase, exoglucanase, β-glucosidase.
2 *Non-cellulosic polysaccharidases*: xylanase, β-xylosidase.
3 *Ligninases*: ligninase, Mg-dependent peroxidase, laccase.

Some ligninase enzyme systems must be functional in a white-rot fungus, but this does not imply a full complement of possible enzyme systems. For example, shiitake does not contain a ligninase while *Pleurotus* contains all three lignin degrading systems (Buswell *et al.*, 1993). Of importance in better understanding the enzymatic capabilities of wood decay mushrooms is that a very high level of lignolytic ability, as in *Pleurotus*, appears to be related to ease of culture and higher efficiency of conversion of wood waste into mushrooms. Enzyme expression can be used both for screening of wild-type cultures and as indicators in selective breeding programmes.

13.5 Future Directions

Environmental and economic pressures will increasingly preclude treating wood wastes as a waste to be abandoned. In countries, especially in Asia, where adequate food production is a concern, only wood decay fungi can readily take wood waste and convert it into a nutritious food. Mushrooms can provide a significant source of protein and also minerals and vitamins (Garcha *et al.*, 1993). While the nutritional value of mushrooms is lower than that of meats, dairy products or nuts, shiitake has a higher nutritional value than corn, potatoes, tomatoes or carrots (Royse *et al.*, 1985). There is also a great deal of recent research, especially in Asia, which demonstrates that wood decay mushrooms can produce a number of beneficial pharmaceutical compounds (Chang, 1993).

In many countries where agricultural land holdings tend to be limited, production of mushrooms on wood waste offers an economic opportunity because a high value crop can be grown on a small amount of land. Mushrooms in general have become much more accepted and desired in recent years as a food item. The technology of cultivating wood decay mushrooms has also made great improvements in the past decade. Cultivation of mushrooms on wood waste should therefore expand greatly in future years.

Acknowledgements

The author wishes to acknowledge the assistance of Daniel Royse for providing the photographs in this chapter and for discussion of relevant literature.

References

ALUM, A. and KHAN, S. M. (1989) Utilization of sugar industry for the production of filamentous protein in Pakistan. *Mushroom Sci.* **12**, 15–22.

BARR, D. P. and AUST, S. D. (1994) Mechanisms white rot fungi use to degrade pollutants. *Environ. Sci. Technol.* **28**, 79–87.

BASSOUS, C., CHAHAL, D. S. and MATHIEU, L. G. (1989) Bioconversion of corn stover into fungal biomass rich in protein with *Pleurotus sajor-caju*. *Mushroom Sci.* **12**, 57–66.

BONONI, V. L. R., MAZIERO, R. and CAPELARI, M. (1991) *Pleurotus ostreatoroseus* cultivation in Brazil. *Mushroom Sci.* **13**, 531–532.

BUSWELL, J. A., CAI, Y. J. and CHANG, S. T. (1993) Fungal and substrate associated factors affecting the ability of individual mushroom species to utilize different lignocellulosic growth substrates. In: Chang, S. T., Buswell, J. A. and Chiu, S. W., eds, *Mushroom Biology and Mushroom Products*, Hong Kong: The Chinese University Press, pp. 141–150.

CHANG, S. T. (1993) Mushroom biology: the impact on mushroom production and mushroom products. In: Chang, S. T., Buswell, J. A. and Chiu, S. W., eds, *Mushroom Biology and Mushroom Products*, Hong Kong: The Chinese University Press, pp. 3–20.

CHANG, S. T. and MILES, P. G. (1991) Recent trends in world production of cultivated mushrooms. *Mushroom J.* **503**, 15–18.

COLLETT, O. (1992) Aromatic compounds as growth substrates for isolates of the brown rot fungus *Lentinus lepideus. Mat. u. Org.* **27**, 67–77.

ETTER, B. E. (1929) New media for developing sporophores of wood-rot fungi. *Mycologia* **21**, 197–203.

GARCHA, H. S., KHANNI, P. K. and SONI, G. L. (1993) Nutritional importance of mushrooms. In: Chang, S. T., Buswell, J. A. and Chiu, S. W., eds, *Mushroom Biology and Mushroom Products*, Hong Kong: The Chinese University Press, pp. 227–236.

HADAR, Y., KEREM, Z. and GORODECKI, B. (1993) Biodegradation of lignocellulosic agricultural wastes by *Pleurotus ostreatus. J. Biotechnol.* **30**, 133–139.

HAN, Y. H., UENG, W. T., CHEN, L. C. and CHENG, S. (1981) Physiology and ecology of *Lentinus edodes* (Berk.) Sing. *Mushroom Sci.* **11**, 623–658.

HIGHLEY, T. L., CLAUSEN, C. A., CROAN, S. C., GREEN, F., ILLMAN, B. L. and MICALES, J. A. (1994) Research on biodeterioration of wood, 1987–1992; I. Decay mechanisms and biocontrol, USDA, Forest Service, Forest Products Laboratory, Research Paper FPL-RP-529, Madison, WI.

HO, M. S. (1989) A new technology, 'plastic bag cultivation method' for growing Shiitake mushroom. *Mushroom Sci.* **12**, 303–307.

ILYAMA, K., STONE, B. A. and MACAULEY, B. J. (1994) Compositional changes in compost during composting and growth of *Agaricus bisporus. Appl. Environ. Microbiol.* **60**, 1538–1546.

ITAVAARA, M. (1989) Comparison of three methods to produce liquid spawn for commercial cultivation of Shiitake. *Mushroom Sci.* **12**, 309–315.

KALBERER, P. R. (1989) The cultivation of Shiitake (*Lentinus edodes*) on supplemented sawdust. *Mushroom Sci.* **12**, 317–335.

KATAGIRI, N., TSUTSUMI, Y. and NISHIDA, T. (1995) Correlation of brightening with cumulative enzyme activity related to lignin biodegradation during biobleaching of kraft pulp by white rot fungi in the solid state fermentation system. *Appl. Environ. Microbiol.* **61**, 617–622.

KAWAI, G., KOBAYASHI, H., FUKUSHIMA, Y. and OHSAKI, K. (1995) Liquid culture induces early fruiting in Shiitake (*Lentinula edodes*). *Mushroom Sci.* **14**, 825–832.

KHAN, S. M. and KHATOON, A. (1989) Oyster mushroom cultivation on soft woods of Swat Valley, Pakistan. *Mushroom Sci.* **13**, 31–34.

KHAN, S. M. and SIDDIQUI, M. A. (1989) Some studies on the cultivation of oyster mushroom (*Pleurotus* spp.) on ligno-cellulosic by products of textile industry. *Mushroom Sci.* **12**, 121–128.

LABORDE, J. (1989) Technologie moderne de production des *Pleurotus. Mushroom Sci.* **13**, 135–153.

LAMAR, R. T. and EVANS, J. W. (1993) Solid phase treatment of a pentachlorophenol contaminated soil using lignin degrading fungi. *Environ. Sci. Technol.* **27**, 2566–2571.

LEE, E. (1996) Shiitake and Maitake mushrooms in Connecticut. *Mushroom News* **44**, 6–11.

LUO, X. C. (1993) Biology of artificial log cultivation of *Auricularia* mushrooms. In: Chang, S. T., Buswell, J. A. and Chiu, S. W., eds, *Mushroom Biology and Mushroom Products*, Hong Kong: The Chinese University Press, pp. 129–132.

MADAN, M., VESUDEVAN, P. and SHARMA, S. (1987) Cultivation of *Pleurotus sajor-caju* on different wastes. *Biol. Wastes* **22**, 241–250.

MARTINEZ-CARRERA, D. (1989) Simple technology to cultivate *Pleurotus* on coffee pulp in the tropics. *Mushroom Sci.* **12**, 169–177.

MILLER, F. C. (1992) Composting as a process based on the control of ecologically selective factors. In: Metting, F. B., ed., *Soil Microbial Ecology*, New York: Marcel Dekker, pp. 515–544.

MILLER, M. W. and JONG, S. C. (1987) Commercial cultivation of Shiitake in sawdust filled plastic bags. In: Wuest, P. J., Royse, D. J. and Beelman, R. B., eds, *Cultivating Edible Fungi*, Amsterdam: Elsevier Science Publishers, pp. 421–426.

MILLER, F. C. and SPEAR, M. (1995) Very large scale commercial trial of a biological control for mushroom blotch disease. *Mushroom Sci.* **14**, 635–642.

NUTALAYA, S. and PATARAGETVIT, S. (1981) Shiitake mushroom cultivation in Thailand. *Mushroom Sci.* **11**, 723–736.

OWENS, E. M., REDDY, C. A. and GRETHLEIN, H. E. (1994) Outcome of interspecific interactions among brown rot and white rot wood decay fungi. *FEMS Microbiol. Ecol.* **14**, 19–24.

PETTIPHER, G. L. (1988) Cultivation of the Shiitake mushroom (*Lentinus edodes*) on lignocellulosic wastes. *J. Sci. Food Agric.* **42**, 195–198.

REDDY, C. A. (1995) The potential for white rot fungi in the treatment of pollutants. *Curr. Opin. Biotechnol.* **6**, 320–328.

ROYSE, D. J. (1992) Recycling of spent shiitake substrate for production of the oyster mushroom, *Pleurotus sajor-caju. Appl. Microbiol. Biotechnol.* **38**, 179–182.

ROYSE, D. J. (1993) Recycling of spent shiitake substrate for the production of the oyster mushroom. *Mushroom News* **41**, 14–20.

ROYSE, D. J. (1995) Speciality mushrooms – cultivation on synthetic substrates in the USA and Japan. *Interdisciplin. Sci. Rev.* **20**, 205–214.

ROYSE, D. J. and BAHLER, C. C. (1986) Effects of genotype, spawn run time, and substrate formulation on biological efficiency of Shiitake. *Appl. Environ. Microbiol.* **52**, 1425–1427.

ROYSE, D. J., SCHISLER, L. C. and DIEHLE, D. A. (1985) Shiitake mushrooms – consumption, production and cultivation. *Interdisciplin. Sci. Rev.* **10**, 329–335.

SAVOIE, J. M., CESBRON, V. and DELPECH, P. (1995) Induction of polyphenol oxidases in the mycelium of *Lentinus edodes. Mushroom Sci.* **14**, 787–793.

SU, C. H., LAI, M. N. and CHAN, M. H. (1993) Hepato-protective triterpenoids from *Ganoderma tsugae* Murrill. In: Chang, S. T., Buswell, J. A. and Chiu, S. W., eds, *Mushroom Biology and Mushroom Products*, Hong Kong: The Chinese University Press, pp. 275–284.

TRIRATANA, S. and TANTIKANJANA, T. (1989) Effects of some environmental factors on morphology and yield of *Lentinus edodes* (Berk.) Sing. *Mushroom Sci.* **12**, 279–289.

TSUNEDA, A. (1994) Shiitake and other edible mushrooms cultivated in Japan: production, biology, and breeding. In: Charalambous, G., ed., *Spices, Herbs and Edible Fungi*, Amsterdam: Elsevier Science, pp. 685–727.

US DEPARTMENT OF AGRICULTURE (1994) *Mushrooms*, Washington, DC: Agricultural Statistics Board.

VAZQUEZ, J. E. S., PALACIOS, G. H. and BADO, L. A. C. (1995a) Progress in the cultivation of the eatable mushroom *Cookenia sulcipes. Mushroom Sci.* **14**, 857–862.

VAZQUEZ, J. E. S., PALACIOS, G. H. and BADO, L. A. C. (1995b) Potential of *Auricularia* sp. in the recycling of agricultural wastes products in the tropics. *Mushroom Sci.* **14**, 877–883.

XIANG, Y. (1991) *Tremella aurantialba* cultivation with substrates of sawdust or agricultural straws. *Mushroom Sci.* **13**, 585–588.

YANG, Q. Y. and JONG, S. C. (1987) Artificial cultivation of the veiled lady mushroom, *Dictyophora indusiata.* In: Wuest, P. J., Royse, D. J. and Beelman, R. B., eds, *Proceedings of the International Symposium on Scientific and Technical Aspects of Cultivating Edible Fungi*, Amsterdam: Elsevier Science Publishers, pp. 437–442.

YANG, G. L, MA, L., WANG, Y. W. and YANG, Y. (1993) Physiology and biochemistry of lignocellulose utilization by *Pholiota nameko*. In: Chang, S. T., Buswell, J. A. and Chiu, S. W., eds, *Mushroom Biology and Mushroom Products*, Hong Kong: The Chinese University Press, pp. 163–168.

ZADRAZIL, F. (1993) Conversion of lignocellulosics into animal feed with white rot fungi. In: Chang, S. T., Buswell, J. A. and Chiu, S. W., eds, *Mushroom Biology and Mushroom Products*, Hong Kong: The Chinese University Press, pp. 151–162.

ZHAO, Z. G. and YANG, X. L. (1983) *Polyporus* (of *Grifola*) *frondosus. Edible Fungi* **5**, 8–9.

Drugs from Plants

AZIZOL ABDUL KADIR

14.1   Introduction

Plants have been used by man since the beginning of human culture for a great variety of purposes, including medicine. The earliest known record of a plant being used in medication is found on an Egyptian papyrus dated about 1550 BC. Since then, plants have provided nearly half of the world's successful drugs, ranging from anticancer drugs from the tiny periwinkle plant, to painkillers from willow bark, and even contraceptives from yams! Plants have provided modern medicine with more diverse and important drugs than any other natural source (Farnsworth and Morris, 1976).

As technology and time progressed some of these natural drugs were synthesized and modified to improve or enhance their properties and as a result these natural products were largely supplanted by their synthetic counterparts. Successes in this field tended to overshadow the pharmaceutical industry's roots in natural products and it was generally thought that ultimately all the plant drugs would be obtained from synthetic sources. However, not all drugs can be commercially produced by synthesis and to this day pharmaceutical chemists still draw on plants when they search for new drug molecules. In the United States alone, pharmaceutical products originating from plants still make up some 25 per cent of prescription drugs.

14.2   The Biotic Resource

Estimates of the total number of higher plant species, both identified and unidentified, range from 250 000 to 500 000 species (Schultes, 1972; Cronquist, 1981) and even to a higher figure of 750 000 (Balandrin *et al.*, 1985). However, most botanists would stick to a conservative figure of 250 000. Whatever the true figure is, one important factor is that only a small percentage of these plants have ever received any more than superficial screening. A great many of those screened comprise plants from the temperate and subtropical regions, due to the fact that phytochemical studies and medical advances in the temperate areas, involving temperate and subtropical plants, have taken place for a much longer period than those for the tropical regions.

The tropical rain forest is hailed to be the most biogenetically diverse of all the forested areas of the world. In a sample plot of 1 hectare of tropical rain forest, up to 100 tree species may be found, compared with only about 10–15, rarely up to 35, at the most, in a temperate forest (Leigh, 1982). It covers only about 7 per cent (9 million km^2) of the earth's land surface (IUCN, 1979), yet it is home to more than half the world's species of flowering plants. Undoubtedly, a vast storehouse of valuable new phytochemicals still awaits discovery. For example, nothing is known about 99 per cent of the flora of Brazil (Balandrin *et al.*, 1985)! Who knows what wonder drug still lies in wait within these dense and dark forest walls?

14.3 The Role of Plants in Drug Discovery and Development

Plants produce a highly individual range of natural products which vary widely from species to species and are mostly structurally distinct from microbial metabolites. There are essentially four basic ways in which plants contribute to modern medicine. First, plants are sources of direct therapeutic agents. For example, the South American jungle liana *Chondodendron tomentosum* is the main source of *d*-tubocurarine, a muscle relaxant much used in surgery. Chemists so far have been unable to produce this drug synthetically in a form which has all the attributes of the natural product. Furthermore, there are incidences where, even when chemical synthesis is possible, it is less costly to harvest the drug from its natural sources. One such example is the hypotensive drug reserpine which is still commercially extracted from *Rauwolfia* species.

Plants are also a starting point for the elaboration of more complex semi-synthetic compounds. Among the most important therapeutic agents used in modern medicine today are steroidal drugs such as the corticosteroids, sex hormones, anabolic agents and oral contraceptives. Although these drugs can be obtained from a number of sources, including total synthesis, steroidal sapogenins obtained from plant species, for example diosgenin, which may be obtained from tubers of various species of *Dioscorea*, constitute one of the major raw materials for the partial syntheses of these drugs. At present the use of diosgenin has decreased by half or less due to the widescale use of stigmasterol and sitosterol which are also obtained from plant sources. Nevertheless diosgenin is still being used and *Dioscorea* species will continue to be an important source for steroidal drugs at least in the developing countries.

Plants are also a source of natural products which serve as models for new, pharmacologically active compounds in the field of drug synthesis. There are several reasons for this. It may be that the plant material is not present in abundance and large scale cultivation is not viable, precluding the direct use of the natural source. Another reason is that in some cases the side effects of a natural product often prevent its use in medicine and can be resolved only by preparation of a synthetic derivative; examples are cocaine, a template for other modern local anaesthetics, and modifications of podophyllotoxin to obtain other antitumour preparations. New and unusual chemical substances found in plants will continue to serve as models for novel synthetic substances and will prove to be increasingly important in the future.

Finally, plants may also act as a natural source of compounds whose side effects are too strong to permit their use as prescription drugs, but which are valuable in research such as in the investigation and characterization of biochemical processes and their mechanisms. This is a very important, though obscure, use which assists in drug discovery and development. Many compounds with anticancer properties have been found to be toxic for use as clinical drugs but nevertheless are widely used and have proved to be very helpful in research.

14.4 Progress in Plant Drug Research

The 1950s coincided with a number of significant events such as the discovery of reserpine, the start of the investigation of the vinca alkaloids, and the development of refined chromatographic procedures and radioactive tracer techniques for biosynthetic studies. These provided considerable impetus to the investigation of drugs from plants. Apart from significant achievements in the field of drug synthesis, during the 40 years that followed, major advances were also continually being made in other areas related to plant drug research.

Much of the progress achieved in plant drug research today has been due to the analytical instruments and methods developed and employed during the last 40 years. Thin layer chromatography (TLC) and liquid chromatography techniques were widely used and remain of considerable importance to this day. Since such techniques were first developed they have undergone tremendous improvements and innovative modifications enabling better separations of mixtures of plant products. Advances in electronics brought more efficiency and sensitivity to spectroscopic techniques, especially nuclear magnetic resonance spectroscopy (NMR), mass spectrometry (MS) and X-ray crystallography. These analytical instruments are now more sophisticated and readily available; they are favoured as indispensable methods for structural determination. The spectral methods have been combined with chromatographic techniques, such as GC/MS, HPLC/MS and HPLC/NMR which permit the direct identification of separated compounds with remarkable ease.

The lack of simple bioassay procedures has been a continuing source of problems for natural products chemists in determining the physiological activity of plant materials, whether in the form of crude fractions or as purified chemical entities. Previously, fairly elaborate assays were used, for example the rat 'Hippocratic' screen. This was followed by the more successful brine shrimp (*Artemia salina*) toxicity assay, and the potato-disc assay which involves observation of the inhibition of crown-gall tumours induced on potato discs by *Agrobacterium tumefaciens* (Smith *et* Townsend) Conn. These methods were found to be rapid, reliable, inexpensive, and may be conveniently applied in-house by natural products chemists. In recent years major advances in bioassay techniques have taken place, in parallel with automated high-throughput screening technology based on the use of microelectronics, robotics and advanced spectroscopic instrumentation. The advent of modern biotechnology has led to the development of 'mode of action' bioassays including immunoassays capable of detecting picogram quantities of potentially useful compounds. Advances have also been made in areas of molecular and biochemical pharmacology which facilitated the development of assays for compounds which can selectively inhibit, or bind to, enzymes and receptors associated with known physiological events. Such integrated systems can screen thousands of samples daily and efficiently pinpoint those with pharmaceutical utility.

Another area relevant to plant drug research is the production of plant material for an adequate supply of the drug for clinical use. As civilization encroaches on forested areas, collection of plant material from the wild becomes less and less feasible. Drug producing plants do not often lend themselves to cultivation easily and agronomic research of drug producing plants has been somewhat limited because these plants were considered to be of relatively minor economic importance. Resorting to synthetic production of these drugs may not be as easy as it sounds as most often the structural complexity inherent in such natural products demands multi-step syntheses, which, although of distinct academic interest, are rarely of practical utility for large-scale industrial production. A solution to this question of increasing material availability and eliminating dependence on the living plant as the source is production using plant-tissue and cell culture techniques. As well as having great potential commercially, these techniques have been useful in the study of·plant biosyntheses and regulation of plant secondary metabolite production. Although there are still limitations to such techniques such as slow growth, expensive media, and the tendency to store desired metabolites in the tissues rather than excrete them into the media, cell suspension cultures seem to be the most appropiate system for the production of secondary products on an economical scale, provided that strategies are developed to shorten fermentation times and increase yields. Currently, certain pharmaceutically important chemicals such as shikonin, digoxin, vinblastine and rosmarinic acid are being successfully produced commercially by cell cultures in large bioreactors.

14.5 Some Significant Plant Drugs

Based on computerized information in the NAPRALERT database on natural products, there are currently about 125 clinically useful presciption drugs worldwide, derived from only 95 species of higher plants (Table 14.1). Obviously it is impossible to review each and every one of the drugs listed in Table 14.1. A few of the more important drugs will therefore be briefly discussed in the following sections. At least 45 of the 125 drugs listed are derived from about 39 plants, originating in and around the tropical rain forests and almost half of these drug-yielding tropical species are Asian plants.

14.5.1 *Drugs for Heart Diseases*

The American foxglove (*Digitalis* species) has been used for medicinal purposes for hundred of years but it was only in the late eighteenth century that it was shown to be effective in the treatment of heart diseases. It is the source of the digitalis drugs such as digitalin (**1**), digoxin, acetyldigitoxin, gitalin, lanatosides A, B, C, etc. These cardiac glycosides encompass compounds which contain a cardenolide linked to one or more glucose-like moieties and have a positive inotropic action on the heart. *D. purpurea* and *D. lanata* are two main sources of the digitalis drugs which are the treatment of choice for arrhythmias and heart failure (Doherty, 1985; Smith, 1988).

Table 14.1 Clinically useful drugs obtained from plants, reproduced with kind permission from N. R. Farnsworth and D. D. Soejarto (1985) Global Importance of Plants; Tables 1 and 2, pp. 43–7, in *The Conservation of Medicinal Plants*, Akerele, Heywood and Synge (eds), Cambridge University Press

Drug	Action/clinical use	Species	Origin
Acetyldigitoxin	Cardiotonic	*Digitalis lanata* Ehrh.	NT
Adoniside	Cardiotonic	*Adonis vernalis* L.	NT
Aescin	Anti-inflammatory	*Aesculus hippocastanum* L.	NT
Aesculetin	Antidysentery	*Fraxinus rhynchophylla* Hance	NT
Agrimophol	Anthelmintic	*Agrimonia eupatoria* L.	NT
Ajmalicine	Circulatory disorders	*Rauvolfia serpentina* (L.) Benth.ex Kurz	T, As
Allantoin	Vulnerary	A number of species	NT
Allyl isothiocyanate	Rubefacient	*Brassica nigra* (L.) Koch	NT
Anabasine	Skeletal muscle relaxant	*Anabasis aphylla* L.	NT
Andrographolide	Antibacillary dysentery	*Andrographis paniculata* Nees	T, As
Anisodamine	Anticholinergic	*Anisodus tanguticus* (Maxim.) Pascher	NT
Anisodine	Anticholinergic	*Anisodus tanguticus* (Maxim.) Pascher	NT
Arecoline	Anthelmintic	*Areca catechu* L.	T, As
Artemisinin	Antimalarial	*Artemisia annua* L.	NT
Asiaticoside	Vulnerary	*Centella asiatica* (L.) Urban	T, As, Am, Af
Atropine	Anticholinergic	*Atropa belladonna* L.	NT
		Hyoscyamus niger L.	NT
Azadirachtin	Insecticide	*Azadirachta indica* Juss.	T, As
Benzyl benzoate	Scabicide	A number of species	NT
Berberine	Antibacterial	*Berberis vulgaris* L.	NT
Bergenin	Antitussive	*Ardisia japonica* Bl.	T, As
Borneol	Antipyretic, analgesic, anti-inflammatory	A number of species	NT

Table 14.1 Continued

Drug	Action/clinical use	Species	Origin
Bromelain	Anti-inflammatory, proteolytic agent	*Ananas comosus* (L.) Mer.	T, Am
Caffeine	CNS stimulant	*Camellia sinensis* (L.) Kuntze	NT
Camphor	Rubefacient	*Cinnamomum camphora* (L.)	T, As
Castor oil	Laxative	*Ricinus communis* (L.)	T, Af
(+)-Catechin	Haemostatic	*Potentilla fragariodes* L.	NT
Chymopapain	Proteolytic, mucolytic	*Carica papaya* L.	T, Am
Cissampeline	Skeletal muscle relaxant	*Cissampelos pareira* L.	T, As
Cocaine	Local anaesthetic	*Erythroxylum coca* Lamk.	T, Am
Codeine	Analgesic, antitussive	*Papaver somniferum* L.	NT
Colchicine amide	Anticancer	*Colchicum autumnale* L.	NT
Colchicine	Antigout	*Colchicum autumnale* L.	NT
Convallatoxin	Cardiotonic	*Convallaria majalis* L.	NT
Curcumin	Choleretic	*Curcuma longa* L.	T, As
Cynarin	Choleretic	*Cynara scolymus* L.	NT
Danthron	Laxative	*Cassia* species	NT
Demecolcine	Anticancer	*Colchicum autumnale* L.	NT
Deserpidine	Antihypertensive, tranquilizer	*Raufolvia canescens* L.	T, Am
Deslanosides	Cardiotonic	*Digitalis lanata* Ehrh.	NT
Digitalin	Cardiotonic	*Digitalis purpurea* L.	NT
Digitoxin	Cardiotonic	*Digitalis lanata* Ehrh. *Digitalis purpurea* L.	NT
Diosgenin	Contraceptive	*Dioscorea* spp	T, As, Am
L-Dopa	Antiparkinsonism	*Mucuna deeringiana* (Bort.) Merr.	T, As

Table 14.1 Continued

Drug	Action/clinical use	Species	Origin
Emetine	Amoebicide, emetic	*Cephaelis ipecacuanha* (Brot.) A. Richard	T, Am
Ephedrine	Bronchodilator	*Ephedra sinica* Stapf	NT
Etoposide	Antitumour agent	*Podophyllum peltatum* L.	NT
Galanthamine	Cholinesterase inhibitor	*Lycoris squamigera* Maxim.	NT
Gitalin	Cardiotonic	*Digitalis purpurea* L.	NT
Glaucarubin	Amoebicide	*Simarouba glauca* D.C.	T, Am
Glaucine	Antitussive	*Glaucium flavum* Crantz	NT
Glaziovine	Antidepressant	*Ocotea glaziovii* Mez	T, Am
Gossypol	Male contraceptive	*Gossypium* species	T, As, Am, Af
Glycyrrhizin	Anti-inflammatory, sweetener	*Glycyrrhiza glabra* L.	NT
Hemsleyadin	Antibacillary dysentery, antipyretic	*Hemsleya amabilis* Diels	NT
Hesperidin	Capillary, antihaemorrhagic	*Citrus* species	NT
Hydrastine	Haemostatic, astringent	*Hydrastis canadensis* L.	NT
Hyoscyamine	Anticholinergic	*Atropa belladonna* L.	NT
		Hyoscyamus niger L.	NT
Kainic acid	Ascaricide	*Digenia simplex* (Wulf.) Agardh	NT
Kawain	Tranquilizer	*Piper mythesticum* Forst.f.	T, As
Khellin	Bronchodilator	*Ammi visnaga* (L.) Lamk.	NT
Lanatosides A, B, C	Cardiotonic	*Digitalis lanata* Ehrh.	NT
α-Lobeline	Tobacco deterrent, respiratory stimulant	*Lobelia inflata* L.	NT

Table 14.1 Continued

Drug	Action/clinical use	Species	Origin
Menthol	Rubefacient	*Mentha* species	NT
Methyl salicylate	Rubefacient	*Gaultheria procumbens* L.	NT
Monocrotaline	Antitumour (topical)	*Crotalaria sessiliflora* L.	T, As
Morphine	Analgesic, antitussive	*Papaver somniferum* L.	NT
Neoandrographolide	Antibacillary dysentery	*Andrographis paniculata* Nees	T, As
Nicotine	Insecticide	*Nicotiana tabacum* L.	T, Am
Nordihydroguaiaretic acid	Antioxidant	*Larrea divaricata* Cav.	NT
Norpseudoephedrine	Bronchodilator	*Ephedra sinica* Stapf	NT
Noscapine	Antitussive	*Papaver somniferum* L.	NT
Ouabain	Cardiotonic	*Strophanthus gratus* Baill.	T, Af
Pachycarpine	Ecbolic	*Sophora pachycarpa* Schrenk ex C.A. Meyer	NT
Palmatine	Antipyretic, detoxicant	*Coptis japonica* Makino	NT
Papaverin	Smooth muscle relaxant	*Papaver somniferum* L.	NT
Phyllodulcin	Sweetener	*Hydrangea macrophylla* (Thunb.) Seringe var. *thunbergii* (Siebold) Makino	NT
Physostigmine	Cholinesterase inhibitor	*Physostigma venenosum* Balf.	T, Af
Picrotoxin	Analeptic	*Anamirta cocculus* (L.) W. & A	T, As
Pilocarpine	Parasympathomimetic	*Pilocarpus jaborandi* Holmes	T, Am
Pinitol	Expectorant		NT
Podophyllotoxin	Antitumour agent	*Podophyllum peltatum* L.	NT
Protoveratrines A, B	Antihypertensive	*Veratrum album* L.	NT
Pseudoephedrine	Bronchodilator	*Digitalis purpurea* L.	NT

Table 14.1 Continued

Drug	Action/clinical use	Species	Origin
Quinidine	Antiarrhythmic	*Cinchona ledgeriana* Moens ex Trimen	T, Am
Quinine	Antimalarial, antipyretic	*Cinchona ledgeriana* Moens ex Trimen	T, Am
Quisqualic acid	Anthelmintic	*Quisqualis indica* L.	T, As
Rebaudioside A	Sweetener	*Stevia rebaudiana* Bertoni	T, As
Rescinnamine	Antihypertensive, tranquilizer	*Rauvolfia serpentina* (L.) Benth. ex Kurz	T, As
Reserpine	Antihypertensive, tranquilizer	*Rauvolfia serpentina* (L.) Benth. ex Kurz	T, As
Rhomitoxin	Antihypertensive	*Rhododendron molle* G. Don	NT
Rorifone	Antitussive	*Rorippa indica* (L.) Hochreut.	T, As
Rotenone	Piscicide	*Lonchocarpus nicou* (Aubl.) DC.	T, Am
Rotundine	Analgesic, sedative, tranquilizer	*Stephania sinica* Diels	NT
Rutin	Capillary antihaemorrhagic	*Citrus* species	NT
Salicine	Analgesic	*Salix alba* L.	NT
Sanguinarine	Dental plaque inhibitor	*Sanguinaria canadensis* L.	NT
Santonin	Anthelmintic	*Artemisia maritima* L.	NT
Scillarin A, B	Cardiotonic	*Urgenia maritima* (L.) Baker	NT
Scopolamine	Sedative	*Datura metel* L.	T, As
Sennosides A, B	Laxative	*Cassia acutifolia* Delile	NT
		Cassia senna L. var. *senna*	NT
Silymarin	Antihepatotoxic	*Silybum marianum* (L.) Gaertn.	NT
Sparteine	Oxytocic	*Cytisus scoparius* (L.) Link	NT
Stevioside	Sweetener	*Stevia rebaudiana* Bertoni	T, As
Strychnine	CNS stimulant	*Strychnos nux-vomica* L.	T, As

Table 14.1 Continued

Drug	Action/clinical use	Species	Origin
Taxol	Antitumour	*Taxus brevifolia* Nutt.	NT
Teniposide	Antitumour	*Podophyllum peltatum* L.	NT
Δ9-Tetrahydrocannabinol	Antiemetic, decrease ocular tension	*Cannabis sativa* L.	NT
(±)-Tetrahydropalmatine	Analgesic, sedative, tranquilizer	*Corydalis ambigua* (Pallas) Cham. & Schlechtdl.	NT
Theobromine	Diuretic	*Theobroma cacao* L.	T, Am
Thymol	Antifungal (topical)	*Thymus vulgaris* L.	NT
Trichosanthin	Abortifacient	*Trichosanthes kirilowii* Maxim.	NT
Tubocurarine	Skeletal muscle relaxant	*Chondodendron tomentosum* R. & P.	T, Am
Valepotriates	Sedative	*Valeriana officinalis* L.	NT
Vinblastine	Antitumour	*Catharanthus roseus* (L.) G. Don	T, Af
Vincamine	Cerebral stimulant	*Vinca minor* L.	NT
Vincristine	Antitumour	*Catharanthus roseus* (L.) G. Don	T, Af
Vasicine	Oxytocic	*Adhatoda vasica* Nees	T, As
Xanthotoxin	Leukoderma, vitiligo	*Ammi majus* L.	NT
Yohimbine	Aphrodisiac	*Pausynistalia yohimbi* (K Schum) Pierre ex Beille	T, Af
Yuanhuacine	Abortifacient	*Daphne genkwa* Sieb. & Zucc	NT
Yuanhuadine	Abortifacient	*Daphne genkwa* Sieb. & Zucc	NT

NT, non-tropical; T, tropical; As, Asia; Am, America; Af, Africa

The major digitalis-producing countries are the USA, UK, The Netherlands, Switzerland and Germany.

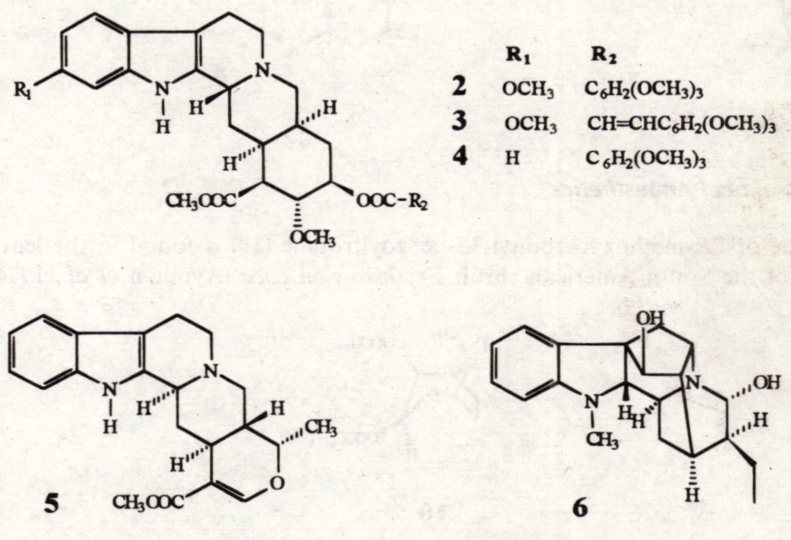

A number of drugs used for cardiovascular disorders are drugs derived from tropical trees. The best example is probably the well-known alkaloids of *Rauwolfia serpentina*, used as antihypertensives and as tranquilizers (Woodsen *et al.*, 1957; Schlittler and Bein, 1967). *R. serpentina* occurs throughout India, Malaysia and Thailand as small trees that grow wild in the humid forests. It is now cultivated in many tropical countries. Because of its highly toxic nature, use of *R. serpentina* has decreased considerably and a sister species, the African serpent wood *R. vomitoria* Afz., has been exploited much more for the world market.

Rauwolfia alkaloids used as drugs include reserpine (2), rescinnamine (3), deserpidine (4), ajmalicine (5) and ajmaline (6). Reserpine is used in combination with diuretics for controlling mild to moderate hypertension. It depletes peripheral noradrenaline stores, resulting in a fall in peripheral resistance and blood pressure as well as bradycardia and CNS depression (Kohli and Mukerji, 1955; Da Prada and Pletscher, 1969; Gilman *et al.*, 1985). It was also formerly used to treat psychotic disorders. Rescinnamine and deserpidine have properties and uses similar to reserpine. Ajmalicine, also known as raubasine, has also been used in conjunction with other agents to treat hypertension and in peripheral and cerebral vascular disorders. Ajmaline has antiarrhythmic acitivity on the heart muscle and is used clinically as a therapeutic agent in cardiac arrhythmia, as well as being used as an antihypertensive and tranquilizer. This compound is found in very large quantities in *R. vomitoria* and has become much more popular as a hypotensive agent than reserpine. Other drugs for cardiovascular disorders derived from tropical species are the cardiotonic, ouabain (**7**) and the antiarrhythmic quinoline alkaloid, quinidine (**8**). Ouabain is an injectable cardiac glycoside, extracted from the seeds of *Strophanthus gratus*. It has a faster onset of action than the usual digitalis digoxin, hence its preferred use over the latter when rapid benefit is required and in emergency situations. Like digitalis, ouabain is used to treat atrial fribrillation (arrhythmia) with an uncontrolled ventricular rate, atrial flutter, supraventricular tachycardia and acute left ventricular failure (Gillis and Quest, 1980). Quinidine is extracted from the bark of *Cinchona ledgeriana* and used as a cardiac depressant or antiarrhythmic (Rosketh and Storstrein, 1963; Stern, 1971). Kawain (**9**) is a naturally occurring pyrone found in the rhizomes of *Piper mythesticum*, a shrub indigenous to islands of the South Pacific (Anon, 1988). Currently produced synthetically, it is used as a tranquilizer and to improve well-being in geriatric patients.

7

8

9

14.5.2 *Local Anaesthetics*

Cocaine or 2*R*-methoxycarbonyl-3*S*-benzoyltropine (**10**) is found in the leaves and barks of the South American shrub *Erythroxylon coca* (Aynilian *et al.*, 1974). It is

10

one of the major coca alkaloids (Clarke, 1977; Wood and Wrigglesworth, 1977). The South American natives have been known to chew the coca plant to stimulate quick recovery from fatigue. The plant is now cultivated in a number of countries including Peru, Bolivia, Colombia, Indonesia and Sri Lanka.

Cocaine is used as a topical local anaesthetic (Ritchie *et al.*, 1970), to relieve pain in cancer patients (Twycross, 1977; Wesson and Smith, 1977; Avorn, 1980) and to relieve pain from cluster headaches (Barre, 1982). Its anaesthetic action comes from its reversible membrane stabilizing effect. It is rapidly absorbed after topical administration and has a vasoconstrictive action, thus enhancing its effectiveness as a local anaesthetic. It also functions as a central nervous system stimulant. Cocaine has been found to cause narcosis and its abuse leads to addiction.

14.5.3 Analgesics

Morphine (**11**) and codeine (**12**) are two well-known opioid analgesic drugs. These alkaloids may be extracted from the dried sap obtained by lancing the unripe seed pods of the opium poppy (*Papaver somniferum*) or by solvent extraction of poppy straw (Casey and Parfitt, 1986). The most important opium-producing countries are India, Turkey, Bulgaria, Yugoslavia, USSR, Australia, France and Spain. To date, morphine is still considered to be the drug of choice for the control of acute and chronic pain of malignant origin such as cancer (Gorlay *et al.*, 1986). Its clinical use is dependent on its interaction with opioid receptors in the brain, spinal cord and gut. It is also employed for treatment of typhoid fever, traumatic shocks and, in combination with atropine sulphate, for relieving renal and intestinal colic and coronary thrombosis. Codeine is less potent than morphine in its pain relief capacity; it is thus used for the control of mild to moderate pain (Beaver *et al.*, 1978). Both morphine and codeine possess antitussive properties and have been used as cough suppressants (Eddy *et al.*, 1969; Hughes, 1978). Codeine is however more acceptable for such uses because morphine tends to increase the incidence of post-operative chest complications by the suppression of a productive cough.

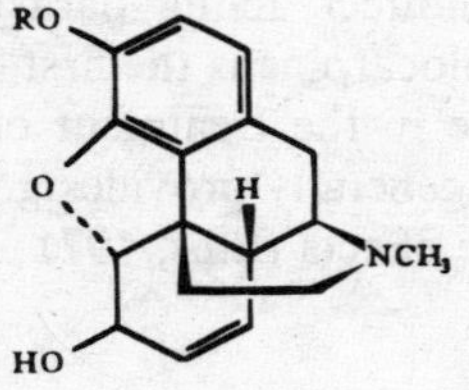

11 R = H

12 R = CH

14.5.4 Antimuscarinics

Antimuscarinic agents are competitive inhibitors of the actions of acetylcholine at the muscarinic receptors of autonomic effector sites innervated by parasympathetic nerves. Plant drugs included in this class are the belladonna alkaloids atropine (**13**), hyosycamine and hyoscine (**14**). Atropine or DL-hyosycamine is the chief alkaloid extracted from the deadly nightshade, *Atropa belladonna, Datura stramonium* and several other Solanaceae plants (Greenblatt and Shader, 1973). *A. belladonna* is indigenous to Western Europe and cultivated in England, Germany, USSR, USA and India.

13　　　　**14**

Atropine may be prepared synthetically or by racemization of the naturally occurring L-hyosycamine. Atropine is the prototype and best known antimuscarinic agent although many of its uses are now superseded by other semi-synthetic anti-muscarinic drugs. It is used primarily for the treatment of stomach spasms and also applied topically to the eye to produce dilatation during ophthalmic examinations. Hyoscine, also known as scopolamine, is a closely related ester of atropine. Although it can be produced synthetically, it is usually obtained by extraction from various members of the Solanaceae, including the well-known herb, *Datura metel*. The drug also has an anticholinergic effect but in contrast to atropine it also has a CNS depressant effect, hence its use as a sedative and to treat motion sickness (Peng *et al.*, 1983).

14.5.5　Miotics

Pilocarpine (**15**) and physostigmine (**16**) are two well-known cholinergic drugs used as miotics in the treatment of glaucoma. Pilocarpine, an alkaloid obtained from the leaves of *Pilocarpus jaborandi*, is a direct-acting muscarinic parasympathomimetic agonist. Physostigmine, an alkaloidal constituent of the calabar bean of the woody vine, *Physostigma venemosum*, is an indirect acting parasympathomimetic agent. Used as the hydrochloride or nitrate, pilocarpine is the first choice when miotics are required to reduce intraocular pressure in the treatment of open-angle glaucoma. This is due to the fact that pilocarpine generally provides good control of intraocular pressure with relatively few adverse effects (Ellis, 1971; Holland, 1974). Physo-

15　　　　**16**

stigmine is not as well tolerated as pilocarpine, hence is rarely used for long-term therapy. However physostigmine has been used for more than a hundred years as an antidote for atropine overdose and more recently in poisoning with tricyclic/ tetracyclic antidepressant drugs (Aquilonius, 1978; Aquilonius and Hedstrand, 1978).

14.5.6 Muscle Relaxants

d-Tubocurarine (**17**) is perhaps the most well-known of the natural skeletal muscle relaxants. The alkaloid is the active principle in 'tubocurare', the arrow poison used by the South American Indians in the Amazon–Orinoco basin (Marini Bettolo, 1981). The compound can be obtained from extracts of the stems and bark of the liana *Chondodendron tomentosum* and several other species of this genus (Dutcher, 1946). Available pharmaceutically as the chloride, it is a competitive neuromuscular blocker, primarily used intravenously to produce skeletal muscle relaxation during surgical procedures (Standaert, 1984). It acts by competing with acetylcholine for receptors on the motor end-plate to produce neuromuscular blockade, seen as flaccid paralysis. A patient is given the drug to reduce the amount of general anaesthetic required to achieve total muscle relaxation, although the resultant paralysis of the respiratory muscles means that artificial ventilation of the patient is necessary.

Papaverine (**18**) is a smooth muscle relaxant and vasodilator (Whipple, 1977). The alkaloid is also extractable from the opium poppy, *Papaver somniferum*, but unlike the other opium derivatives, it is not habit forming. Its use has been largely replaced by drugs with more specific actions such as α-adrenergic blockers and calcium slow channel antagonists but it may still have a place in the treatment of vascular spasms.

17

18

14.5.7 Bronchodilators

The ephedra alkaloids obtained from the Chinese plant 'Ma Huang' (*Ephedra sinica*), ephedrine (**19**), pseudoephedrine and norseudoephedrine, are sympathomimetic agents with direct and indirect effects on adrenergic receptors (Weiner, 1985). These drugs are now produced synthetically. Ephedrine is used for the treatment of nasal congestion and as a bronchodilator for treating the symptoms of

19

asthma. Pseudoephedrine, its naturally occurring stereoisomer, and nor-pseudoephedrine are also used for the same purposes but possess less potent pharmacological properties. The two drugs are widely used as constituents in over-the-counter remedies for treating symptoms of the common cold. They have also been found to show significant CNS excitatory and pressor effects which makes them undesirable for use in hypertensive patients. Ephedrine is no longer a drug of choice in the treatment of asthma as more selective drugs with less cardiac and CNS stimulation effect are available.

14.5.8 Antineoplastic Agents

Cancer is a serious life-threatening disease for modern society today. Among the most prevalent forms of cancer are breast, colorectal, lung, ovarian, prostate and uterine. Many different structural classes of plant secondary metabolites have been found to be cytotoxic but only a few are used clinically.

The most frequently used anticancer drugs are the dimeric indole alkaloids, vinblastine (**20**) and vincristine (**21**), which have been available since the 1960s. First discovered from the Madagascar periwinkle *Vinca rosea* (*Catharanthus roseus*), they have become the two most important clinically useful anticancer agents from any plant source. The drugs are still extracted from natural sources. Most of the *C. roseus* used for production of the two drugs is grown under cultivation in India, Madagascar, Israel and the USA.

Vinblastine and vincristine are used, either singly or in combination therapy, in the management of malignant diseases, particularly lymphomas and sarcomas

20 R = CH₃
21 R = CHO

(Rosenberg, 1979; Bramwell *et al.*, 1985; Volberding *et al.*, 1985), Hodgkin's disease (De Vita *et al.*, 1978; Sutcliffe *et al.*, 1978) and the leukaemias (Frei and Sallan, 1978). These vinca alkaloids exert their biological effects by binding specifically with the protein tubulin, inhibiting its assembly into microtubules with resultant dissolution of the mitotic spindle which eventually leads to cell death (Owellen *et al.*, 1976). Although not a first-line drug, vinblastine and vincristine have also been used in combination with other anticancer drugs for breast cancer (Yap *et al.*, 1980; Steiner *et al.*, 1983).

22

Podophyllotoxin (**22**) is the active principle in podophyllin, the alcoholic extract of the dried rhizomes and root of the North American May apple, *Podophyllum peltatum*. The drug has applications in dermatology where it is an effective therapy for anogenital warts (Perez-Figarado and Baden, 1976), and possibly nasal papillomas (Bennett and Grist, 1985) as well as psoriasis (Lassus and Rosen, 1986). Like the vinca alkaloids, podophyllotoxin is also a microtubule inhibitor. This compound and its congeners have however been found to attack both normal and cancerous cells. The toxic side-effects of these lignans have limited their application as drugs in cancer chemotherapy, except for etoposide and teniposide which are two semi-synthetic derivatives of podophyllotoxin. These two antitumour agents do not exhibit any effect on intracellular microtubules but induce breaks in single and double stranded DNA, through their interactions with topoisomerase II which is a critical enzyme in DNA replication (Liu, 1983; Chen *et al.*, 1984; Long *et al.*, 1984; Ross *et al.*, 1984).

14.5.9 Antiprotozoals

Protozoal infections are responsible for a number of major diseases including malaria, amoebiasis, leishmaniasis, giardiasis and trypanosomiasis. These diseases affect millions of people worldwide, both in the developing and developed part of the world. The AIDS epidemic has resulted in an increase in infections due to *Cryptosporidium parvum* which causes severe diarrhoea and *Pneumocystis carinii* which results in pneumonia. In the last century there have been two major antiprotozoal drugs obtained from higher plants: the antimalarial drug, quinine (**23**), and the amoebicidal drug, emetine (**24**).

23

24

Quinine, a quinolinemethanol, was used in the treatment of malaria long before the malaria parasite was even discovered. It is extracted from the bark of various species of *Cinchona*, in particular *C. ledgeriana*, where it occurs together with its dextrorotatory stereoisomer quinidine and two other main alkaloids, cinchonidine and cinchonine. The main *Cinchona* producing countries are Indonesia, Zaire, Tanzania, Kenya, Rwanda, Sri Lanka, Bolivia, Colombia, Costa Rica and India.

Quinine is used primarily for the treatment of severe and complicated *Plasmodium falciparum* induced cases of malaria, especially when the cases are resistant to synthetic chemotherapy (Bunnag and Harinasuta, 1987; Delacollette *et al.*, 1987; Salako, 1987; Warrell, 1987; Wernsdorfer, 1987; White, 1987). The chemical structure of quinine has also served as a template molecule for the design and development of several other antimalarial drugs including chloroquine. At one time quinine was superseded by synthetic antimalarials which have fewer side-effects. However resistance of *P. falciparum* to antimalarial chemotherapy favours the continued use of quinine.

The other three *Cinchona* alkaloids have also been shown to be effective in the treatment of falciparum malaria (Dawson, 1932). Only quinidine however, which is actually superior to quinine in its antimalarial effect but more likely to cause cardiac toxicity and hypersensitivity, has been recommended for oral or parenteral use in the event of quinine unavailability (Anon, 1990).

Emetine is a long recognized effective therapy for invasive amoebiasis, widely employed in developing countries for treating amoebic dysentery (Wilmot, 1962). Used in the form of the hydrochloride salt, the alkaloid is obtained from ipecac which is the dried root of the Brazilian plant *Cephaelis ipecahuanha* (Brossi *et al.*, 1971). It may also be semi-synthetically obtained by methylation of cephaeline, the other main constituent of ipecac. Ipecac root products (particularly ipecac syrup) are widely used as emetics in cases of poisoning. Emetine has a direct lethal action on the protozoan *Entamoeba histolytica* in tissues, including bowel, invaded by the organism (Knight, 1980). It has no effect however on amoebae confined to the lumen of the bowel. High doses of emetine are toxic to humans. It has severe adverse effects, exhibiting toxicity to heart, liver, kidneys and gastrointestinal tract. It is currently superseded by the synthetic metronidazole and its use is restricted to initial treatment of severe amoebic abcesses. Another plant drug, glaucarubin (**25**), which is a quassinoid-type degraded triterpene and is extracted from the American *Simarouba glauca*, has also been used in the oral treatment of amoebic dysentery.

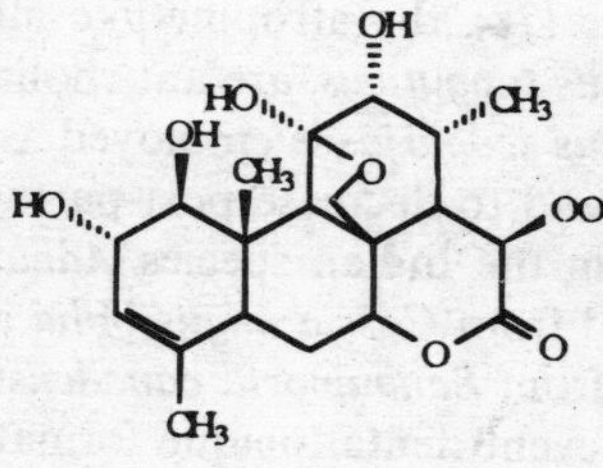

25

14.5.10   *Other Miscellaneous Drugs from Plants*

There are a myriad of other diseases and medical conditions for which drugs derived from plants are prescribed. For example, berberine (**26**) is a widely used drug for the treatment of bacillary dysentery, a disease caused by bacteria of the genus *Shigella*. It is an abundant alkaloid extractable from *Berberis vulgaris* and many other plant sources of the families Annonaceae, Ranunculaceae (mostly herbaceous), Berberidaceae, Menispermaceae and Papaveraceae. It has good antibacterial activity and is poorly absorbed from the gastrointestinal tract when administered orally.

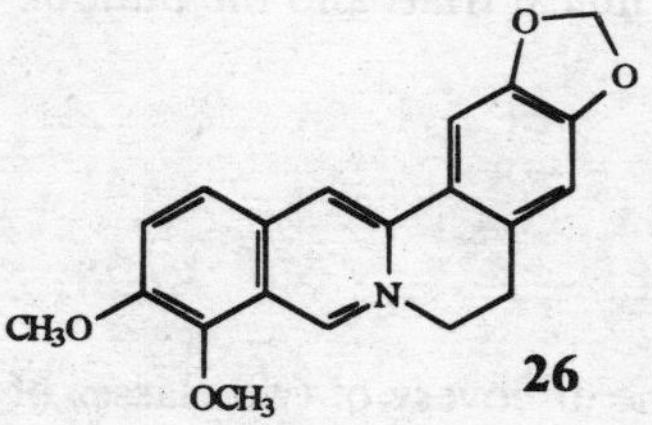

26

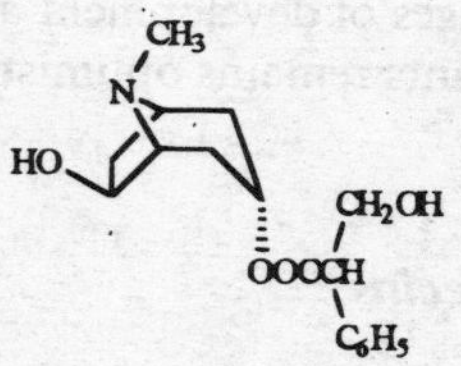

27

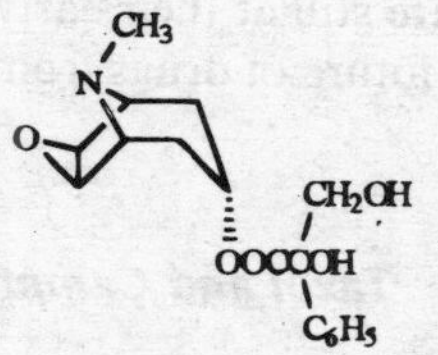

28

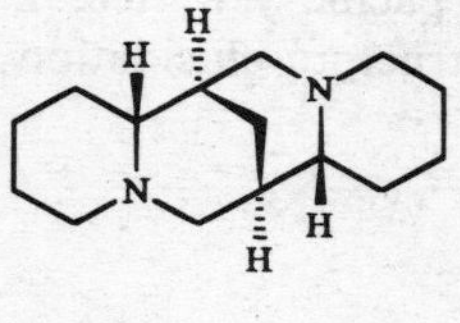

29

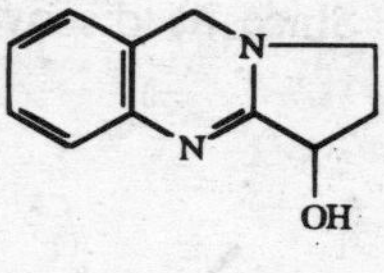

30

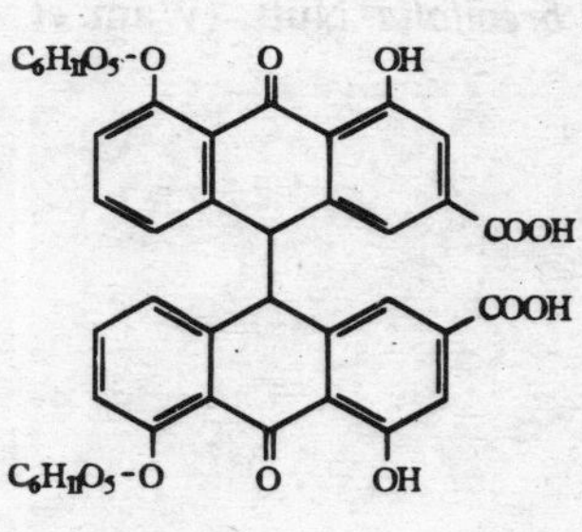

31

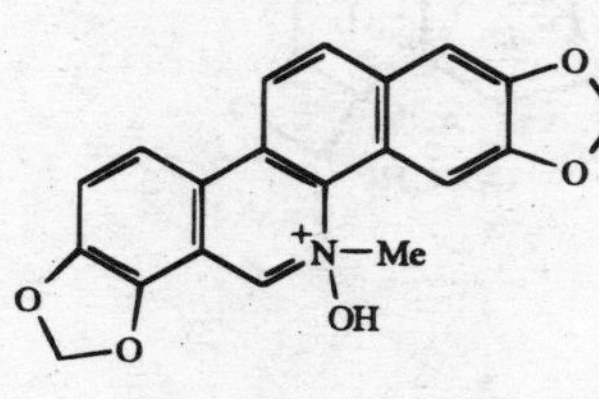

32

Anisodamine (**27**) and anisodine (**28**), the atropine-like alkaloids isolated from the Chinese painkiller plant, *Anisodus tanguticus*, are anticholinergic agents. The quinolizidine sparteine (**29**) from *Cytisus scoparius* is employed as an oxytocic in childbirth to initiate uterine contractions and to decrease post-partum haemorrhage. Vasicine (**30**), a quinazoline alkaloid from the Indian species *Adhatoda vasica*, has a similar use. The sennosides (**31**), derived from *Cassia angustifolia* and *C. acutifolia*, are used as laxatives. Sanguinarine (**32**) from *Sanguinaria canadensis*, is used in dental preparations due to its ability to prevent dental plaque formation. For numerous other examples refer to Table 14.1.

14.6 Recently Discovered Plant-based Drugs

Since the 1950s, spurred by the discoveries of the rauwolfia and vinca alkaloids, numerous plant species have been studied for their chemical constituents and biological properties. In 1985 NAPRALERT listed almost 200 000 compounds which have been identified from natural sources, 70 per cent of which were plants (Loub *et al.*, 1985). This number may well have increased significantly since then. Success in the development of new, commercially viable drugs has however been modest. Since the last major breakthrough of vinblastine and vincristine, the only other drug that the world has heralded with equal excitement has been taxol, which is also an antitumour agent. The discovery of the antimalarial, artemisinin, was also of particular importance, especially for tropical countries. Nevertheless many other potential drugs are still at their early stages of development and clinical trials and the outlook for the future of drugs from plants remains optimistic.

14.6.1 *Taxol and Camptothecins*

Current advances in cancer chemotherapy are due to the discovery of two classes of natural products, the taxoids and the camptothecins. Taxol or paclitaxel (**33**) was first discovered in 1971, from the bark of the northwest pacific yew tree, *Taxus brevifolia* Nutt. (Wani *et al.*, 1971). Since its discovery, structure elucidation and

33

biological activity more than 20 years ago, taxol has finally been accepted as an anticancer agent, culminating in its recent FDA approval for use in the treatment of breast and ovarian cancers, two of the most difficult forms of cancer to treat (Chabner, 1991). Taxol's mode of action is by targeting microtubule formation. Unlike other antimicrotubule agents, including the vinca alkaloids, which block microtubule production, however, taxol promotes tubulin polymerization and stabilizes microtubules against depolymerization (Rowinsky and Donehower, 1991; Gelmon, 1994). Much remains to be learned about the clinical use of taxol, such as its activity towards other forms of cancer, dose–response relationships, effectiveness in combination with other drugs in combination therapy, etc. Nevertheless, based on just the ovarian and breast cancer utilization, worldwide need of taxol has been estimated to be hundreds of kilograms per year. The yield of taxol from its rare and slow-growing natural source is very low, less than 0.02 per cent dry weight (Wani *et al.*, 1971). If the current trend of promising activities against other cancer forms continues, larger supplies and alternative sources of taxol will be needed and many mature trees would have to be felled in order to supply sufficient drug for clinical use (Cragg *et al.*, 1993). Although the semi-synthesis of the compound has been achieved (Holton, 1990), it will probably still not answer the supply problem. Efforts at solving this problem through tissue culture and microbial fermentation are currently underway and show great promise of success (Stierle *et al.*, 1995).

Camptothecins, the second group of natural antineoplastic compounds, are derivatives of the isoquinoline alkaloid camptothecin (**34**), first isolated from the Chinese tree *Camptotheca acuminata*. The alkaloid has been subjected to limited clinical trials but its effects have not paralleled those seen in animal studies. The compound was found to have a serious side-effect, causing bleeding in the bladder and kidneys. Better results have however been observed in clinical trials of the relatively new camptothecin derivatives, topotecan (**35**) and irinotecan (**36**) (Cragg *et al.*, 1993). Whereas several families of topoisomerase II inhibitors, such as anthracyclines, epipodophyllotoxins, acridines, and ellipticines, are known and widely used in clinical practice, the camptothecins are the unique representative of selective topoisomerase I inhibitors with clinical applications (Slichenmeyer *et al.*, 1993; Potmesil, 1994).

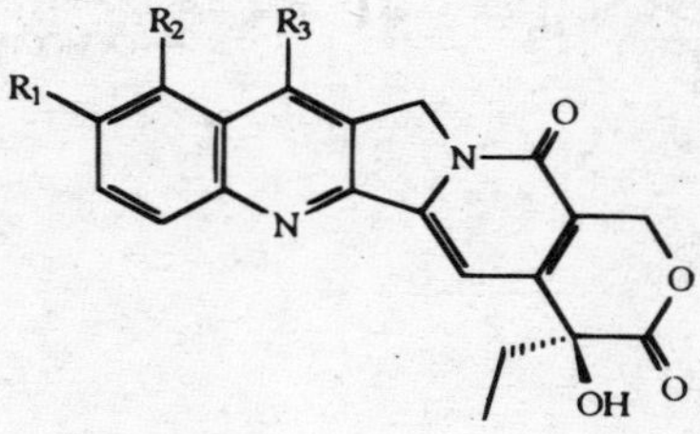

	R1	R2	R3
34	H	H	H
35	H	CH$_2$NH(CH$_3$)$_2$	OH
36	CH$_2$CH$_3$	H	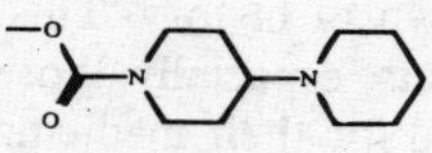

14.6.2 *Artemisinin*

Until the recent discovery of artemisinin (**37**) from the Chinese antimalarial plant, *Artemisia annua* (Qinghao), there had, for many years, been no further antiprotozoal drugs from higher plants. Also known as qinghaosu, this is an unusual endoperoxide sesquiterpene lactone (Ranasinghe *et al.*, 1993). Artemisinin has been found to be an effective antimalarial drug against both chloroquinine-resistant and chloroquinine-sensitive strains of *Plasmodium falciparum* as well as against cerebral malaria (Trigg, 1989). Artemisinin is currently in clinical use and in some parts of south-east Asia it seems to be the only effective drug against infections from multi-drug resistant *P. falciparum*.

The significant biological activity, novel chemical structure and its low yield from natural sources have prompted efforts directed at its synthesis. In the past few years, a number of derivatives have been developed and found to be clinically active against malaria, including multi-drug resistant *P. falciparum*. For example, reduction of the lactone carbonyl of artemisinin yields dihydroartemisinin (**38**) from which the semi-synthetic products (**39–41**) have been prepared. The World Health Organization and Rhone-Poulenc Rorer have in fact collaborated in the development of artemeter (**39**). Registration of these products has been approved in six African countries and applications filed in another eight countries. The current interest in artemisinin is considerable and it is possible that other antimalarial drugs based on its structure will be developed in the future, in a similar way to quinine being used as a template molecule for the development of chloroquine and mefloquine.

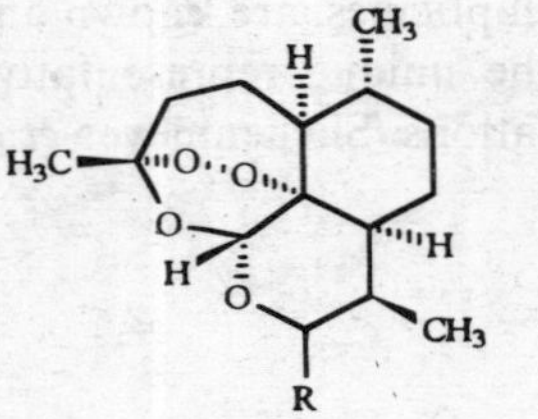

37 R = O
38 R = OH
39 R = ——OMe
40 R = ——OEt
41 R = ——OCHOCH₂CH₂COONa

14.7 Summary and Conclusions

It is apparent that the plant kingdom is a rich source of biologically active natural products and no doubt it will continue to serve mankind in the future just as it has done since the dawn of history. This is likely in view of the fact that only a small proportion of plants, especially those of the tropical rain forests, have been thoroughly investigated for their medicinal potential. Plants are useful in their crude or advanced forms as drugs, and biologically active compounds from plants can serve as templates for the synthesis of modern drugs. Methods of plant drug research have improved tremendously over the years, and much more has been learned about the basics of plant metabolism, plant analysis and even plant production. Looking ahead to the future, there are still many diseases such as AIDS, certain cancers, Parkinson's disease, muscular dystrophy, and cystic fibrosis which require improved and/or satisfactory cures. It is hoped that significant new plant drugs and new methods of producing them will continue to be discovered and developed.

References

ANONYMOUS (1988) Kava. *Lancet* **ii**, 258–259.

ANONYMOUS (1990) Practical chemotherapy of malaria: report of a WHO scientific group. *WHO Tech. Rep. Ser.* **805**.

AQUILONIUS, S. M. (1978) Physostigmine in the treatment of drug overdose. In: Jenden, D. J., ed., *Cholinergic Mechanisms and Psychopharmacology*, New York: Plenum Press, pp. 817–825.

AQUILONIUS, S. M. and HEDSTRAND, U. (1978) The use of physostigmine as an antidote in tricyclic antidepressant intoxication. *Acta Anaes. Scand.* **22**, 40–45.

AVORN, J. L. (1980) The role of cocaine in treating intractable pain in terminal disease. In: Jeri, F. R., ed., *Cocaine*, Lima, Peru: Pacific Press.

AYNILIAN, G. H., DUKE J. A., GRENTER W. A. and FARNSWORTH N. R. (1974) Cocaine content of *Erythroxylon* species, *J. Pharmaceut. Sci.* **63**, 1938.

BALANDRIN, M. F., KLOCKE, J. A., WURTELE, E. S. and BOLLINGER, W. H. (1985) Natural plant chemicals: sources of industrial and medicinal materials. *Science* **228**, 1154–1160.

BARRE, F. (1982) Cocaine as an abortive agent in cluster headaches. *Headache* **22**, 353–356.

BEAVER, W. T., WALLENSTEIN, S. L., ROGERS, A. and HOUDE, R. W. (1978) Analgesic studies of codeine and oxycodone in patients with cancer. I. Comparison of oral with intramuscular oxycodone with intramuscular morphine and codeine. *J. Pharmacol. Exp. Ther.* **207**, 101–108.

BENNETT, R. G. and GRIST, W. J. (1985) Nasal papillomas: successful treatment with podophyllin. *Southern Med. J.* **78**, 224–225.

BRAMWELL, V. H. C., CROWTHER, D., DEAKIN, D. P., SWINDELL, R. and HARRIS, M. (1985) Combined modality management of local and disseminated adult soft tissue sarcomas; a review of 257 cases seen over 10 years at the Christie Hospital and Holt Radium Institute, Manchester. *Br. J. Cancer* **51**, 301–318.

BROSSI, A., TEITEL, S. and PARRY, G. V. (1971) The ipecac alkaloids. In: Manske, R. H. F. and Holmes, H. L., eds, *The Alkaloids. XIII*, New York: Academic Press, pp. 189–212.

BUNNAG, D. and HARINASUTA, T. (1987) Quinine and quinidine in malaria in Thailand. *Acta Leidensia* **55**, 99–113.

CASEY, A. F. and PARFITT, R. T. (1986) *Opioid Analgesics. Chemistry and Receptors*, New York: Plenum Press, pp. 9–10.

CHABNER, B. A. (1991) *Principles and Practice of Oncology: PPO Updates* **5**, 1.

CHEN, G. L., YANG, L., ROWE, T. C., HALLIGAN, B. D., TEWEY, K. M. and LIU, L. F. (1984) Nonintercalative antitumour drugs interfere with the breakage-reunion reaction of mammalian DNA topoisomerase II. *J. Biol. Chem.* **259**, 560–566.

CLARKE, R. L. (1977) The tropane alkaloids. In: Manske, R. H. L. and Holmes, H. L., eds, *The Alkaloids. XVL*, New York: Academic Press, pp. 83–180.

CRAGG, G. M., SCHEPARTZ, S. A., SUFFNESS, M. and GREVER, M. R. (1993) The taxol supply crisis. New NCI policies for handling the large-scale production of novel natural products anticancer and anti-HIV agents. *J. Nat. Prod.* **56**, 1657–1668.

CRONQUIST, A. (1981) *An Integrated System of Classification of Flowering Plants*, New York: Columbia University Press.

DA PRADA, M. and PLETSCHER, A. (1969) Storage of exogenous monoamines and reserpine in 5-hydroxytryptamine organelles of blood platelets. *Eur. J. Pharmacol.* **7**, 45–48.

DAWSON, W. T. (1932) Antimalarial value of *Cinchona* alkaloids other than quinine. *Southern Med. J.* **25**, 529–533.

DELACOLLETTE, C., CIBANGUKA, J., BUREGEA, H. and NKERA, J. (1987) Compared response to chloroquine, Fansidar, quinine and quinidine slow release of infections with *Plasmodium falciparum* in the Kivu region of Zaire. *Acta Leidensia* **55**, 159–162.

DE VITA V. T. JR, LEWIS, B. J., ROZENCWEIG, M. and MUGGIA, F. M. (1978) The chemotherapy of Hodgkin's diseases; past experiences and future directions. *Cancer*, **42**, 979–990.

DOHERTY, J. E. (1985) Clinical use of digitalis glycosides: an update. *Cardiology* **72**, 225–254.

DUTCHER, J. D. (1946) Curare alkaloids from *Chondrodendron tomentosum* Ruiz and Pavon. *J. Am. Chem. Soc.* **68**, 419–424.

EDDY, N. B., FRIEBEL, H., HOHN, K. J. and HALBACH, H. (1969) Codeine and its alternates for pain and cough relief. *Bull. WHO* **40**, 639–719.

ELLIS, P. P. (1971) Pilocarpine therapy. *Survey Ophthalmol.* **16**, 165–169.

FARNSWORTH, N. R. and MORRIS, R. W. (1976) Higher plants: the sleeping giants for drug development. *Am. J. Pharm.* **148**, 46–52.

FREI III, E. and SALLAN, S. E. (1978) Acute lymphoblastic leukemia: treatment. *Cancer (Philad.)* **42**, 828.

GELMON, K. (1994) The taxoids: paclitaxel and docetaxel. *Lancet* **344**, 1267–1272.

GILLIS, R. A. and QUEST, J. A. (1980) The role of central nervous system in cardio-vascular effects of digitalis. *Pharmacol. Rev.* **31**, 19–97.

GILMAN, A. G., GOODMAN, L. S., RALL, T. W. and MURAD, F. (eds) (1985) *The Pharmacological Basis of Therapeutics*, 7th edition, New York: Macmillan, pp. 207–210.

GORLAY, G. K., CHERRY, D. A. and COUSINS, M. J. (1986) A comparative study of the efficacy and pharmacokinetics of oral methadone and morphine in the treatment of severe pain in patients with cancer. *Pain* **25**, 297–312.

GREENBLATT, D. J. and SHADER, R. I. (1973) Drug therapy: anticholinergics. *N. Engl. J. Med.* **288**, 1215–1219.

HOLLAND, M. G. (1974) Autonomic drugs in ophthalmology: some problems and promises. Section I: directly and indirectly acting parasympathomimetic drugs. *Ann. Ophthalmol.* **6**, 447–461.

HOLTON, R. A. (1990) European Patent Application, EP O 400 971 A2.

HUGHES, D. T. D. (1978) Diseases of the respiratory system: cough suppressants, expectorants and mucolytic agents. *BMJ* 1202–1203.

IUCN (1979) Categories, objectives and criteria for protected areas. Annex to General Assembly Paper GA 78/24. In: *Proceedings of IUCN Fourteenth Technical Meeting*, Morges: IUCN.

KNIGHT, R. (1980) The chemotherapy of amoebiasis. *J. Antimicrob. Chemother.* **6**, 577–593.

KOHLI, J. D. and MUKERJI, B. (1955) Comparative activity of reserpine and total alkaloids of Rauvolfia. *Curr. Sci.* **24**, 198–199.

LASSUS, A. and ROSEN, B. (1986) Response of solitary psoriatic plaques to experimental application of podophyllotoxin. *Dermatologica* **172**, 319–322.

LEIGH, E. G. (1982) Introduction: why are there so many kinds of tropical trees? In: Leigh, E. G., Rand, A. S. and Windsor, D. M., eds, *The Ecology of a Tropical Forest*, Washington DC: Smithsonian Institution.

LIU, L. F. (1983) DNA topoisomerases-enzymes that catalyse the breaking and rejoining of DNA. *CRC Crit. Rev. Biochem.* **15**, 1–24.

LONG, B. H., MUSIAL, S. T. and BRATTAIN, M. G. (1984) Comparison of cytotoxicity and DNA breakage activity of congeners of podophyllotoxin including VP16-213 and VM-26: a quantitative structure-activity relationship. *Biochemistry* **23**, 1183–1188.

LOUB, W. D., FARNSWORTH, N. R., SOEJARTO, D. D. and QUINN, M. L. (1985) NAPRALERT: computer handling of natural product research data. *J. Chem. Inf. Computer Sci.* **25**, 99–103.

MARINI BETTOLO, G. B. (1981) Recent advances in the research on curare. *Udit (Verhandelingen van de Koninglijke Academie voor Geneeskunde van Belgie) XLIII* 3, 185–212.

OWELLEN, R. J., HARTKE, C. A., DICKERSON, R. M. and HAINS, F. Q. (1976) Inhibition of tubulin-microtubule polymerization by drugs of the vinca alkaloid class. *Cancer Res.* **36**, 1499–1502.

PENG, J. Z., JIN, L. R., CHEN, X. Y. and CHEN, Z. X. (1983) Central effects of anisodamine, atropine, anisodine and scopolamine after intraventricular injection. *Chung-Kuo Yao Li Hsueh Pao* **4**, 81–87.

PEREZ-FIGARADO, R. A. and BADEN, H. P. (1976) The pharmacology of podophyllum. *Prog. Dermatol.* **10**, 1–4.

POTMESIL, M. (1994) Camptothecins: from bench research to hospital wards. *Cancer Res.* **54**, 1431–1439.

RANASINGHE, A., SWEATLOCK, J. D. and COOKS, R. G. (1993) A rapid screening method for artemisinin and its congeners using MS/MS: search for new analogues in *Artemisia annua*. *J. Nat. Prod.* **56**, 552–563.

RITCHIE, J. M., COHEN, P. J. and DRIPPS, R. D. (1970) Cocaine, prococaine and other local anaesthetics. In: Goodman, L. S. and Gilman, A., eds, *The Pharmacological Basis of Therapeutics*, New York: Macmillan, pp. 367–377.

ROSENBERG, S. A. (1979) Current concepts in cancer: non-Hodgkin's lymphoma – selection of treatment on the basis of histologic type. *N. Eng. J. Med.* **301**, 924–928.

ROSKETH, R. and STORSTREIN, O. (1963) Quinidine therapy of chronic auricular fibrillation. *Arch. Intern. Med.* **111**, 184–189.

ROSS, W., ROWE, T., GLISSON, B., YALOWICH, J. and LIU, L. (1984) Role of topoisomerase II in mediating epipodophyllotoxin induced DNA cleavage. *Cancer Res.* **44**, 5857–5860.

ROWINSKY, E. K. and DONEHOWER, R. C. (1991) The clinical pharmacology and use of antimicrotubule agents in cancer chemotherapeutics. *Pharma. Ther.* **52**, 35–84.

SALAKO, L. A. (1987) Quinine and malaria: the African experience. *Acta Leidensia*, **55**, 167–180.

SCHLITTLER, E. and BEIN, H. J. (1967) Rauwolfia alkaloids. In: Schlittler E., ed., *Antihypertensive Agents*, New York: Academic Press, pp. 191–221.

SCHULTES, R. E. (1972) The future of plants as sources of new biodynamic compounds. In: Swain, T., ed., *Plants in the Development of Modern Medicine*, Cambridge, MA; Harvard University Press, pp. 103–124.

SLICHENMEYER, W. J., ROWINSKY, E. K., DONEHOWER, R. C. and KAUFMAN, S. H. (1993) The current status of camptothecin analogues as antitumour agents. *J. Nat. Cancer Inst.* **85**, 271–291.

SMITH, T. W. (1988) Digitalis: mechanisms of action and clinical use. *N. Engl. J. Med.* **318**, 358–365.

STANDAERT, F. G. (1984) Pharmacology of the neuromuscular junction. In: Brumback, R. A. and Gerst, J., eds, *The Neuromuscular Junction*, Mt Kisco, NJ: Futura.

STEINER, R., STEWART, J. F., CARTWELL, B. M. J., MINTON, M. J., KNIGHT, R. R. and RUBENS, R. I. (1983) Adriamycin alone or in combination with vincristine in the treatment of advanced breast cancer. *Eur. J. Clin. Oncol.* **19**, 1553–1557.

STERN, S. (1971) Treatment and prevention of cardiac arrhythmias with propranolol and quinidine. *Br. Heart J.* **33**, 522–525.

STIERLE, A., STROBEL, G., STIERLE, D., GROTHAUS, P. and BIGNAMI, G. (1995) The search for taxol-producing microorganism among the endophytic fungi of the Pacific yew, *Taxus brevifolia. J. Nat. Prod.* **58**, 1315–1324.

SUTCLIFFE, S. B., WRIGLEY, P. F. M. and PETO, J. (1978) MVPP chemotherapy regimen for advanced Hodgkin's diseases. *BMJ* **1**, 679.

TRIGG, P. (1989) Qinhhaosu (artemisinin) as an antimalarial drug. In: Wagner, H., Hikino, H. and Farnsworth, N. R., eds, *Economic and Medicinal Plant Research*, Vol. 3, London: Academic Press, p. 20.

TWYCROSS, R. G. (1977) Value of cocaine in opiate-containing elixirs. *BMJ* **2**, 1348.

VOLBERDING, P. A., ABRAMS, D. I. and CONANT, M. (1985) Vinblastine therapy for kaposis sarcoma in the acquired deficiency syndrome. *Ann. Intern. Med.* **103**, 333–338.

WANI, M. C., TAYLOR, H. C., WALL, M. E., COGGON, P. and McPHAIL, A. T. (1971) Plant antitumour agents. VI. The isolation and structure of taxol, a novel anti-leukemic and antitumour agent from *Taxus brevifolia. J. Am. Chem. Soc.* **93**, 2325–2327.

WARRELL, D. A. (1987) Clinical management of severe falciparum malaria. *Acta Leidensia* **55**, 99–113.

WEINER, N. (1985) Norephedrine, epinephrine and the sympathomimetic amines. In: Goodman, L. S. and Gilman, A. G., eds, *The Pharmacological Basis of Therapeutics*, 7th edtion, New York: Macmillan, pp. 145–180.

WERNSDORFER, W. H. (1987) Quinine in health care in the tropics. *Acta Leidensia* **55**, 197–208.

WESSON, D. R. and SMITH, D. E. (1977) Cocaine: its use for central nervous system stimulation including recreational and medical uses. In: Peterson, R. C. and Stillman, R. C., eds, *Cocaine: 1977, NIDA Research Monograph*, **13**, pp. 137–152.

WHIPPLE, G. H. (1977) Papaverine as an antiarrhythmic agent. *Angiology* **28**, 737–749.

WHITE, N. J. (1987) The pharmacokinetics of quinine and quinidine in malaria. *Acta Leidensia*, **55**, 65–76.

WILMOT, A. J. (1962) *Clinical Amoebiasis*, Oxford: Blackwell Scientific Publications.

WOOD, H. C. S. and WRIGGLESWORTH, R. (1977) Tropane alkaloids. In: Coffey, S., ed., *Rodd's Chemistry of Carbon Compounds. IVB*, New York: Elsevier, pp. 205–235.

WOODSEN, R. E., YOUNGKEN, H. W., SCHLITTLER, E. and SCHNEIDER, J. A. (1957) *Rauwolfia: Botany, Pharmacognosy, Chemistry and Pharmacology*, Boston: Little Brown & Co.

YAP, H. Y., BLUMENSCHEIN, G. R., KEATING, M. J., HORTOBAGYI, G. N., TASHIMA, C. K. and LOO, T. L. (1980) Vinblastine given as a continuous five-day infusion in the treatment of refractory-advanced breast cancer. *Cancer Treat. Rep.* **64**, 270–283.

The Role of Biological Metal Chelators in Wood Degradation and in Xenobiotic Degradation

BARRY GOODELL AND JODY JELLISON

15.1   Introduction

In this chapter we focus on a review of previous work which supports the hypothesis of the role of phenolate chelating compounds produced by brown-rot wood decay fungi in generating reactive oxygen species to initiate wood deterioration. The role of these chelating compounds in the degradative process and their interaction with other metabolites and metals is explored. This chapter begins with a brief review of brown-rot degradative processes and the production of reactive oxygen species, the availability of transition metals in the environment, and the production of chelators by microorganisms. We then progress into a review of research which has been conducted specifically on chelators isolated from the brown-rot fungus *Gloeophyllum trabeum*, and how these compounds may potentially be employed in processes adapted to industrial biotechnology.

15.2   Brown-Rot Degradation of Wood and the Involvement of Non-Enzymatic Systems

Brown-rot fungi cause a rapid depolymerization of cellulose and a modification of lignin components (Cowling, 1975; Kirk, 1975) during the wood degradation process. It is now recognized that cellulose degrading enzymes produced by fungi are too large to penetrate the non-modified wood cell wall and that some smaller agent must be involved (Flournoy *et al.*, 1991). Previous research has suggested that the action of organic acids alone or simple Fenton chemistry does not adequately explain the natural process of brown-rot decay (Cowling, 1961). The involvement of low molecular weight metabolites in brown-rot attack has been implicated (Enoki *et al.*, 1989; Fekete *et al.*, 1989; Goodell, 1989; Jellison *et al.*, 1991b; Chandhoke *et al.*, 1992; Goodell *et al.*, 1995; Hirano *et al.*, 1995; Goodell *et al.*, 1997). Immuno-electron microscopic studies have confirmed that degradative enzymes and metabolites from both white- and brown-rot fungi, including a variety of cellulases, ligninase, manganese peroxidases, laccase, and others, are not capable of penetrating the wood cell wall in early decay stages (Daniel *et al.*, 1989, 1990; Srebotnik and Messner, 1991; Kim *et al.*, 1991, 1993; Blanchette, 1996).

To account for the unexplained depolymerization of the wood cell wall, various research groups have focused on possible mechanisms employing low molecular weight agents which could potentially penetrate and initiate depolymerization of the cell wall constituents. Research by Koenigs (1975) and others (Highley, 1980; Schmidt *et al.*, 1981) on the action of Fenton reagents against cellulose and wood cell wall components suggested that iron and hydrogen peroxide were involved in the production of highly reactive hydroxyl radicals which could initiate the depolymerization of cellulose in wood. Several more recent hypotheses, including those involving oxalic acid, have been proposed (Green *et al.*, 1991; Shimada *et al.*, 1992). Oxalate has been postulated to play a role in direct acid attack on the wood cellulose and hemicellulose. It has also been suggested that oxalate may function to reduce iron III to iron II which then reacts with H_2O_2 to yield $OH^{\cdot}$. However, Hyde and Wood (1995) and others (Zepp *et al.*, 1992; Sedlak and Hoigné, 1993; Sulzberger and Laubscher, 1995) have observed that oxalate does not reduce ferric iron except as a light-dependent reaction. Therefore, oxalate does not appear to function as a direct catalyst of Fenton type chemical reactions in wood.

The attack of wood by brown-rot fungi is both physically and chemically similar to the action of iron(II) and hydrogen peroxide (Fenton's reagents) in wood (Kirk *et al.*, 1991). This suggests therefore that not only soluble iron, but also soluble reduced iron, or perhaps another suitable transition metal such as manganese, must be present for brown-rot wood degradation processes to occur. In alkaline and calcareous soils free ferric iron is often limiting and is not available for plant or microbial activity (Hartwig and Loepper, 1993). In aqueous environments relatively insoluble (hydr)oxide forms of iron predominate (Cotton and Wilkinson, 1976; Winterbourn, 1991) and very small amounts of soluble iron would be available (approx. 10^{-38} M Fe(III)) in aerobic environments (van der Helm and Winkelmann, 1994). In addition, very limited amounts of soluble iron can be extracted from most natural iron containing minerals (Page, 1993). Total iron levels in the wood (as well as levels of manganese and other transition metals) are usually significantly lower than those found in the soil and are often below 2 μM concentration in non-degraded wood (Jellison *et al.*, 1992, 1993). These metals would be expected to be found either bound to components in the wood cell wall, or alternatively would be found as insoluble oxy(hydr)oxide forms. Soluble metal concentrations in wood therefore would be expected to be much lower than the total metal concentrations previously reported in wood.

Low levels of available iron in the environment affect the manner in which most microorganisms adapt, and high-affinity iron binding compounds known as siderophores (normally repressible by high iron concentrations in the medium) have been found to be produced under iron-limiting conditions by most aerobic and facultative anaerobic microorganisms examined (Höfte, 1993; Guerinot, 1994). The production of metal-chelating compounds by most microorganisms therefore is essential to the sequestration of transition metals required for life processes. Many forms of microbially produced chelators have been isolated but the basic forms include phenolate (or catecholate) derivatives, hydroxamic acid derivatives, and mixed function derivative compounds. With few exceptions, chelators previously isolated from the fungi have been hydroxamic acid derivatives (Winkelmann and Winge, 1994). Recently however, wood degrading basidiomycetes were found to produce phenolate chelators (Fekete *et al.*, 1989) and phenolate siderophores have also been reported from *Trichoderma* species (Srinivasan *et al.*, 1995). For an additional review of chelators isolated from wood decay fungi see Goodell *et al.* (1997). Further information about chelator structures, and mechanisms for metal sequestration and release can be found in Neilands *et al.* (1987), Lodge (1993), Emery (1987) and Winkelmann and Winge (1994).

15.3 Review of Research on Chelators Isolated from the Brown-Rot Fungus *Gloeophyllum trabeum*

15.3.1 *Isolation and Characterization of Phenolate Chelators*

The general procedures used to produce and isolate phenolate derivative chelators have been described previously (Chandhoke, 1991). The brown-rot fungus *Gloeophyllum trabeum* (Pers.:Fr.) Murr., is grown in 200 ml of liquid culture medium in 500 ml Erlenmeyer flasks as described by Highley (1973) except that manganese is omitted and the concentration of iron is 20 μM. Stationary cultures are incubated in the dark at 37°C for 4–5 weeks. Cultures are then monitored for the production of iron-chelating compounds using the Chrome Azural-S (CAS) assay (Fekete *et al.*, 1989) and are then harvested by filtration of the mycelium through Whatman No. 2 filter paper. The concentrated filtrate is then placed in an Amicon 400 ultrafiltration unit with a nominal cut-off of 1000 daltons (Amicon YM1 membrane). The filtrate is acidified with HCl to pH 3.0 and extracted with an equal volume of ethyl acetate three times. The ethyl acetate phase is dried and finally resuspended in deionized water. Chelators are identified as catecholate phenolics using Arnow (1937) and Rioux assays (Ishimaru and Loper, 1993; Easwaran, 1994); these assays are also used to determine the concentration of active phenolate chelator present in a sample relative to a dihydroxybenzoic (DHBA) acid standard. Previous work by Chandhoke (1991), Easwaran (1994) and Goodell *et al.* (1997) has shown that the chelators isolated are benzene ring derivative compounds with many having phenolate or phenolate derivative character. Many microorganisms are known to produce more than one siderophore species (Emery, 1980; Ecker *et al.*, 1982) and it is therefore not unexpected that more that one chelator species would be produced by *G. trabeum*. The possible additional production of hydroxamic acid or mixed function chelators by *G. trabeum* has not been investigated. Because multiple chelator fractions are isolated (Goodell *et al.*, 1997) with the procedure cited above, and these are not fractionated before use in assays, the term Gt chelator used throughout the remainder of this chapter is understood as referring to the <1000 dalton, ethyl acetate soluble phenolate chelator fraction unless otherwise noted. In some of the subsequent work presented, where specified, purified DHBA has also been used in reaction mixtures instead of the Gt chelator. DHBA is similar in structure to many of the chelator fractions identified, and functions similarly, in most assays, as outlined in subsequent sections of this chapter.

Reduction Potential

Previous work has shown that the Gt chelator can alter the redox potential of iron to favour its reaction with other potential reactants (Lu *et al.*, 1994; Goodell *et al.*,

1997). This is demonstrated by monitoring the reduction of ferric iron in the presence of ferrozine, an Fe(II) chelator used in the detection of reduced iron (Stookey, 1970; Gibbs, 1976; Sørensen, 1982). In a typical assay, freshly prepared ferric iron (anhydrous Fe_2Cl_3) is mixed in a buffered solution with 2.0 mM ferrozine. An aliquot of Gt chelator is then added to start the reaction. When the Gt chelator is present in excess at pH values below neutrality, all available iron is rapidly reduced and then bound by the ferrozine indicator. Increasing the amount of Fe(III) in the presence of excess chelator increases the rate of reduction of iron in solution as indicated by an increase in absorbance of the ferrozine at 562 nm. Addition of the chelator 2,3-DHBA to iron produced data similar to that of the Gt chelator (Figure 15.1). With both Gt chelator and DHBA the rate of reaction is increased in lower pH environments as these environments are more favourable for the reduction of iron. At pH values of 7 and above, the reduction of iron is greatly limited.

15.3.2 Solubilization and Phase Transfer of Iron

When an aqueous solution of ferric iron complexes with oxygen such as occurs in natural aerated environments, oxygenated forms of iron (oxy(hydr)oxy species) are produced. A variety of iron oxy(hydr)oxy species exist in nature and these have been previously reviewed (Cotton and Wilkinson, 1976; Winterbourn, 1991). Most oxygenated forms of iron have extremely low solubility. In experiments designed to simulate the sequestration of iron from natural environments, suspension mixtures of iron (hydr)oxides have been produced (Goodell *et al.*, 1997) which permit reactions to be carried out in aqueous solution.

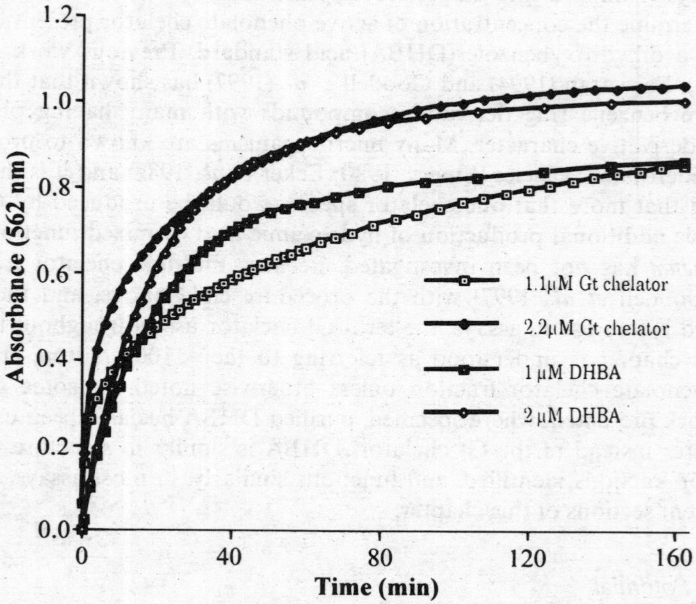

Figure 15.1 Comparison of Gt chelator and 2,3-dihydroxybenzoic acid in the reduction of ferric iron. Solutions contained 30 μM Fe (III) with 2.5 mM ferrozine reagent in pH 4.5 acetate buffer

Previous research in our laboratories has shown that the Gt chelator and chelators like 2,3-DHBA cannot readily sequester oxygenated forms of iron. Only freshly prepared aqueous solutions of ferric iron are readily chelated. When oxalate is added to oxygenated forms of iron first however, the Gt chelator readily sequesters the iron from the oxalate in a phase transfer reaction and alters the redox potential of the iron so that it can then be chelated by the ferrous iron indicator ferrozine (Figure 15.2). Lowering the pH of the environment is known to create more favourable conditions for reduction of iron and it has been observed that ferric iron reduction by the chelator occurs more readily at lower pH values. However, simple reduction of pH with buffer solutions does not increase the capacity of the Gt chelator to sequester and reduce oxygenated forms of iron. Interestingly, the phase transfer of iron from oxalate to the Gt chelator did not occur when the pH of the reaction mixture was reduced below pH 4. This suggests that the phase transfer of iron is a pH-dependent phenomenon and would therefore be subject to changes in pH controlled largely by the fungus in wood undergoing fungal degradation. The mechanism for this is reviewed later in this chapter.

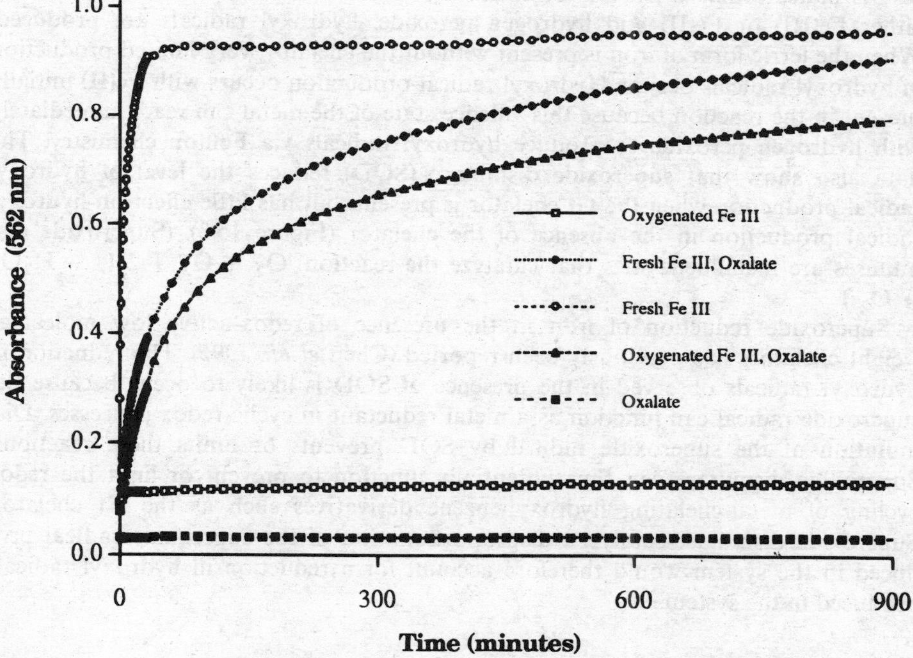

Figure 15.2 Freshly prepared Fe(III) solutions (30 μM) compared with oxygenated Fe(III) solutions (30 μM) with the Gt chelator in a ferrozine assay in pH 4.4 acetate buffer. For oxalate treated samples, iron solutions were first incubated with oxalate (at a ratio of 1 mM Fe(III):1.5 mM oxalate in aqueous solution) for more than 5 hours until the colour in the oxygenated iron solutions had cleared. Final reaction mixtures contained 2 μM of Gt chelator and 250 μM of ferrozine reagent (reprinted from Goodell, B., Jellison, J., Liu, J. *et al.* (1997). Low molecular weight chelators and phenolic compounds isolated from wood decay fungi and their role in the fungal biodegradation of wood: *Journal of Biotechnology* (**53**, 133–162), with kind permission of Elsevier Science-NL, Sara Burgerharrstraat 25, Ioss KV Amsterdam, Netherlands)

Oxalate, a metabolite produced by wood-degrading fungi, is also a well-known chelator and a reagent commonly used in the household to remove iron and rust stains from fabrics and other substrates. It is interesting that the brown-rot fungi have perhaps used a similar mechanism to solubilize oxygenated forms of iron as a first step in wood degradation processes long before humanity learned to use oxalate for solubilizing iron for other purposes.

15.3.3 Hydroxyl Radical and Superoxide Radical Production

Previous experiments with both oxidized and reduced forms of iron have shown that production of reactive oxygen species such as the hydroxyl radical and superoxide radical are affected by the presence of the Gt chelator and hydrogen peroxide in reaction mixtures. Studies were also carried out to determine the effect of oxalic acid as a solubilizing and chelating agent on the production of reactive oxygen species in the reaction mixtures tested. In addition, the role of the superoxide radical was explored by using superoxide dismutase in some tests (Lu *et al.*, 1994). Our results indicate that when the Gt chelator is present in the reaction mixture with either Fe(III) or Fe(II) and hydrogen peroxide, hydroxyl radicals are produced. When the ferric form of iron is present without the chelator, very limited production of hydroxyl radicals occurs. Hydroxyl radical production occurs with Fe(II) initially present in the reaction because this valence state of the metal can react immediately with hydrogen peroxide to produce hydroxyl radicals via Fenton chemistry. The data also show that superoxide dismutase (SOD) reduces the level of hydroxyl radical production when the Gt chelator is present, but has little effect on hydroxyl radical production in the absence of the chelator (Figure 15.3). (Superoxide dismutases are metalloenzymes that catalyze the reaction: $O_2^- + O_2^- + 2H^+ \rightarrow H_2O_2 + O_2$.)

Superoxide reduction of iron in the presence of redox-active low molecular weight chelators has previously been reported (Chen *et al.*, 1995). The reduction in hydroxyl radicals observed in the presence of SOD is likely to occur because the superoxide radical can function as a metal reductant in cyclic redox processes. Dismutation of the superoxide radical by SOD prevents or limits these reactions. Superoxide dismutase can thus potentially function to prevent or limit the redox cycling of metal-chelating hydroxybenzene derivatives such as the Gt chelator. Superoxide dismutase-catalyzed disproportionation of the superoxide radical produced in the system would therefore account for a reduction in hydroxyl radicals produced in the system.

15.3.4 Immunological Studies

The enzyme-linked immunosorbent assay (ELISA) is a serological technique which has been previously used for the detection and quantification of fungal metabolites in wood (Goodell and Jellison, 1988; Jellison and Goodell, 1988). Polyclonal antisera produced by HPLC-purified phenolate iron-binding chelator fractions have been used to detect and quantify phenolate chelators in fungally degraded wood and in wood-free fungal cultures. The chelator fraction typically is purified from the

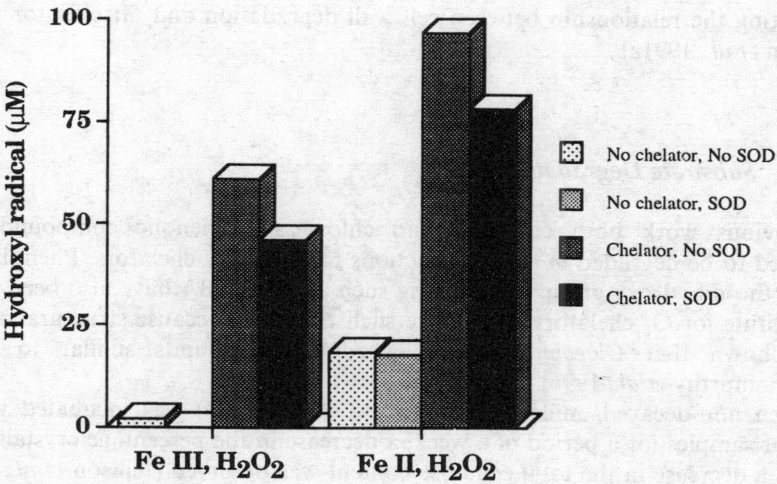

Figure 15.3 The effect of superoxide dismutase (SOD, 450 units) on hydroxyl radical production in the presence of ferric or ferrous iron (50 μM) and hydrogen peroxide (1 mM) in pH 4.0 acetate buffer. Gt chelator, when added was at 15 μM concentration. Reactions were carried out over a 3-hour reaction period at ambient temperature (reprinted from Goodell, B., Jellison, J., Liu, J. *et al.* (1997). Low molecular weight chelators and phenolic compounds isolated from wood decay fungi and their role in the fungal biodegradation of wood. *Journal of Biotechnology* (**53**, 133–163), with kind permission of Elsevier Science-NL, Sara Burgerharrstraat 25, loss KV Amsterdam, Netherlands)

extracellular filtrate of *G. trabeum* grown in liquid culture as outlined above. Antigen injection, production of antibodies, and antisera purification follow procedures outlined in previous literature (Goodell and Jellison, 1986; Jellison and Goodell, 1988; Easwaran, 1994).

The Gt chelator antibody displays strong reactivity against extracts of *G. trabeum* and against *G. trabeum* infected spruce (*Picea rubens*) wood. The reactions did not occur when non-infected control wood was tested against the antibody. The antibody made to the Gt chelator also reacted with spruce wood infected with the brown-rot fungus *Postia placenta*, suggesting that similar compounds are found in wood degraded by at least one other brown-rot fungus. The antigenic compounds in the samples could be of fungal origin and/or modified lignin breakdown products but appear to be present only in the samples of brown-rot degraded wood and not in the non-infected control wood. It should also be noted that especially in the *G. trabeum* degraded wood samples, the serological data suggest a high initial concentration of these compounds in the degraded wood which is maintained through more advanced decay stages.

In addition to the ELISA assays, antibodies produced to the Gt chelator have also been used to label wood degraded by *G. trabeum* in an immunogold labelling procedure (Jellison *et al.*, 1991a). The Gt chelator was found localized in the plasma membrane and cell wall of the fungus and internally within the hyphal cytoplasm. Labelling within the wood cell wall was observed throughout, but was seen in higher concentration in more highly degraded (lower electron dense) regions of the cell wall confirming the presence of the Gt chelator in the degraded wood and illustrating the relationship between cell wall degradation and Gt chelator activity (Jellison *et al.*, 1991a).

15.3.5 Substrate Degradation

In previous work both cellulosic and chlorinated phenolic compounds were observed to be degraded *in vitro* in reactions mediated by chelators. Phenolic acids with ortho-dihydroxy groups on the ring such as 2,3-DHBA have also been used as a substitute for Gt chelators to mediate such reactions, because structural analyses have shown that *Gloeophyllum* spp. produce compounds similar to DHBA (Krishnamurthy *et al.*, 1996).

When non-decayed, milled, poplar wood (*Populus* sp.) was incubated with Gt chelator samples for a period of a week, a decrease in the percentage crystallinity as well as a decrease in the total cellulose content was observed (Jellison *et al.*, 1991b). Wood samples decayed by the brown-rot fungus *Postia placenta* also displayed similar reductions. Calculations of estimated cellulose micelle width of Gt chelator-treated and non-treated poplar wood showed a small decrease in average micelle width in the treated wood, consistent with an attack and removal of the outer layers of the cellulose crystal. Additional work has shown that the addition of oxalic acid to the chelator–iron system with milled poplar wood causes a further decrease in cellulose crystallinity values (Hayashi, Jellison and Ishihara, unpublished) possibly due to the role of oxalate in increasing iron availability.

Related work on depolymerization of cellulose and on xylan has also been carried out. Both α-cellulose and xylan were exposed to the Gt chelator or to DHBA in the presence of iron with hydrogen peroxide in a buffered solution. Analysis of the samples was then performed using gel permeation chromatography (GPC), indicating that the Gt chelator in the presence of iron and hydrogen peroxide will degrade cellulose. The samples treated with Gt chelator, iron and hydrogen peroxide had a greater proportion of cellulose with a lower DP (degree of polymerization). Similar results have been obtained using xylans incubated with DHBA, iron and hydrogen peroxide using viscosity measurements (Anon, 1990) as a measure of the change in DP. It was assumed that the hydroxyl radicals generated in the chelator-mediated system were responsible for the degradation and depolymerization of the cellulosic materials.

Because hydroxyl radicals have been reported to degrade a wide variety of hydrocarbon substrates including pesticides and other xenobiotics, it was of interest to determine whether pentachlorophenol, a chlorinated wood preservative commonly found as a soil contaminant around old wood-treatment plants, could be degraded by the Gt chelator system. Pentachlorophenol (PCP) is commonly regarded as one of the most difficult to degrade xenobiotics when found as a contaminant in soil and water environments. When PCP is incubated with a mixture of ferric iron, hydrogen peroxide and either the Gt chelator or DHBA, the PCP is observed to degrade over several hours until the hydrogen peroxide in the system is depleted. In aerobic environments suitable for fungal growth, only forms of ferric iron would be available for reaction. In laboratory experiments however, ferrous iron may be used, which permits immediate reaction with hydrogen peroxide to form hydroxyl radicals and a return of the iron to its oxidized (ferric) valence state. Subsequent

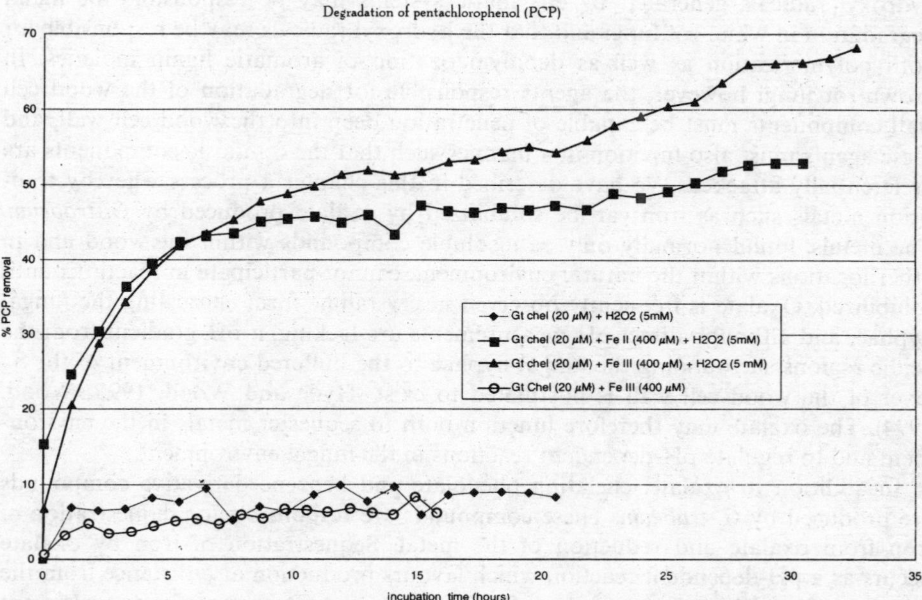

Figure 15.4   Degradation of pentachlorophenol (PCP) mediated by Gt chelator. The reaction mixtures contained 25 μM PCP in pH 4.0 aqueous buffer. All solutions were incubated at room temperature in the dark. Samples with all three components (Gt chelator, iron and hydrogen peroxide) were pulsed with hydrogen peroxide after initial reaction (Krishnamurthy and Goodell, unpublished)

reduction of this ferric iron by the Gt chelator (or DHBA) then permits additional reactions with the hydrogen peroxide to occur (Figure 15.4). Previous researchers have reported that the degradation of non-polar organic molecules, including a variety of aromatic and hydrocarbon compounds, by Fenton's reagent alone is limited (Pignatello, 1993; Carberry and Sang, 1994). The chelator mediated system increases the rate of degradation of PCP beyond that of Fenton's reagent chemicals alone, but the reason for this is still being explored. Since the publication of research on the chelator mediated PCP degradation system (Goodell *et al.*, 1997), the degradation of other xenobiotics has been explored, and compounds such as trichlorophenol have also been shown to be readily degraded by chelator mediated systems (unpublished data).

15.4   Summary

Brown-rot fungi are characterized by their ability to attack rapidly and selectively the hemicellulose and cellulose of the wood cell wall. The lignin in the wood cell wall is typically modified by cleavage of phenyl-propane groups from the ring and by demethylation and demethoxylation, but the ring itself remains and is not metabolized to a significant extent by most brown-rot fungi. It has been postulated that hydroxyl radicals generated by enzymatic systems may be responsible for lignin degradation in white-rot fungi and that the hydroxyl radicals may be responsible for both polymerization as well as depolymerization of aromatic lignin moieties. In brown-rot fungi however, the agents responsible for degradation of the wood cell wall components must be capable of penetration deep into the wood cell wall, and these agents must also function in a manner such that the cellulosic components are preferentially attacked. We have described in this chapter a process whereby transition metals such as iron can be solubilized by oxalate produced by *G. trabeum*. The metals, found normally only as insoluble compounds within the wood and in other locations within the natural environment, cannot participate in reactions until solubilized. Oxalate is frequently observed in crystalline form encrusting the fungal hyphae, and although direct pH measurements are lacking, a pH gradient from the acidic regions surrounding the fungal hyphae to the buffered environment of the S_2 layer of the wood cell wall is postulated to exist (Hyde and Wood, 1995; Wood, 1994). The oxalate may therefore function both to sequester metals in the environment and to regulate pH-dependent reactions in the fungal environment.

In addition to oxalate, chelating phenolate and benzene derivative compounds are produced by *G. trabeum*. These compounds are responsible for sequestration of iron from oxalate and reduction of the metal. Sequestration of iron by oxalate occurs as a pH-dependent reaction which favours production at a distance from the fungal hyphae. Thus the phase transfer of iron to the Gt chelator and subsequent reduction of the metal would also occur at a distance from the hyphae. This helps to ensure that the fungus is not damaged by this reactive metal species through interaction with hydrogen peroxide or other oxidants present in the environment immediately surrounding the hyphae. Other research has shown that hydrogen peroxide is produced only as pH-dependent autoxidation of phenolic compounds with limited peroxide produced in low pH environments (Pueyo and Ariza, 1993; Stich and Anders, 1989). This may also be a mechanism which the fungi employ to help reduce radical generation in close proximity to the fungal hyphae.

That chelators are produced by degradative fungi is not surprising. Most microorganisms studied to date have been found to produce chelators and in addition they maintain systems for the acquisition of the metals sequestered by the chelators. It is interesting to note that other chelator-producing organisms do not degrade substrates in their environment via production of hydroxyl radicals (Srinivasan *et al.*, 1995). We cannot rule out the possibility that other organisms such as pathogenic fungi and bacteria could employ a mechanism such as the one we have proposed. Such a system is dependent not only on the production of chelators, but on the production of chelators with specific capability to reduce metals. In addition, the microorganisms would have to be capable of producing oxalate or some similar organic acid capable of solubilizing and sequestering insoluble metal oxides, and hydrogen peroxide to permit reaction of the reduced potential metal made available by the chelator. The organism must also have the capability to regulate the pH of its environment to favour both metal reduction and hydroxyl radical production. The chelators produced by *G. trabeum* do not appear to be particularly unique as they have similar structures to chelators produced by other microorganisms. However, this fungus and other brown-rot fungi appear to employ a different strategy for use of the chelators produced, and we hypothesize that they have adapted their metabolism through the production and regulation of oxalate, hydrogen peroxide, and perhaps other metabolites to generate free radicals efficiently, allowing for the degradation of the cellulosic portions of the wood cell wall.

The hypothesis presented provides evidence to support a mechanism for production of hydroxyl radicals in wood undergoing degradation by *G. trabeum*. In the proposed mechanism, although the fungus would be protected from random generation of reactive oxygen species in the environment of the fungal hyphae, hydroxyl radicals would still be randomly produced within the wood cell wall. This may be desirable in some instances, but it is possible that binding of metals specifically to the cellulosic components of the cell wall allows hydroxyl generating reactions to occur in a site specific manner similar to that which has been proposed in the cleavage of nucleic acids in animal systems (Halliwell and Aruoma, 1993). Such a mechanism, in combination with the short half-life of the hydroxyl radical, would help to explain the rapid depolymerization of cellulose within the wood cell wall while lignin derivatives remain relatively protected from degradation. Additional work in our laboratories is underway to explore this aspect of the hypothesis.

Acknowledgements

The authors would like to acknowledge the technical assistance of B. Doyle, Y. Shao, and A.-S. Hansen in portions of this research and would also like to acknowledge current and former students and collaborators including G. Daniel, A. Paszczynski, V. Chandhoke, J. Liu, L. Jun, V. Easwaran and S. Krishnamurthy. Portions of this review have been published in the articles cited in the text of this paper.

References

ANONYMOUS (1990) *Standard Test Method for Intrinsic Viscosity of Cellulose*, ASTM D 1795-90, American Society for Testing and Materials, pp. 310–314.

ARNOW, L. E. (1937) Colorimetric determination of components of 3,4-dihydroxyphenyalaninetyrosine mixative. *J. Biol. Chem.* **118**, 531–537.

BLANCHETTE, R. A. (1996) Cell wall alterations in loblolly pine wood decayed by the white-rot fungus *Ceriporiopsis subvermispora*. *J. Biotech.* (in press).

CARBERRY, J. B. and SANG, H. L. (1994) Enhancement of pentachlorophenol biodegradation by Fenton's reagent partial oxidation. In: Tedder, D. W. and Pohland, F. G., eds, *Emerging Technologies in Hazardous Waste Management*, Vol. IV, ACS Symp. Ser. 554, pp. 197–223.

CHANDHOKE, V. (1991) Iron-binding compounds produced by the brown-rot fungus *Gloeophyllum trabeum*. PhD Thesis, University of Maine, Orono, Maine, USA.

CHANDHOKE, V., GOODELL, B., JELLISON, J. and FEKETE, F. A. (1992) Oxidation of KTBA by iron-binding compounds produced by the wood-decaying fungus *Gloeophyllum trabeum*. *FEMS Microbiol. Lett.* **90**, 236–266.

CHEN, Y., MILES, A. M. and GRISHAM, M. B. (1995) Pathophysiology and reactive oxygen metabolites. In: Ahmad, S., ed., *Oxidative Stress and Antioxidant Defenses in Biology*, New York: Chapman and Hall, International Thompson Publishing, pp. 62–95.

COTTON, F. A. and WILKINSON, G. (1976) The chemistry of iron. In: Cotton, F. A. and Wilkinson, G., eds, *Basic Inorganic Chemistry*, New York: John Wiley & Sons, pp. 24–29.

COWLING, E. B. (1961) Comparative biochemistry of the decay of sweetgum sapwood by white-rot and brown-rot fungi. Tech. Bull. 1258, Washington, DC: USDA.

COWLING, E. B. (1975) Physical and chemical constraints in the hydrolysis of cellulose and lignocellulose materials. In: Wilke, C. R., ed., *Cellulose as a Chemical and Energy Resource, Symposium 5, Biotechnology and Bioengineering*, New York: John Wiley and Sons.

DANIEL, G., NILSSON, T. and PETTERSON, B. (1989) Intra- and extracellular localization of lignin peroxidase during the degradation of solid wood and wood fragments by *Phanerochaete chrysosporium* by using transmission electron microscopy and immuno-gold labelling. *Appl. Environ. Microbiol.* **55**, 871–881.

DANIEL, G., PETTERSON, B. and VOLC, J. (1990) Use of immunogold cytochemistry to detect Mn(II)-dependent and lignin peroxidase in wood degraded by the white rot fungi *Phanerochaete chrysosporium* and *Lentinula edodes*. *Can. J. Bot.* **68**, 920–933.

EASWARAN, V. (1994) The purification and partial characterization of iron-binding compounds produced by *Gloeophyllum trabeum*. MS thesis, December 1994, University of Maine.

ECKER, D. J., LANCASTER JR, J. R. and EMERY, T. (1982) Siderophore iron transport followed by electron paramagnetic resonance spectroscopy. *J. Biol. Chem.* **257**, 8623–8626.

EMERY, T. (1987) Reductive mechanisms of iron assimilation. In: Winkelmann, G., van der Helm, D. and Neilands, J. B., eds, *Iron Transport in Microbes, Plants and Animals*, New York: BVCH Publishers, pp. 235–250.

EMERY, T. (1980) Malonichrome, a new iron chelate from *Fusarium roseum. Biochim. Biophys. Acta* **629**, 382–390.

ENOKI, A., TANAKA, H. and FUSE, G. (1989) Relationship between degradation of wood and production of H_2O_2-producing one-electron oxidases in brown-rot fungi. *Wood Sci. Technol.* **23**, 1–12.

FEKETE, F. A., CHANDHOKE, V. and JELLISON, J. (1989) Iron-binding compounds produced by wood-decaying basidiomycetes. *Appl. Environ. Microbiol.* **55**, 2720–2722.

FLOURNOY, D. S., KIRK, T. K. and HIGHLEY, T. (1991) Wood decay by brown-rot fungi: changes in pore structure and cell wall volume. *Holzforsch* **45**, 383–388.

GIBBS, C. R. (1976) Characterization and application of ferrozine iron reagent as a ferrous iron indicator. *Anal. Chem.* **48**, 1197–1201.

GOODELL, B. (1989) The potential of biotechnology applications in the forest products industry. In: Schniewind, A., ed., *Advances in Materials Science and Engineering, Encyclopedia of Wood and Wood-Based Materials*, Oxford: Pergamon Press.

GOODELL, B. and JELLISON, J. (1986) *Detection of a Brown-Rot Fungus using Serological Assays*, Document IRG/WP 1305, International Research Group on Wood Preservation Series, Stockholm, Sweden.

GOODELL, B. and JELLISON, J. (1988) Serological detection of wood decay fungi. *For. Prod. J.* **38**, 59–62.

GOODELL, B., LIU, J., JELLISON, J., LU, J., PASZCZYNSKI, A. and FEKETE, F. (1996) Chelation activity and hydroxyl radical production mediated by low molecular weight phenolate compounds isolated from *Gloeophyllum trabeum*. In: Srebotnik, E. and Messner, K., eds, *Biotechnology in the Pulp and Paper Industry*, Vienna: Facultas-Universitätsvewrlag.

GOODELL, B., JELLISON, J., LIU, J., DANIEL, G., PASZCZYNSKI, A., FEKETE, F., KRISHNAMURTHY, S., JUN, L. and XU, G. (1997) Low molecular weight chelators and phenolic compounds isolated from wood decay fungi and their role in the fungal biodegradation of wood. *J. Biotech.* **53**, 133–162.

GREEN, F., LARSEN, M. J., WINANDY, J. E. and HIGHLEY, T. L. (1991) Role of oxalic acid in incipient brown-rot decay. *Mater. Org.* **26** 191–213.

GUERINOT, M. L. (1994) Microbial iron transport. *Ann. Rev. Microbiol.* **48**, 743–772.

HALLIWELL, B. and ARUOMA, O. I. (1993) *DNA and Free Radicals*, West Sussex, England: Ellis Horwood, Simon and Schuster.

HARTWIG, R. C. and LOEPPER, R. H. (1993) Evaluation of soil iron. In: Barton, L. L. and Hemming, B. C., eds, *Iron Chelation in Plants and Soil Microorganisms*, New York: Academic Press, pp. 465–482.

HIGHLEY, T. (1973) Influence of carbon source on cellulase activity of white- and brown-rot fungi. *Wood Fiber* **5**, 50–58.

HIGHLEY, T. (1980) Degradation of cellulose by *Poria placenta* in the presence of compounds that affect hydrogen peroxide. *Mater. Org.* **15**, 81–90.

HIRANO, T., TANAKA, H. and ENOKI, A. (1995) Extracellular substance from the brown-rot fungus *Tyromyces palustris* that reduces molecular oxygen to hydroxyl radicals and ferric iron to ferrous iron. *Mokuzai Gak.* **41**, 334–341.

HÖFTE, M. (1993) Production and characteristics of metal chelators. In: Barton, L. L. and Hemming, B. C., eds, *Iron Chelation in Plants and Soil Microorganisms*, San Diego: Academic Press, pp. 3–27.

HYDE, S. M. and WOOD, P. M. (1995) *A Model for Attack at a Distance from the Hyphae Based on Studies with the Brown Rot* Coniophora puteana, Document IRG/WP 95-10104, International Research Group on Wood Preservation Series, Stockholm, Sweden.

ISHIMARU, C. and LOPER, J. (1993) Biochemical and genetic analysis of siderophores produced by plant-associated *Pseudomonas* and *Erwinia* species. In: Barton, L. L. and Hemming, B., eds, *Iron Chelation in Plants and Soil Microorganisms*, San Diego: Academic Press, pp. 27–73.

JELLISON, J. and GOODELL, B. (1988) Immunological detection of decay in wood. *Wood Sci. Technol.* **22**, 293–297.

JELLISON, J., CHANDHOKE, V., GOODELL, B. and FEKETE, F. (1991a) The isolation and immunology of iron-binding compounds produced by *Gloeophyllum trabeum. Appl. Microbiol. Biotechnol.* **35**, 805–809.

JELLISON, J., CHANDHOKE, V., GOODELL, B., FEKETE, F., HAYASHI, N., ISHIHARA, M. and YAMAMOTO, K. (1991b) *The Action of Siderophores Isolated from* Gloeophyllum trabeum *on the Structure of Crystallinity of Cellulose*, Document IRG/WP 1479, International Research Group on Wood Preservation Series, Stockholm, Sweden.

JELLISON, J., SMITH, K. and SHORTLE, W. (1992) *Cation Analysis of Wood Degraded by White and Brown Rot Fungi*, Document IRG/WP 1552, International Research Group on Wood Preservation Series, Stockholm, Sweden.

JELLISON, J., CONNOLLY, J., SMITH, K. and SHORTLE, W. (1993) *A Comparison of Inductively Coupled Plasma Spectroscopy and Neutron Activation Analysis for the Determination of Cation Concentrations in Wood*, Document IRG/WP 10048-93, International Research Group on Wood Preservation Series, Stockholm, Sweden.

KIM, Y. S., GOODELL, B. and JELLISON, J. (1991) Immuno-electron microscopic localization of extracellular metabolites in spruce wood decayed by brown-rot fungus *Postia placenta. Holzforsch.* **45**, 389–393.

KIM, Y. S., GOODELL, B. and JELLISON, J. (1993) Immunogold labelling of extracellular metabolites from the white-rot fungus *Trametes versicolor. Holzforsch.* **47**, 25–28.

KIRK, T. K. (1975) Effects of a brown-rot fungus *Lenzites trabea* on lignin in spruce wood. *Holzforsch.* **29**, 99–107.

KIRK, T. K., IBACH, R., MOZUCH, M. D., CONNER, A. H. and HIGHLEY, T. (1991) Characteristics of cotton cellulose depolymerized by a brown-rot fungus, by acid, or by chemical oxidants. *Holzforsch.* **45**, 239–244.

KOENIGS, J. W. (1975) Hydrogen peroxide and iron: a microbial cellulolytic system? In: Wilke, C. R., ed., *Cellulose as a Chemical and Energy Resource. Symposium 5, Biotechnology and Bioengineering*, New York: John Wiley & Sons, **5**, 151–159.

KRISHNAMURTHY, U., HIGHLEY, T. L. and BRUCE, A. (1996) The role of siderophore production in the biological control of wood decay fungi by *Trichoderma* spp., *Interna-*

tional Biodeterioration and Biodegradation Symposium.

LODGE, J. S. (1993) Enzymatic reduction of iron in siderophores. In: Barton, L. L. and Hemming, B. C., eds, *Iron Chelation in Plants and Soil Microorganisms*, San Diego: Academic Press, pp. 241–251.

LU, J., GOODELL, B., LIU, J., ENOKI, A., JELLISON, J., TANAKA, H. and FEKETE, F. (1994) *The Role of Oxygen and Oxygen Radicals in the One-electron Oxidative Reactions Mediated by Low Molecular Weight Chelators Isolated from* Gloeophyllum trabeum, Document IRG/WP/1086, International Working Group on Wood Preservation Series, Stockholm, Sweden.

NEILANDS, J. B., KONOPKA, K., SCHWYN, B., COY, M., FRANCIS, R. T., PAW, P. W. and BAGG, A. (1987) Comparative biochemistry of microbial iron assimilation. In: Winkelmann, G., van der Helm, D. and Neilands, J. B., eds, *Iron Transport in Microbes, Plants and Animals*, New York: VCH Publishers, pp. 3–33.

PAGE, W. J. (1993) Growth conditions for the demonstration of siderophores and iron-repressible outer membrane proteins in soil bacteria, with an emphasis on free-living diazotrophs. In: Barton, L. L. and Hemming, B. C., eds, *Iron Chelation in Plants and Soil Microorganisms*, San Diego: Academic Press, pp. 76–110.

PIGNATELLO, J. J. (1993) Degradation of pesticides by ferric reagents and peroxide in the presence of light, United States Patent No. 5 232 484.

PUEYO, C. and ARIZA, R. R. (1993) Role of reactive oxygen species in the mutagenicity of complex mixtures of plant origin. In: Halliwell, B. and Aruoma, O. I., eds, *DNA and Free Radicals*, West Sussex, England: Ellis Horwood, Simon and Schuster, pp. 275–291.

SCHMIDT, C. J., WHITTEN, B. K. and NICHOLAS, D. D. (1981) A proposed role for oxalic acid in nonenzymatic wood decay by brown-rot fungi. *Proc. Am. Wood Preserv. Assoc.* **77**, 157–164.

SEDLAK, D. L. and HOIGNÉ, J. (1993) The role of copper and oxalate in the redox cycling of iron in atmospheric waters. *Atmos. Env.* **27A**, 2173–2185.

SHIMADA, M., AKAMATSU, Y., MA, D. B. and TAKAHASHI, M. (1992) New biochemical aspects of oxalic acid production and decomposition by wood destroying fungi. *5th International Conference on Biotechnology in the Pulp and Paper Industry*, 27–30 May, Kyoto, Japan: UNI Publishers, pp. 273–278.

SØRENSEN, J. (1982) Reduction of ferric iron in anaerobic, marine sediment and interaction with reduction of nitrate and sulfate. *Appl. Environ. Microbiol.* **43**, 319–324.

SREBOTNIK, E. and MESSNER, K. (1991) Immunoelectron microscopical study of the porosity of brown-rot degraded wood. *Holzforsch.* **45**, 95–101.

SRINIVASON, U., HIGHLEY, T. L. and BRUCE, A. (1995) The role of siderophore production in the biological control of wood decay fungi by *Trichoderma* spp. In: Bousher, A., Chandra, M. and Edyvean, R., eds, *Biodeterioration and Biodegradation*, Vol. 9, Rugby: Institute of Chemical Engineering, pp. 226–230.

STICH, H. F. and ANDERS, F. (1989) The involvement of reactive oxygen species in oral cancers of betel quid/tobacco chewers. *Mutat. Res.* **214**, 47–61.

STOOKEY, L. L. (1970) Ferrozine – a new spectrophotometric reagent for iron. *Anal. Chem.* **42**, 779–782.

SULZBERGER, B. and LAUBSCHER, H. (1995) Reactivity of various types of iron (III) (hydr)oxides towards light-induced dissolution. *Mar. Chem.* **50**, 103–115.

VAN DER HELM, D. and WINKELMANN, G. (1994) Hydroxamates and polycarboxylates as iron transport agents (siderophores) in fungi. In: Winkelmann, G. and Winge, D. R., eds, *Metal Ions in Fungi*, New York: Marcel Dekker, pp. 39–98.

WINKELMANN, G. and WINGE, D. R. (1994) *Metal Ions in Fungi*, New York: Marcel Dekker.

WINTERBOURN, C. (1991) Free radical biology of iron. In: Dreosti, I. E., ed., *Trace Elements, Micronutrients, and Free Radicals*, Clifton, NJ: Humana Press.

WOOD, P. M. (1994) Pathways for production of Fenton's reagent by wood-rotting fungi

FEMS Microbiol. Rev. **13**, 313–320.

ZEPP, R. G., FAUST, B. C. and HOIGNE, J. (1992) Hydroxyl radical formation in aqueous reactions (pH 3–8) of iron (II) with hydrogen peroxide: the photo-Fenton reaction. *Environ. Sci. Technol.* **26**, 313–319.

Biological Control of Wood Decay

ALAN BRUCE

16.1 Introduction

Biological control may generally be described as the introduced use of organisms (macro or micro) or their products to keep in check the numbers or activities of particular pest species. While this definition is probably the most commonly accepted one, it is nevertheless rather limited and does not include control strategies which involve manipulation of the substrate, environment, or inherent factors in the pest, all of which are considered as biological control methods by many researchers. Common to all definitions however is that biological control is the manipulation by mankind of factors which are fundamental to natural control within ecosystems. As such therefore, biological control has been portrayed as a much more environmentally acceptable approach when compared with control strategies involving chemicals which by definition involve the artificial introduction of toxic chemicals not naturally found in the particular ecosystem.

Biological control in agricultural systems is now a well established and accepted technology for the control of a wide range of pests including nematodes (Mankau, 1980), weeds (Julien, 1987) insects (Mackauer *et al.*, 1990), as well as microbial plant pathogens (Cook and Baker, 1983; Hornby, 1990). The technology is not as advanced, however, for the biological protection of processed wooden products even though successful systems have been studied or commercially used for the protection of growing trees and stumps against basidiomycete fungi including *Heterobasidium annosum* (Rishbeth, 1963; Ricard, 1970), *Chondostereum purpureum* (Mercer and Kirk, 1984a,b), *Phellinus weirii* (Nelson *et al.*, 1985), *Ophiostoma ulmi* (Atkins and Fairhurst, 1987), *Phellinus tremulae* (Hutchinson *et al.*, 1994) and *Armillaria luteobubalina* (Nelson *et al.*, 1995).

16.2 Particular Requirements for Biological Control of Wood Deterioration

The principle reasons for the development of biological control research in the wood preservation field are the same as those which have driven similar developments in agriculture, i.e. the need to develop more environmentally acceptable treatments at a time when the legislation governing the use of toxic chemicals has become increasingly restrictive. The change in focus towards more environmentally safe preservatives in the UK is nicely illustrated in a recent government report on the use of timber in construction (Anon, 1995) which highlights the need to 'develop improved, environmentally benign, protection strategies to enhance the durability of wood components'.

The strategies developed and experiences gained from biological control systems in agriculture have undoubtedly provided an excellent database for researchers in the wood preservation field. In both situations the objective has been to develop efficient, non-toxic control systems which are cost-effective when compared with the currently used preservatives, and have minimal environmental impact.

Prevention of wood deterioration however provides a number of particular challenges for the biological control researcher, which are not encountered in the agricultural situation.

16.2.1 Diversity of Target

Wood can be attacked by a range of microbial as well as insect and invertebrate deteriogens and the particular organism responsible for the decay will be dependent on either the wood type or the environmental conditions in which the wood is used (Zabel and Morrell, 1992). One of the attractive features of biological control agents in agricultural systems is their target specificity which decreases the likelihood of adverse reactions against other members of the ecosystem. Since most crop species, due to their constitutive and inducible defences, are attacked by a relatively narrow range of pathogens, development of a target specific control system is feasible. Although some wood types contain chemical extractives which will confer resistance against wood decay fungi (Dinwoodie and Desch, 1996), most are non-durable and subject to attack by a wide range of fungi, thereby necessitating a broader spectrum biological control agent. In addition, changing environmental conditions will also influence the range of deteriogens which must be controlled, unlike the situation in agriculture where the host plant is the major determinant of pest type.

16.2.2 Length of Protection

Most agricultural crops only need to be protected from disease during their growing season which in many cases will be less than a year. This means that any control agent need only be effective during this time; reapplication on a regular basis is good business for manufacturers as it results in annual sales of the biological control product. Wood, depending on its intended use, may require to be protected for varying lengths of time ranging from a few months to 40 years field exposure. This has major implications for any biological control system including: formulation considerations, for example whether continued viability of the organism must be maintained; whether any active metabolites are stable and will be permanently retained in the wood; suitability of delivery systems and/or need for reapplication; and continued tolerance by the control agent of fluctuating biotic and abiotic conditions.

16.2.3   *Efficacy of the System*

Wood is often used as a structural material and therefore an essential prerequisite of any treatment system is that it does not compromise the strength of the timber. This limits the range of potential control agents to those organisms which will only use the non-structural materials, i.e. starches and sugars in the ray parenchyma and other storage tissues of the timber, and not themselves attack any of the structural polymers of the timber. Additionally the protection they provide must be 100 per cent effective since even in the early stages of colonization (for example by brown-rot fungi) significant strength losses can occur due to depolymerization of the cellulose before any substantial weight losses are generated. This is different from agricultural systems where the success of the control agent is measured against the increase or decrease in crop yield compared with the use of chemical biocides. It may be justified to use a biological control agent which is marginally less efficient than its chemical counterpart if its use is associated with cost savings or is environmentally more acceptable than the chemical which it replaces.

16.2.4   *Formulation and Delivery System Technology*

As highlighted above, biological control agents for timber applications may need to be effective over long time periods without damaging the structure of the wood. This may necessitate the development of formulations with sufficient extraneous nutrients to allow survival and growth in a wooden substrate which itself is devoid of nutrients, particularly nitrogen. Similarly isolates must be tolerant of the chemical extractives present in wood (Philp *et al.*, 1995).

Method and timing of application is generally a critical determinant of the success of any biological control system and in agricultural systems most control agents are delivered as seed dressings, spray inoculation or by drenching, i.e. in a similar manner to traditional chemical biocides. Wood preservatives however are often applied by pressure processes under high temperature conditions. Treatment technology and development of suitable delivery systems therefore provide an additional challenge to biological control researchers. Biological agents do however, unlike chemicals, have the ability to colonize a substrate by their growth and this may, in suitable systems, be used to advantage. While ingenious surface application systems have been efficient in allowing control agents to colonize freshly felled timber (Schoeman *et al.*, 1994b) it may be necessary to inoculate finished wooden structures by alternative methods. Bruce and King (1986a) applied a commercial pelleted product into boreholes to colonize the groundline region of creosoted distribution poles and recent unpublished work undertaken at the Scottish Institute for Wood Technology has shown that *Trichoderma* conidiospores can be successfully applied to wooden stakes in a pressure treatment cylinder. Pressure operating conditions were those which would normally be used for the chemical treatment of spruce and pine lumber.

16.3   Target Areas for Biological Control in Wood

Wood is a variable substrate which is used for a variety of purposes such as building construction, composites manufacture and pulp and paper products, and a variety of finished materials including furniture, posts, pilings and distribution poles. As such it is often attacked by a wide range of deteriogens at any one time and therefore it is unlikely that a single biological control system will be suitable for all situations just as no one chemical preservative is useful for all applications. As in agriculture, most biological control treatments are likely to be applied as prophylactic treatments to prevent colonization and subsequent decay by the deteriogens. In some situations however, for example decay in buildings, it may be essential to apply a remedial treatment against established decay. With this in mind Freitag *et al.* (1991) suggested the terms 'bioprotection' and 'biocontrol' to differentiate between prophylactic and remedial wood treatment systems, respectively.

Freitag *et al.* (1991) and Bruce (1992) have reviewed the use of biological control for wood protection against microorganisms and particularly fungi, but as far as the author is aware, no biological control systems have yet been reported for the control of wood deterioration by insects or marine borers. Bacteria, though major contributors to decay of waterlogged wood, are much less important than fungi as agents of wood degradation (Eaton and Hale, 1993) and as such do not represent a significant target for biological control systems. Most research to date has therefore been targeted against either sapstain/bluestain or decay fungi.

16.3.1 Sapstain Biocontrol

Moulds and staining fungi (including bluestain) cause surface growth and deep sapwood discoloration, respectively, and are among the primary colonizers of freshly felled timber prior to drying (Zabel and Morrel, 1992; Eaton and Hale, 1993). While neither category causes any significant structural damage (they use only non-structural materials in the wood) and only minimal strength losses (Blanchette *et al.*, 1992), they are nevertheless significant biodeteriogens; due to their pigmented mycelia, the aesthetic value of the timber is lost and its usefulness for pulp and paper manufacture significantly reduced. Bluestain fungi can discolour timber shortly after felling or in service; in the UK however this term is more commonly used to describe colonization and growth of these pigmented fungi in finished wooden structures such as windows and doors.

As primary colonizers of freshly felled lumber, sapstain fungi are an ideal target for biological control systems. Lack of previous colonizers means that bioprotectant agents can become readily established in the substrate and generally the wood only requires to be protected for a reasonably short time during storage prior to the wood being seasoned. A major limitation however is that the control agent itself must not add any coloration to the wood. This has led to a number of researchers examining the use of non-pigmented or hyaline strains of selected fungal isolates. Horvath *et al.* (1995) have reported the development of mutant strains of *Trichoderma* with reduced sporulation and production of non-pigmented spores for general biocontrol uses, while Behrendt *et al.* (1995a,b) and Croan (1996) have examined the use of non-pigmented strains of *Ophiostoma piliferum* for control of staining in chip samples.

Some of the earlier studies on sapstain biocontrol examined bacteria and culture filtrate from species including *Bacillus subtilis* (Bernier *et al.*, 1986; Siefert *et al.*, 1987; Florence and Sharma, 1990) and *Pseudomonas cepacia* (Benko 1988, 1989). Various authors have continued to examine bacteria (Benko and Highley, 1990a,b;

Croan and Highley, 1991a, 1992, 1994; Highley *et al.*, 1991; Kreber and Morrell, 1993; Morrell and Sexton 1993; Morrell and Velicheti, 1995; Jin and Morrell, 1996). Other workers however have concentrated on the use of fungal species as bio-protectants. As early as 1973, Klingstrom and Johansson reported the activity of various *Scytalidium* spp. against a wood stain fungus, *Leptographium lundbergii*, on agar and in wood, while Stranks (1976) noted that fungal produced antibiotics such as scytalidin, hyalodendrin and cryptosporiopsin were capable of inhibiting blue-stain in pine sapwood. Since these early studies many authors have reported the use of various other fungal antagonists of bluestain including *Trichoderma* spp. (Schoeman *et al.*, 1994b; Jin and Morrell, 1996); *Ophiostoma piliferum* (Blanchette *et al.*, 1992; Behrendt *et al.*, 1995a,b; Croan, 1996); *Talaromyces flavus* (Croan and Highley, 1991b, 1993); *Stachybotrys cylindrospora* (Hiratsuka *et al.*, 1994); *Lecythophora hoffmannii* (Chakravarty and Hiratsuka, 1994); *Sporomiella similis* (Chakravarty *et al.*, 1994); basidiomycetes and/or their products (Benko and Henningson, 1986; Croan and Highley, 1991b,c, 1993); ranges of fungi (Siefert *et al.*, 1988; Kreber and Morrell, 1993) and yeasts (Walker *et al.*, 1995).

In most of the above cases the intention is to use the biological agent as a prophylactic treatment to prevent stain. Croan and Highley (1996) however have recently examined the use of fungal metabolites from *Bjerkandera adusta* and *Talaromyces flavus* to de-stain existing sapstained wood (i.e. as a remedial treatment).

An alternative strategy to control sapstain development is to target specific key enzymes of sapstain fungi. Breuil *et al.* (1995) identified and characterized (Abraham and Breuil, 1996) a specific subtilisin-like serine proteinase enzyme produced by the staining fungus *Ophiostoma piceae*. Characterization of the cleavage specificity of this enzyme (Abraham *et al.*, 1995a), and knowledge of the factors likely to affect the autolysis of such enzymes (Abraham and Breuil, 1995) may lead to more specific control systems for sapstain fungi (Abraham *et al.*, 1995b).

16.3.2 Decay Biocontrol

The protection of wood from decay fungi presents a greater challenge to biological control than that of sapstain biocontrol. While prior colonization and removal of available soluble nutrients may help biocontrol agents to prevent sapstain, basidiomycete fungi are also able to utilize the structural elements of the wood and therefore any control agent cannot be expected successfully to exclude decay fungi solely on the basis of competition for nutrients.

Many authors have reported the antagonism of various wood decay fungi by biological control agents in cultural studies using either agar or wood block test methods (Klingstrom and Beyer, 1965; Toole, 1971; Hulme and Shields, 1972a; Klingstrom and Johansson, 1973; Morton and Eggins, 1976; Bruce and King, 1983; Morris *et al.*, 1986; Gramss, 1987; Bettucci *et al.*, 1988; Highley and Ricard, 1988; Murmanis *et al.*, 1988a; Giron and Morrell, 1989; Highley, 1989; Benko and Highley, 1990c; Dawson-Andoh and Morrell, 1990; Freitag and Morrell, 1990; Morrell and Sexton, 1990; Doi and Yamada, 1991; Srinivasan *et al.*, 1993; Ejechi and Obuekwe, 1994, 1996; Highley, 1994; Schoeman *et al.*, 1994c; Score and Palfreyman, 1994; Palfreyman *et al.*, 1995; Tucker and Bruce, 1995; Highley *et al.*, 1996; Rattray *et al.*, 1996; Tucker *et al.*, 1996).

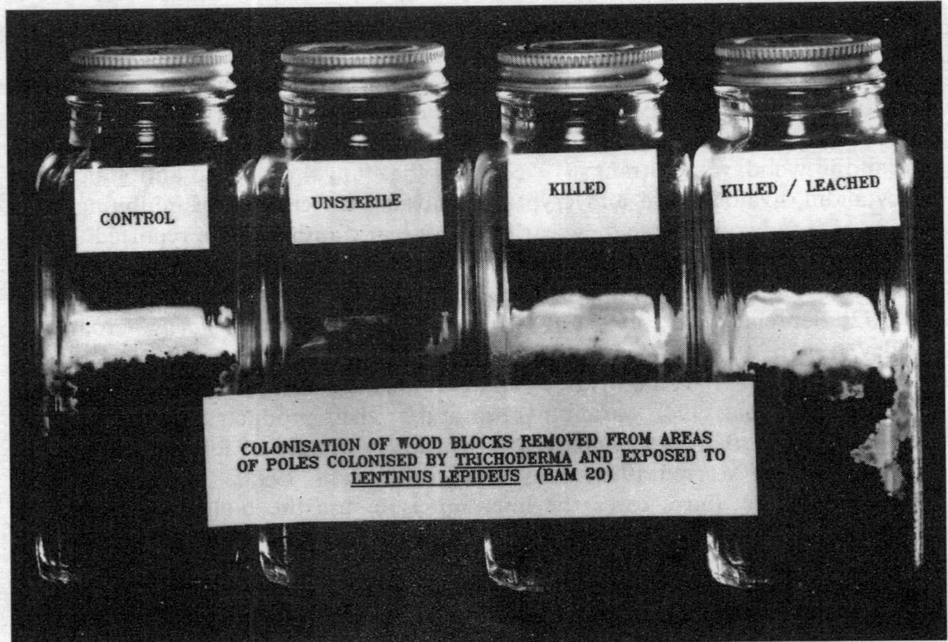

Figure 16.1 Protection of wood blocks from attack by *Lentinus lepideus* in AWPA standard soil block test (AWPA 1413, 1977). 'Control' represents unsterile wood from internal regions of creosoted poles uninoculated with *Trichoderma* (30.5 per cent weight loss after 12 weeks). 'Unsterile' is wood from regions of pole interiors known to be colonized by *Trichoderma* (1.3 per cent weight loss). 'Killed' is the same as 'Unsterile' but after steam sterilization (29.5 per cent weight loss). 'Killed/leached' is the same as 'Unsterile' after steam sterilization and leaching (30.1 per cent weight loss), (extracted from Bruce *et al.*, 1991)

Despite this wealth of research effort few workers have reported on field performance of biological control agents. Processed wood is used in many applications and thereby subjected to varying environmental conditions which will determine the range and variety of decay fungi against which the wood must be protected. This represents a challenge for biological control agents which have been tested most often in pure culture systems against single target fungi. It is perhaps not too surprising therefore that where field testing of biological control agents has taken place it has been directed at particular wooden products such as distribution poles (see below) or against specific target organisms such as the dry rot fungus *Serpula lacrymans* (Score, 1996, unpublished data).

The earliest field tests of biocontrol agents for wood decay were carried out on birch logs (Shields and Atwell, 1963; Hulme and Shields, 1972b), while in the late 1960s and early 1970s internal decay in wooden distribution poles became the focus of attention for biological control systems. Ricard and Bollen (1968) and Ricard *et al.* (1969) examined *Scytalidium* and *Trichoderma* respectively to control decay in Douglas fir poles and subsequent field work has concentrated on these two fungal genera and particularly *Trichoderma* spp. Further field trials of *Trichoderma* in poles (Ricard, 1975, 1976; Morris *et al.*, 1984; Bruce and King, 1986a,b; Bruce *et al.*, 1990; Morris *et al.*, 1992) continued with variable levels of control reported. Bruce and King (1986b) reported some decrease in the level of decay of poles artificially inoculated with *Lentinus lepideus* when *Trichoderma* spp. were applied either before or after the decay fungus. The level of control was found to be variable and was adversely affected by the level of resident fungi already present in pole interiors. This failure of the *Trichoderma* spp. to colonize throughout the groundline regions of the poles did not improve even after subsequent exposure periods (Bruce *et al.*, 1990). Wood removed from *Trichoderma*-colonized regions of poles 7 years after pole inoculation was however resistant to attack by selected basidiomycete fungi (Bruce *et al.*, 1991) when tested using standard soil block test systems (see Figure 16.1). This result is significant since it indicates that provided control agents can properly colonize throughout a wooden structure, they can biologically protect the material even after extended time periods.

Other authors have considered that *Trichoderma* spp. may be better suited to provide short-term pre-seasoning bioprotection of freshly felled logs from decay (Schoeman and Dickinson, 1992, 1993a) and successfully applied *Trichoderma* spores to the wood by incorporating them in the chain-saw oil (Schoeman *et al.*, 1995). Schoeman *et al.* (1994a) also examined the use of a *Trichoderma viride* isolate in conjunction with aqueous disodium octoborate and reported good control using this integrated control system.

Scytalidium and *Trichoderma* spp. have received the greatest attention as control agents in field trials since *Scytalidium* had previously been shown to produce diffusible antibiotics with antifungal activity (Strunz *et al.*, 1972; Klingstrom and Johansson, 1973; Overeem and Mackor, 1973) whereas *Trichoderma* isolates have long been recognized for their antagonistic traits (Dennis and Webster, 1971a,b,c). *Trichoderma* spp. are also generally tolerant of other wood preservatives (Wallace *et al.*, 1992) which makes them attractive for use in integrated control strategies with chemicals.

16.4 Future Considerations

The future development of biological control systems for wood protection or treatment will ultimately depend on how they measure up against traditional chemical preservatives. New biological systems must perform as well as chemical preservatives under field conditions often over extended periods of service of the treated wood product; be competitive in terms of product cost; be easy to apply, store and handle; and satisfy the same level of stringent testing and regulatory control which is required during the development of any new chemical wood preservative. Only when a biocontrol agent has satisfied all the above will the wood preservative industries happily embrace the technology.

16.4.1 Test Methodologies

Many different approaches have been developed to screen potential biocontrol agents. While agar systems have most commonly been used as a primary screening method various authors have adapted their screening systems to be more reflective of wood as a substrate. Freitag and Morrell (1990) developed a wood sandwich method and Schoeman *et al.* (1994c) combined pine discs and agar media to produce a bilayer assay for screening purposes and also introduced a computer assisted system to aid assessment during screening of large numbers of isolates (Schoeman and Dickinson, 1993b). Tucker and Bruce (1995) developed an agar medium containing the carbon:nitrogen balance and major amino acids composition present in Scots pine sapwood and found that isolates selected using this medium gave excellent protection from decay when subsequently assessed in wood (Tucker *et al.*, 1996).

Ultimately however, a control agent must be tested in solid wood. It would seem reasonable to attempt, if appropriate, to use standard test methods which are currently used for this purpose to evaluate chemical preservatives. Tucker *et al.* (1997) showed that standard American and European test methods designed to establish the concentrations of chemicals required to protect wood could also be successfully used, with only slight modification, to evaluate the efficacy of biocontrol agents for wood protection. Another important consideration that has been addressed by some researchers is that of isolate identity and subsequent product quality assurance. Since the efficiency of most biological control agents is strain dependent it is essential that the end user has surety with the product. Schlick *et al.* (1994a, 1994b) showed that DNA and PCR (polymerase chain reaction) fingerprinting methods could be used to identify individual strains of *Trichoderma* for biocontrol. In addition to improving consumer confidence, development of specific identification processes also allows product protection for the manufacturer through patenting of identifiable isolates. Further work is however required to provide field performance data and to develop suitable delivery systems before control agents can be accepted as viable alternatives to chemical preservatives.

16.4.2 *Modes of Action*

Further research is also required to establish the specific mechanisms of action of biocontrol agents against wood decay fungi. Many antagonistic mechanisms have been attributed to the control of plant pathogens by *Trichoderma* spp. and these can be categorized as follows: soluble metabolites (Dennis and Webster, 1971a; Dreyfuss *et al.*, 1976; Fujiwara *et al.*, 1982; Bodo *et al.*, 1985; Taylor, 1986; Claydon *et al.*, 1991; Ghisalberti *et al.*, 1992; Ordentlich *et al.*, 1992; Corley *et al.*, 1994); inhibitory volatiles (Dennis and Webster, 1971b; Claydon *et al.*, 1987); mycoparasitism via lytic enzymes (Chet *et al.*, 1981; Harman *et al.*, 1981; Chet and Elad, 1982; Elad *et al.*, 1982); and production of siderophores (Anke *et al.*, 1991). In contrast fewer studies have been undertaken into the specific mechanisms of action of these and other fungi against wood biodeteriogens.

Srinivasan *et al.* (1992) and Srinivasan (1993), however, examined a range of antagonistic mechanisms of *Trichoderma* against wood decay fungi and assessed how each was influenced by media type. Bruce *et al.* (1995) showed that the presence of the cell walls of wood decay fungi stimulated the production of chitinase and laminarinase by *Trichoderma* isolates and indeed found that the concentration of these enzymes produced was dependent on the species of wood decay fungal cell wall used. Other workers have also examined mycoparasitsm and lytic enzymes as a mechanism of biocontrol of wood decay fungi (Hutterman and Cwielong, 1982; Murmanis *et al.*, 1988a; Liu and Morrell, 1996).

The inhibitory effect of *Trichoderma* volatiles against wood decay fungi was first reported by Bruce *et al.* (1984) and recent work has identified some of the volatile organic compounds (VOCs) produced by these organisms (Bruce *et al.*, 1996). As with most other antagonistic traits the range of VOCs produced is isolate specific and is dependent on the growing conditions. Recent work has, however, attempted to identify those VOCs which may be responsible for the volatile inhibition of the wood decay fungi (Wheatley *et al.*, 1997).

Bruce and Highley (1991) showed that the culture filtrate from *Trichoderma* spp. could significantly reduce the growth of a wide range of wood decay fungi and that white-rot fungi were affected to a lesser extent than brown-rot organisms. Canessa and Morrell (1996) reported that a *T. harzianum* isolate induced increased laccase production in *Trametes versicolor*; this may account for the greater selective action of *Trichoderma* isolates against brown-rots. Other workers have reported the effect of metabolite production by biocontrol agents on various wood decay fungi (Croan and Highley, 1991a; Ananthapadmanabha *et al.*, 1992; Sexton and Morrell, 1992; Burgel *et al.*, 1994; Doi and Mori, 1994). Srinivasan *et al.* (1993) showed that germination of basidiospores could also be inhibited by soluble metabolites from *Trichoderma* spp. Little work has been undertaken however to identify the metabolites responsible although Horvath *et al.* (1995) concluded that Trichorzianines, associated with the process of conidiogenesis, were responsible for the inhibition of wood decay fungi by a *T. harzianum* isolate.

It is known that iron plays an important role in the biological degradation of wood both as an essential component of the extracellular haem enzymes involved in white-rot decay (Paszczynski *et al.*, 1988) and possibly in brown-rot organisms during non-enzymic catalysis of cellulose degradation (Murmanis *et al.*, 1988b). Competition for iron between wood decay fungi and biocontrol agents is therefore likely to be a significant mechanism to prevent wood decay. Srinivasan *et al.* (1995) showed that *Trichoderma* spp. produced both hydroxymate and phenolate siderophores and that these are implicated in the biological control of wood decay fungi in agar test systems.

It is clear from the above that, although some interesting and exciting work is currently being undertaken into the factors which determine the outcome of fungal interactions, more work needs to be undertaken in this area if biological control systems are to compete with chemical preservatives as bioprotectants or biocontrol agents of wood decay.

References

ABRAHAM, L. and BREUIL, C. (1995) Factors affecting autolysis of a subtilisin-like serine protease secreted by *Ophiostoma piceae* and identification of the cleavage site. *Biochim. Biophys. Acta* **1245**, 76–84.

ABRAHAM, L. and BREUIL, C. (1996) Isolation and characterisation of a subtilisin-like serine protease secreted by the sap-staining fungus *Ophiostoma piceae*. *Enzyme Microb. Technol.* **18**, 133–140.

ABRAHAM, L., CHOW, D. T. and BREUIL, C. (1995a) Characterization of the cleavage specificity of a subtilisin-like serine protease from *Ophiostoma piceae* by liquid-chromatography, mass-spectroscopy and tandem MS. FEBS *Lett.* **374**, 208–210.

ABRAHAM, L., BRADSHAW, D. E., BYRNE, A., MORRIS, P.I. and BREUIL, C. (1995b) *Targetting Fungal Proteases to Prevent Sapstain on Wood*, Document No. IRG/WP/95-10097, International Research Group on Wood Preservation.

ANANTHAPADMANABHA, H. S., NAGAVENI, H. C. and SRINIVASAN, V. V. (1992) *Control of Wood Biodeterioration by Fungal Metabolites*, Document No. IRG/WP/1527-92, International Research Group on Wood Preseration.

ANKE, H., KINN, J., BERQUIST, K. E. and STERNER, O. (1991) Production of siderophores by strains of the genus *Trichoderma*. Isolation and characterisation of the new lipophilic coprogen derivative palmitoyl coprogen. *Biometals* **4**, 176–180.

ANON 1995 *Timber 2005: A Research and Innovation Strategy for Timber in Construction*, Department of Environment, UK.

ATKINS, P. D. and FAIRHURST, C. P. (1987) Biological control of Dutch Elm disease (DED) in northern England, 2nd International Workshop on *Trichoderma* and *Gliocladium*, University of Salford, UK.

BEHRENDT, C. J., BLANCHETTE, R. A. and FARRELL, R. L. (1995a) An integrated approach, using biological and chemical control, to prevent blue stain in pine logs. *Can. J. Bot.* **73**, 613–619.

BEHRENDT, C. J., BLANCHETTE, R. A. and FARRELL, R. L. (1995b) Biological control of blue-stain in wood. *Phytopathology* **85**, 92–97.

BENKO, R. (1988) *Bacteria as Possible Organisms for Biological Control of Blue Stain*, Document No. IRG/WP/ 1339, International Research Group on Wood Preservation.

BENKO, R. (1989) *Biological Control of Blue Stain on Wood with* Pseudomonas cepacia *6253: Laboratory and Field Test*, Document No. IRG/WP/1380, International Research Group on Wood Preservation.

BENKO, R. and HENNINGSON, B. (1986) *Mycoparasitism of some White Rot Fungi on Blue Stain Fungi in Culture*, Document No. IRG/WP/1339, International Research Group on Wood Preservation.

BENKO, R. and HIGHLEY, T. L. (1990a) Selection of media on screening interaction of wood attacking fungi and antagonistic bacteria. I. Interaction on agar. *Mat. u. Org.* **25**, 161–171.

BENKO, R. and HIGHLEY, T. L. (1990b) Selection of media on screening interaction of wood attacking fungi and antagonistic bacteria. II. Interaction on wood. *Mat. u. Org.* **25**, 173–180.

BENKO, R. and HIGHLEY, T. L. (1990c) *Evaluation of Bacteria for Biological Control of Wood Decay*, Document No. IRG/WP/1426, International Research Group on Wood Preservation.

BERNIER, Jr, R., DESROCHER, M. and JURASEK, L. (1986) Antagonistic effects between *Bacillus subtilis* and wood staining fungi. *J. Inst. Wood Sci.* **10**, 214–216.

BETTUCCI, L., LUPOS, S. and SILVAS, S. (1988) Growth control of wood rotting fungi by nonvolatile metabolites from *Trichoderma* spp. and *Gliocladium virens*. *Cryptogamie Mycol.* **9**, 157–165.

BLANCHETTE, R. A., FARRELL, R. L. and BURNES, T. A. (1992) Biological control of pitch in pulp and paper production by *Ophiostoma piliferum*. *Tappi* **75**, 102–106.

BODO, B., REBUFFAT, S., HAJJI, S. and DAVOUS, D. (1985) Structure of Trichorazianine A IIIc, an antifungal peptide from *Trichoderma harzianum*. *J. Am. Chem. Soc.* **107**, 6011–6017.

BREUIL, C., YAGODNIK, C. and ABRAHAM, L. (1995) Staining fungi growing in softwood produce proteinases and aminopeptidases. *Mat. u. Org.* **29**, 15–25.

BRUCE, A. (1992) *Biological Control of Wood Decay*, Document No. IRG/WP/1531-92, International Research Group on Wood Preservation.

BRUCE, A. and HIGHLEY, T. L. (1991) Control of growth of wood decay basidiomycetes by *Trichoderma* spp. and other potentially antagonistic fungi. *For. Prod. J.* **41**, 63–67.

BRUCE, A. and KING, B. (1983) Biological control of wood decay by *Lentinus lepideus* (Fr.) produced by *Scytalidium* and *Trichoderma* residues. *Mat. u. Org.* **18**, 171–181.

BRUCE, A. and KING, B. (1986a) Biological control of decay in creosoted distribution poles: I. Establishment of immunizing commensal fungi in poles. *Mat. u. Org.* **21**, 1–13.

BRUCE, A. and KING, B. (1986b) Biological control of decay in creosoted distribution poles: II. Control of decay in poles by immunizing commensal fungi. *Mat. u. Org.* **21** 165–179.

BRUCE, A., AUSTIN, W. J. and KING, B. (1984) Control of growth of *Lentinus lepideus* by volatiles from *Trichoderma. Trans. Br. Mycol. Soc.* **82**, 423–428.

BRUCE, A., FAIRNINGTON, A. and KING, B. (1990) Biological control of decay in creosoted distribution poles: III. Control of decay in poles by immunizing commensal fungi after extended incubation periods. *Mat. u. Org.* **25**, 15–28.

BRUCE, A., KING, B. and HIGHLEY, T. L. (1991) Decay resistance of wood removed from poles biologically treated with *Trichoderma. Holzforschung* **45**, 307–311.

BRUCE, A., SRINIVASAN, U., STAINES, H. J. and HIGHLEY, T. L. (1995) Chitinase and laminarinase production in liquid culture by *Trichoderma* spp. and their role in the biocontrol of wood decay fungi. *Intern. Biodet. Biodeg.* **35**, 337–353.

BRUCE, A., KUNDZEWICZ, A. and WHEATLEY, R. (1996) Influence of culture age on the volatile organic compounds produced by *Trichoderma aureoviride* and associated inhibitory effects on selected wood decay fungi. *Mat. u. Org.*, **30**, 79–94.

BURGEL, J., HORVATH, E. M., HASCHKA, J. and MESSNER, K. (1994) *Biological Control with* Trichoderma harzianum *in Relation to the Formation of Spores and the Production of Soluble Metabolites,* Document No. IRG/WP/94-10073, International Research Group on Wood Preservation.

CANESSA, E. A. and MORRELL, J. J. (1996) *Effect of* Trichoderma harzianum *on the Induction of Laccase by* Trametes versicolor *on Ponderosa Pine Sapwood,* Document No. IRG/WP/96-10177, International Research Group on Wood Preservation.

CHAKRAVARTY, P. and HIRATSUKA, Y. (1994) Evaluation of *Lecythophora hoffmannii* as a potential biological control agent against a blue stain fungus on *Populus tremuloides. J. Plant Dis. Protect.* **101**, 74–79.

CHAKRAVARTY, P., TRIFONOV, L., HUTCHINSON, L. J., HIRATSUKS, Y and AYER, W. A. (1994) Role of *Sporomiella similis*, as a potential bioprotectant of *Populus tremuloides* wood against the blue stain fungus *Ophiostoma piliferum. Can. J. For. Res.* **24**, 2235–2239.

CHET, I. and ELAD, Y. (1982) Prevention of plant infection by biological means. In: *La Selection des Plantes*, Bordeaux (France), Colloq. l'INRA, **11**, 195–204.

CHET, I., HARMAN, G. E. and BAKER, R. (1981) *Trichoderma hamatum*, its hyphal interaction with *Rhizoctonia solani* and *Phythium* spp., *Microb. Ecol.* **7**, 29–38.

CLAYDON, N., ALLEN, M., HANSON, J. R. and AVENT, A. G. (1987) Antifungal alkyl pyrones of *Trichoderma harzianum. Trans. Br. Mycol. Soc.* **88**, 503–513.

CLAYDON, N., HANSON, J. R., TRUNEH, A. and AVENT, A. G. (1991) Harziolide, a butenolide metabolite from cultures of *Trichoderma harzianum. Phytochemistry* **30**, 3802–3803.

COOK, R. J. and BAKER, K. F. (1983) *The Nature and Practice of Biological Control of Plant Pathogens*, St Paul, MN: American Phytopathological Society.

CORLEY, D. G., MILLER-WIDEMANN, M. and DURLEY, R.C. (1994) Isolation and structure of harzianum A: a new trichothecene from *Trichoderma harzianum. J. Nat. Prod.* **57**, 422–425.

CROAN, S. C. (1996) *Biological Control of Sapstain Fungi in Wood,* Document No. IRG/WP/ 96-10158, International Research Group on Wood Preservation.

CROAN, S. C. and HIGHLEY, T. L. (1991a) *Antifungal Activity in Metabolites from* Streptomyces rimosus, Document No. IRG/WP/1440, International Research Group on Wood Preservation.

CROAN, S. C. and HIGHLEY, T. L. (1991b) Biological control of the blue stain fungus *Ceratocystis coerulescens* with fungal antagonists. *Mat. u. Org.* **25**, 255–266.

CROAN, S. C. and HIGHLEY, T. L. (1991c) *Control of Sapwood Inhabiting Fungi by Fractionated Extracellular Metabolites from* Coniophora puteana, Document No. IRG/WP/

1494, International Research Group on Wood Preservation.

CROAN, S. C. and HIGHLEY, T. L. (1992) *Biological Control of Sapwood-inhabiting Fungi by Living Bacterial Cells of* Streptomyces rimosus *as a Bioprotectant*, Document No. IRG/WP/1564-92, International Research Group on Wood Preservation.

CROAN, S. C. and HIGHLEY, T. L. (1993) *Controlling the Sapstain Fungus*, Ceratocystis coerulescens *by Metabolites Obtained from* Bjerkandera adusta *and* Talaromyces flavus, Document No. IRG/WP/93-10024, International Research Group on Wood Preservation.

CROAN, S. C. and HIGHLEY, T. L. (1994) Biological control of sapwood-inhabiting fungi by metabolites from *Streptomyces rimosus*, In: Llewellyn, G. C. *et al.*, eds, *Biodeterioration Research 4*, New York: Plenum Press.

CROAN, S. C. and HIGHLEY, T. L. (1996) Fungal removal of wood sapstain caused by *Ceratocystis coerulescens. Mat. u. Org.* **30**, 45–56.

DAWSON-ANDOH, B. and MORRELL, J. J. (1990) *Effects of Chemical Pretreatment of Douglas-fir Heartwood: Efficacy of Potential Bioprotection Agents*, Document No. IRG/WP/1440, International Research Group on Wood Preservation.

DENNIS, C. and WEBSTER, J. (1971a) Antagonistic properties of species groups of *Trichoderma*, I. Production of non-volatile antibiotics. *Trans. Br. Mycol. Soc.* **57**, 25–39.

DENNIS, C. and WEBSTER, J. (1971b) Antagonistic properties of species groups of *Trichoderma*, II. Production of volatile antibiotics. *Trans. Br. Mycol. Soc.* **57**, 41–48.

DENNIS, C. and WEBSTER, J. (1971c) Antagonistic properties of species groups of *Trichoderma*, III. Hyphal interaction. *Trans. Br. Mycol. Soc.* **57**, 363–369.

DINWOODIE, J. M. and DESCH, H. E. (1996) *Timber: Its Structure, Properties and Utilization*, 7th edition, London: Macmillan.

DOI, S. and MORI, M. (1994) *Antifungal Properties of Metabolites Produced by* Trichoderma *Isolates from Sawdust Media of Edible Fungi against Wood Decay Fungi*, Document No. IRG/WP/94-10051, International Research Group on Wood Preservation.

DOI, S. and YAMADA, A. (1991) *Antagonistic Effect of* Trichoderma *spp. against* Serpula lacrymans *in the Soil Treatment Test*, Document No. IRG/WP/1473, International Research Group on Wood Preservation.

DREYFUSS, M., HARRI, E., HOFMANN, H., KOBEL, H., PACHE, W. and TSCHERTER, H. (1976) Cyclosporin A and C, new metabolites from *Trichoderma polysporum* (Link ex Pers.) Rifai. *Eur. J. Appl. Microbiol.* **3**, 125–133.

EATON, R. A. and HALE, M. D. C. (1993) *Wood Decay, Pests and Protection*, London: Chapman & Hall.

EJECHI, B. O. and OBUEKWE, C. O. (1994) The effect of mixed cultures of some microfungi and basidiomycetes on the decay of three tropical timbers. *Intern. Biodet. Biodeg.* **33**, 173–185.

EJECHI, B. O. and OBUEKWE, C. O. (1996) The influence of pH on biodeterioration of a tropical timber exposed to mixed cultures of biocontrol and wood rotting fungi. *J. Phytopathol.* **144**, 119–123.

ELAD, Y., CHET, I. and HENIS, Y. (1982) Degradation of plant pathogenic fungi by *T. harzianum. Can. J. Microbiol.* **28**, 719–725.

FLORENCE, E. J. M. and SHARMA, J. K. (1990) *Botryodiplodia theobromae* associated with blue staining in commercially important timbers of Kerala and its possible biological control. *Mat. u. Org.* **25**, 193–199.

FREITAG, M. and MORRELL, J. J. (1990) Wood sandwich tests of potential biological control agents for basidiomycetous decay fungi. *Mat. u. Org.* **25**, 63–70.

FREITAG, M., MORRELL, J. J. and BRUCE, A. (1991) Biological protection of wood: status and prospects. *Biodeter. Abstr.* **5**, 1–13.

FUJIWARA, A., OKUDU, T., MASUDA, S., SHIOMI, S., MIYAMOTO, C., SEKINE, Y., TAZOE, M. and FUJIWARA, M. (1982) Fermentation, isolation and characterisation of isonitrile antibiotics. *Agric. Biol. Chem.* **46**, 1803–1809.

GHISALBERTI, E. L., HACKLESS, D. C. R., ROWLAND, C. and WHITE, A. H. (1992) Harziandione, a new class of diterpene from *Trichoderma harzianum. J. Nat. Prod.* **55**, 1690–1694.

GIRON, M. Y. and MORRELL, J. J. (1989) Interactions between microfungi isolated from fumigant treated Douglas-fir heartwood and *Poria placenta* or *Poria carbonica. Mat. u. Org.* **24**, 39–49.

GRAMSS, G. (1987) The colonisation of timber by wood-decay fungi as a dynamic interaction with the microbial wood substrate contaminants. *Mat. u. Org.* **22**, 271–287.

HARMAN, G. E., CHET, I. and BAKER, R. (1981) Factors affecting *T. hamatum* applied to seeds as a biocontrol agent. *Phytopathology* **71**, 569–572.

HIGHLEY, T. L. (1989) *Antagonism of* Scytalidium lignicola *against Wood Decay Fungi*, Document No. IRG/WP/1392, International Research Group on Wood Preservation.

HIGHLEY, T. L. (1994) *Effect of* Scytalidium lignicola *on Decay Resistance and Strength of Wood*, Document No. IRG/WP/94-10061, International Research Group on Wood Preservation.

HIGHLEY, T. L. and RICARD, J. (1988) Antagonism of *Trichoderma* spp. and *Gliocladium virens* against wood decay fungi. *Mat. u. Org.* **23**, 157–169.

HIGHLEY, T. L., BENKO, R. and CROAN, S. C. (1991) *Laboratory Studies on Control of Sapstain and Mould on Unseasoned Lumber*, Document No. IRG/WP/1493, International Research Group on Wood Preservation.

HIGHLEY, T. L., ANANTHAPADMANABHA, H. S. and HOWELL, C. R. (1996) *Antagonistic Properties of* Gliocladium virens *against Wood Attacking Fungi*, Document No. IRG/WP/96-10162, International Research Group on Wood Preservation.

HIRATSUKA, Y., CHAKRAVARTY, P., MIAO, S. and AYER, W. A. (1994) Potential for biological protection against blue stain in *Populus tremuloides* with a hyphomycetous fungus, *Stachybotrys cylindrospora, Can. J. For. Res.* **24**, 157–179.

HORNBY, D. (1990) *Biological Control of Soil-borne Plant Pathogens*, UK: CAB International.

HORVATH, E. M., BURGEL, J. L. and MESSNER, K. (1995) The production of soluble antifungal metabolites by the biocontrol fungus *Trichoderma harzianum* in connection with the formation of conidiospores. *Mat. u. Org.* **29**, 1–14.

HULME, M. A. and SHIELDS, J. K. (1972a) Interaction between fungi in wood blocks, *Can. J. Bot.* **50**, 1421–1427.

HULME, M. A. and SHIELDS, J. K. (1972b) Effect of primary fungal infection upon secondary colonisation of birch bolts. *Mat. u. Org.* **7**, 177–184.

HUTCHINSON, L. J., CHAKRAVARTY, P., KAWCHUCK, L. M. and HIRATSUKA, Y. (1994) *Phoma etheridgei* sp. *nova* from black gills and cankers of trembling Aspen (*Populus tremuloides*) and its potential role as a bioprotectant against the Aspen decay pathogen *Phellinus tremulae. Can. J. Bot.* **72**, 1424–1431.

HUTTERMAN, A. and CWIELONG, P. (1982) The cell wall of *Fomes annosus* (*Heterobasidium annosum*) as a target for biological control. I. Direct lysis of the cell wall by lytic enzyme from *Trichoderma harzianum. Eur. J. For. Path.* **12**, 238–245.

JIN, Z. W. and MORRELL, J. J. (1996) Bioprotection of bamboo against fungal mould and stain. *Mat. u. Org.* **30**, 57–62.

JULIEN, C. (1987) *Biological Control of Weeds: A World Catalogue of Agents and their Target Weeds*, UK: CAB International.

KLINGSTROM, A. E. and BEYER, L. (1965) Two new species of *Scytalidium*, with antagonistic properties to *Fomes annosus* (Fr.) Cke. *Svensk Botanik Tidskrift.* **59**, 30–36.

KLINGSTROM, A. E. and JOHANSSON, S. M. (1973) Antagonism of *Scytalidium* isolates against wood decay fungi. *Phytopathology*, **63**, 473–479.

KREBER, B. and MORRELL, J. J. (1993) Ability of selected bacterial and fungal bioprotectants to limit stain in ponderosa pine sapwood. *Wood Fiber Sci.* **25**, 23–34.

LIU, J. and MORRELL, J. J. (1996) *The Role of Chitinase in Bioprotectant Activity against Wood Staining Fungi*, Document No. IRG/WP/96-10175, International Research Group on Wood Preservation.

MACKAUER, M., EHLER, L. E. and ROLAND, J. (1990) *Critical Issues in Biological Control*, Intercept, Andover, Hants.

MANKAU, R. (1980) Biological control of nematode pests by natural enemies. *Ann. Rev. Phytopathol.* **18**, 415–440.

MERCER, P. C. and KIRK, S. A. (1984a) Biological treatments for the control of decay in tree wounds. I. Laboratory tests. *Ann. Appl. Biol.* **104**, 211–219.

MERCER, P. C. and KIRK, S. A. (1984b) Biological treatments for the control of decay in tree wounds. II. Field tests. *Ann. Appl. Biol.* **104**, 221–229.

MORRELL, J. J. and SEXTON, C. M. (1990) Evaluation of a biocontrol agent for controlling Basidiomycete attack of Douglas-fir and southern pine. *Wood Fibre Sci.* **22**, 10–21.

MORRELL, J. J. and SEXTON, C. M. (1993) Fungal staining of ponderosa pine sapwood: effects of wood preconditioning and bioprotectants. *Wood Fibre Sci.*, **25**, 322–325.

MORRELL, J. J. and VELICHETI, R. K. (1995) *Effect of* Pseudomonas cepacia *on the Activity of a Mixture of Wood Staining Fungi on Ponderosa Pine Sapwood*, Document No. IRG/WP/95-10107, International Research Group on Wood Preservation.

MORRIS, P. I., DICKINSON, D. J. and LEVY, J. F. (1984) The nature and control of decay in creosoted electricity poles, Rec. BWPA Ann Conv., 42–53.

MORRIS, P. I., SUMMERS, N. A. and DICKINSON, D. J. (1986) *The Leachability and Specificity of the Biological Protection of Timber using* Scytalidium *sp. and* Trichoderma *spp.*, Document No. IRG/WP/1302, International Research Group on Wood Preservation.

MORRIS, P. I., DICKINSON, D. J. and CALVER, B. (1992) *Biological Control of Internal Decay in Scots Pine Poles: A Seven Year Experiment*, Document No. IRG/WP/1529-92, International Research Group on Wood Preservation.

MORTON, L. H. G. and EGGINS, H. O. W. (1976) Studies of interactions between wood inhabiting microfungi. *Mat. u. Org.* **11**, 197–214.

MURMANIS, L. L., HIGHLEY, T. L. and RICARD, J. (1988a) Hyphal interaction of *Trichoderma harzianum* and *Trichoderma polysporum*, with wood decay fungi. *Mat. u. Org.* **23**, 271–279.

MURMANIS, L. L., HIGHLEY, T. L. and PALMER, J. G. (1988b) The action of isolated brown rot cell free culture filtrate, H_2O_2-Fe^{2+}, and the combination of both on wood. *Wood Sci. Technol.* **22**, 59–66.

NELSON, E. E., GOLDFARB, G. and THIES, W. G. (1985) Colonisation of *Phellinus weirii* infested stumps by *Trichoderma viride*. I. Effect of isolate and inoculum base. *Eur. J. For. Path.* **15**, 425–431.

NELSON, E. E., PEARCE, M. H. and MALAJCZUK, N. (1995) Effects of *Trichoderma* spp. and ammonium sulfamate on establishment of *Armillaria luteobubalina* on stumps of *Eucalyptus diversicolor*. *Mycol. Res.* **99**, 957–962.

ORDENTLICH, A., WIESMAN, Z., GOTTLIEB, H. E., COJOCARU, M. and CHET, I. (1992) Inhibitory furanone produced by the biocontrol agent *Trichoderma harzianum*. *Phytochemistry* **31**, 485–486.

OVEREEM, J. C. and MACKOR, A. (1973) Scytalidic acid, a novel compound from *Scytalidium* species. *Recueil* **92**, 349–359.

PALFREYMAN, J. W., WHITE, N. A., BUULTJENS, T. E. J. and GLANCY, H. (1995) The impact of current research on the treatment of infestations by the dry rot fungus *Serpula lacrymans*. *Intern. Biodet. Biodeg.* **35**, 369–395.

PASZCZYNSKI, A., CRAWFORD, R. L. and BLANCHETTE, R. A. (1988) Delignification of wood chips and pulps by using natural and synthetic porphyrin: models of fungal decay. *Appl. Environ. Microbiol.* **54**, 62–68.

PHILP, R. W., BRUCE, A. and MUNRO, A. G. (1995) The effect of water soluble Scots pine (*Pinus sylvestris* L.) and Sitka spruce (*Picea sitchensis* (Bong.) Carr.) heartwood and sapwood extracts on the growth of selected *Trichoderma* species. *Intern. Biodet. Biodeg.* **35**, 355–367.

RATTRAY, P., McGILL, G. and CLARKE, D. D. (1996) *Antagonistic Effects of a Range of Fungi to* Serpula lacrymans, Document No. IRG/WP/10156, International Research Group on Wood Preservation.

RICARD, J. (1970) *Biological Control of* Fomes annosus *in Norway Spruce* (Picea abies) *with Immunizing Commensals*, Studia Forestalia Suecica 84, Stockholm, Sweden.

RICARD, J. (1975) Biological control of decay in Douglas-fir poles: seven years experience. *Eur. J. For. Path.* **5**, 175–177.

RICARD, J. (1976) Biological control of decay in creosote treated poles. *J. Inst. Wood Sci.* **7**, 6–9.

RICARD, J. and BOLLEN, W. B. (1968) Inhibition of *Poria carbonica* by *Scytalidium* sp., an imperfect fungus isolated from Douglas-fir poles. *Can. J. Bot.* **46**, 643–647.

RICARD, J., WILSON, M. M. and BOLLEN, W. B. (1969) Biological control of decay in Douglas-fir poles. *For. Prod. J.* **19**, 41–45.

RISHBETH, J. (1963) Stump protection against *Fomes annosus*. III Inoculation with *Peniophora gigantea. Ann. Appl. Biol.* **52**, 63–77.

SCHLICK, A., KUHLS, K., MEYER, W., LIECKFELD, E., BORNER, T. and MESSNER, K. (1994a) *Application of DNA Fingerprinting Methods to Identify Biocontrol Strains of Fungi Imperfecti*, Document No. IRG/WP/94-10068, International Research Group on Wood Preservation.

SCHLICK, A., KUHLS, K., MEYER, W., LIECKFELDT, E., BORNER, T. and MESSNER, K. (1994b) Fingerprinting reveals gamma-ray induced mutations in fungal DNA: implications for identification of patent strains of *Trichoderma harzianum. Curr. Genet.* **26**, 74–78.

SCHOEMAN, M. and DICKINSON, D. J. (1992) *Shorter-term Biological Control of Wood Decay in Pre-seasoning Pine Roundwood as an Alternative to Chemical Methods*, Document No. IRG/WP/1555-92, International Research Group on Wood Preservation.

SCHOEMAN, M. and DICKINSON, D. J. (1993a) *Field Studies Investigating the Efficacy of Biological Treatments in Preventing Decay of Freshly Felled Pine*, Document No. IRG/WP/93-10022, International Research Group on Wood Preservation.

SCHOEMAN, M. and DICKINSON, D. J. (1993b) *Computer Assisted Ranking of Potential Biocontrol Fungi Based on Data from Laboratory Screening Trials*, Document No. IRG/WP/93-10023, International Resesarch Group on Wood Preservation.

SCHOEMAN, M., DICKINSON, D. J. and WEBBER, J. F. (1994a) *Disodium Octaborate Tetrahydrate, Alone and in Conjunction with a Selected Isolate of* Trichoderma viride, *Reduces Decay of Freshly Felled Pine Independent of the Effect of Weathering*, Document No. IRG/WP/94-10054, International Research Group on Wood Preservation.

SCHOEMAN, M., WEBBER, J. F. and DICKINSON, D. J. (1994b) Chain-saw application of *Trichoderma harzianum* Rifai to reduce fungal deterioration of freshly felled pine logs. *Mat. u. Org.* **28**, 243–250.

SCHOEMAN, M., WEBBER, J. F. and DICKINSON, D. J. (1994c) A rapid method for screening potential biological control agents of wood decay. *Eur. J. For. Path.* **24**, 154–159.

SCHOEMAN, M., WEBBER, J. F. and DICKINSON, D. J. (1995) Efficacy of *Trichoderma* protection applied in chain-saw oil or by direct application in water. *Mat. u. Org.* **29**, 305–309.

SCORE, A. J. and PALFREYMAN, J. W. (1994) Biological control of the dry rot fungus *Serpula lacrymans* by *Trichoderma* species – the effect of complex and synthetic media on interaction and hyphal extension rates. *Intern. Biodet. Biodeg.* **33**, 115–128.

SEXTON, C. M. and MORRELL, J. J. (1992) *Effects of* Trichoderma harzianum *on Enzyme Activity and Oxalic Acid Production of* Gleophyllum trabeum *in Ponderosa Pine Sapwood*

Blocks, Document No. IRG/WP/1550-92, International Research Group on Wood Preservation.

SHIELDS, J. K. and ATWELL, E. A. (1963) Effect of a mold, *Trichoderma viride,* on decay of birch by four storage-rot fungi. *For. Prod. J.* **13**, 262–265.

SIEFERT, K. A., HAMILTON, W. E., BREUIL, C. and BEST, M. (1987) Evaluation of *Bacillus subtilis* C186 as a potential biocontrol of sapstain and mould on unseasoned lumber. *Can. J. Microbiol.* **33**, 1102–1107.

SIEFERT, K. A., BREUIL, C., ROSSINGNOL, L., BEST, M. and SADDLER, J. H. (1988) Screening microorganisms with the potential for biological control of sapstain on unseasoned lumber. *Mat. u. Org.* **23**, 81–95.

SRINIVASAN, U. (1993) A study of the mechanisms of antagonism by the biocontrol fungi *Trichoderma* against wood decay basidiomycetes, PhD thesis, Dundee Institute of Technology.

SRINIVASAN, U., STAINES, H. J. and BRUCE, A. (1992) Influence of media type on antagonistic modes of *Trichoderma* spp. *Mat. u. Org.* **27**, 303–321.

SRINIVASAN, U., HIGHLEY, T. L., CROAN, S. C. and BRUCE, A. (1993) *Antagonistic Effect of* Trichoderma *spp. against Basidiospores,* Document No. IRG/WP/93-10027, International Research Group on Wood Preservation.

SRINIVASAN, U., HIGHLEY, T. L. and BRUCE, A. (1995) The role of siderophore production in the biological control of wood decay fungi by *Trichoderma* spp. In: Bousher, A., Chandra, M. and Edyvean, R., eds, *Biodeterioration and Biodegradation 9*, Rugby: Institute of Chemical Engineering, pp. 226–230.

STRANKS, D. W. (1976) Scytalidin, hyalodendrin, cytosporiopsin – antibiotics for preventing blue stain in white pine sapwood. *Wood Sci.* **9**, 110–112.

STRUNZ, G. M., KAKUSHIMA, M. and STILLWELL, M. A. (1972) Scytalidin: a new fungitoxic metabolite produced by *Scytalidium* sp. *J. Chem. Soc.* **18**, 2280–2283.

TAYLOR, A. (1986) Some aspects of the chemistry and biology of the genus *Hypocrea* and its anamorphs *Trichoderma* and *Gliocladium. Proc. Nova Scotia Inst. Sci.* **36**, 27–58.

TOOLE, E. R. (1971) Interaction of mold and decay fungi on wood in laboratory tests. *Phytopathology* **61**, 124–125.

TUCKER, E. J. B. and BRUCE, A. (1995) *Preliminary Agar Screening Studies as the First Stage in the Development of a Biological Control System against Basidiomycetes in Ground Contact Situations,* Document No. IRG/WP/95-10111, International Research Group on Wood Preservation.

TUCKER, E. J. B., BRUCE, A. and STAINES, H. J. (1996) *Protection of Wood Blocks Treated with* Trichoderma *Isolates Selected on the Basis of Preliminary Agar Screening Studies,* Document No. IRG/WP/96-10154, International Research Group on Wood Preservation.

TUCKER, E. J. B., BRUCE, A. and STAINES, H. J. (1997) Application of modified international wood preservative chemical testing standards for assessment of biocontrol treatments. *Intern. Biodet. Biodeg.* **39**, 189–198.

WALKER, G. M., MCLEOD, A. H. and HODGSON, V. J. (1995) Interactions between killer yeasts and pathogenic fungi. *Microb. Lett.* **127**, 213–222.

WALLACE, R. J., EATON, R. A., CARTER, M. A. and WILLIAMS, G. R. (1992) *The Identification and Preservative Tolerance of Species Aggregates of* Trichoderma *Isolated from Freshly Felled Timber,* Document No. IRG/WP/1553-92, International Research Group on Wood Preservation.

WHEATLEY, R., HACKETT, C., BRUCE, A. and KUNDZEWICZ, A. (1997) Effect of substrate composition on the production and inhibitory activity against wood decay fungi of volatile organic compounds from *Trichoderma* spp. *Intern. Biodet. Biodeg.* **39**, 199–206.

ZABEL, R. A. and MORRELL, J. J. (1992) *Wood Microbiology Decay and its Prevention,* London: Academic Press.

Biological Control of Forest Pests: A Biotechnological Perspective

SUBBA REDDY PALLI AND ARTHUR RETNAKARAN

17.1 Introduction

Since 1962, with the publication of *Silent Spring* by Rachel Carson highlighting the adverse effects of pesticides on birds, there has been renewed interest in biological pest control. Biological control is based on two ecological principles: that one organism can be used to control another organism; and that some of the organisms that can be used to control other organisms have a limited host range. Since this approach is close to nature's own way of controlling an organism, it has found widespread acceptance with the public. Biological control is perceived, for good reasons, as being a highly selective and effective method that is long lasting. At the same time, some organisms have been used for practical or commercial purposes such as using yeast to make bread or beer since ancient times, and this use has been defined as 'biotechnology'. Beginning in the 1970s, biotechnology has taken a dramatically different turn. With the knowledge gained in recombinant DNA technology, it became possible to alter the genetic make-up of an organism by inserting foreign genes, deleting genes, or modifying the functions of genes. The integration of traditional biological control with modern biotechnology has resulted in some exciting new developments, some of which will be outlined in this chapter.

Modern biotechnology is an immensely powerful tool and has found applications in almost all walks of life, promising improved agricultural and forest products, health care, and crime detection, to mention a few. Unfortunately, research in biotechnology has become controversial, mainly because it is difficult for the general public to comprehend this field since it is mired in complex jargon. It has been cited as being a threat to environmental safety, violating natural laws, and religious zealots have even labelled it as trivializing the meaning of life! In this chapter we will examine some of the recent advances in pest control using biotechnology and attempt to allay some of the concerns of the public.

17.2 Pests of Forests

Broadly speaking, there are three major groups of forest pests: pathogenic microorganisms, competing vegetation, and insect pests. Historically many of these pests of economic importance have been managed by using chemical control agents such as chemical fungicides, herbicides and pesticides. While biological control methods have been attempted, there was no sustained interest until recently when some of the non-target effects of the chemical pesticides became apparent. After a cursory examination of the biological control of competing weeds and pathogenic organisms, much of the discussion will be focused on the control of insect pests.

17.3   Vegetation Management

Phytophagous biological control organisms have been tested as weed control agents for many decades; the organisms used have been either insects or microorganisms, including fungi, nematodes or mites (Goeden, 1988). Testing for host specificity has been a major hurdle. In Australia, the common heliotrope *Heliotropium europaeum* was successfully controlled with the fungus *Uromyces heliotropii* and tests were conducted on 96 plants of the region for microscopic and macroscopic effects (Hasan *et al.*, 1992). Rather than continuing to depend on chemical herbicides, North American forestry is rapidly changing to 'ecosystem management', which is an effort to combine ecological principles, sustainable forests, and land stewardship ethics (Wagner, 1992). Since chemical herbicides are generally inexpensive and effective there has been a reluctance to invest in research on the use of alternative herbicides. Public belief that forested landscapes are among the only remaining unspoiled ecosystems has more recently lead to a search for other types of weed control agents including mycoherbicides (Wall *et al.*, 1992). Some mycoherbicides have been successfully developed in recent years (Te Beest *et al.*, 1992). *Chondrostereum purpureum* is a common pathogen that invades the xylem vessels through fresh wounds and causes silverleaf disease in various hardwood shrubs and trees (Spiers and Hopcroft, 1988). Although *C. purpureum* is not host specific, it prefers broadleaf trees. The fungus lives in living tissues or in trees that have been dead for less than 2 years. The airborne basidiospores infect tree wounds or stumps.

Biotechnological approaches have been used to engineer herbicide resistance genes into the trees to protect them against herbicides that are sprayed to control competing vegetation (Mazur and Falco, 1989). A sulphonyl urea herbicide, chlorosulfuron, was shown to inhibit cell division. Biochemical studies showed that it inhibits acetolactate synthase (ALS), which is required for the synthesis of isoleucine, leucine and valine. Using the ALS gene as a probe, the ALS mutant gene from herbicide resistant plants was isolated. The resistance was conferred by selective amino acid substitution. When such a resistant ALS gene was cloned into the tobacco plant, the transgenic tobacco plant was found to be resistant to sulphonyl urea herbicides. A similar approach was used to isolate a glyphosate resistant gene that was transferred into the crop plant. In the case of phosphoinothricin, a gene that produces a detoxifying enzyme was found to be better than a resistant glutamine synthase gene. In the case of atrazine and bromoxynil, the herbicide binding protein or Q_β protein (32-kDa membrane protein in the chloroplast) was involved in the resistance mechanism. The resistant variety of the Q_β protein has some amino acid substitutions (Mazur and Falco, 1989).

17.4   Plant Pathogens

Many species of bacteria, fungi and nematodes are pathogenic to plants in general. The molecular mode of action of phytopathogenic bacteria is being actively studied and the results of these investigations may pave the way for developing resistant varieties by recombinant DNA technology. *Erwinia*, *Pseudomonas* and *Xanthomonas* have genes that code for cellulases (endo-β-1,4-glucanase) but their role in pathogenicity has not been established. Proteases, production and export of extracellular enzymes, polysaccharides, plant growth substances, toxins and unknown products have all been implicated in pathogenicity and genes for many of them have been cloned. *Agrobacterium tumefaciens*, the soil bacterium that causes crown gall in dicotyledonous plants, has been by far the most studied and is extensively used for gene transfer. During the disease process, a segment of this bacterium's DNA (T-DNA) is transferred to the host plant and becomes integrated into the plant genome. The T-DNA originates from a 200-kb plasmid and foreign genes can be inserted into this DNA for transfer into the host plant (Daniels *et al.*, 1988).

The molecular biology of several fungal pathogens of plants has been well studied. A resistance gene may be cloned by a shot-gun method and thereby inserted into a host plant (Ellis *et al.*, 1988). The large genome size and, consequently, the size of the cloned DNA fragments however requires the screening of more than 10 000 transformed plants, which is an enormous task. Another method is the insertion of a transposable DNA sequence (transposable element) which in essence functions as a tag and can carry a resistance gene (Ellis *et al.*, 1988).

17.5 Nematodes

Resistance to phytopathogenic nematodes has been reported and has been traced to an H_1 resistance gene in certain cultivars of potato. While at this time this resistance gene has been shown for the potato cyst nematode, *Globodera rostochiensis*, similar studies can be extended to others that are pathogenic to trees such as the pinewood nematode. The excellent genetic information available for the free-living nematode *Caenorhabditis elegans* will serve as an elegant backdrop to characterizing the biochemical mechanism of resistance in parasitic forms such as the potato cyst nematode. Since the H_1 resistance gene has been shown to follow a Mendelian pattern, it lends itself to genetic manipulation (Bakker *et al.*, 1993).

17.6 Phytoalexins

Phytoalexins are low molecular weight, antimicrobial agents that are synthesized by plants in response to a pathogen or stress factor. Many compounds from the invading fungus or bacterium, such as oligoglucans, ethylene, chitosan oligomers, and polypeptides serve as elicitors for phytoalexin production in the host. The accumulation of phytoalexins at the infection site inhibits the growth of the fungus or bacterium and serves as a defence mechanism. The ability to produce phytoalexins confers a degree of resistance against the pathogen. Work that will lead to gene expression for phytoalexin accumulation and disease control is underway (Kuc, 1995).

17.7 Viral Pathogens of Plants

Viral pathogens of plants are well known and transmission of many of them by sucking insects such as aphids and leafhoppers has been extensively studied. One of the novel ways of providing protection against the pathogenic viruses is similar to vaccination. The coat protein of the virus protects the host plant from viral infection. The exact mechanism for this protection has not yet been elucidated. For example, the cucumber mosaic virus (CMV) infects 775 species of plants from 85 different families and causes severe viral disease. The coat protein (CP) gene of this virus was inserted into tobacco leaf discs using *Agrobacterium tumefaciens*. Of the 22 transformant lines, 15 showed resistance to CMV (Liang *et al.*, 1994). Such a technology may be extended to trees for protection against viral pathogens.

17.8　Insect Pests

Chemical and biological control of insect pests using parasites and predators dates back to the 1950s and has been extensively reviewed (Sweetman, 1958; Croft, 1990; Strand and Obrycki, 1996). Control of forest insect pests with the introduction of various egg parasites (*Trichogramma* sp.), external parasites (Tachinids), and internal parasites (Ichneumonids, Chalcids, etc.), has been attempted with various degrees of success, but in general, the results have not been spectacular. However, success stories of classical examples such as the introduction of the Australian vedalia lady beetle in 1899 to control the cottony-cushion scale in Californian orange groves reinforces the need to continue this approach. One of the areas that needs particular attention, especially in forestry, is the environmental impact of classical biological control agents. In the words of Howarth, 'Absence of evidence is not evidence of absence' (McEvoy, 1996).

Microbial pathogens such as fungi, bacteria and viruses that cause diseases in insects have been tested extensively as biological control agents (Federici and Maddox, 1996). Among the fungi, more than 50 genera contain species that are pathogenic to insects and are candidates for biological control (McCoy *et al.*, 1988). They have widely varied life cycles and host range and are transmitted primarily through either spores (sexual forms) or conidia (asexual forms). When a spore or a conidium lands on a susceptible host, it adheres to the cuticle, germinates and digests its entry into the insect by means of its germ tube. The fungal mycelia soon overpower and mummify the insect. The conidia or spores are released and the life history starts all over again. *Beauveria bassiana* is an imperfect fungus that infects over 100 insect species from many orders; it has been used extensively in Russia and China for insect control. Identification of the various isolates (pathotypes) that are morphologically similar is being done with molecular techniques to resolve taxonomic problems (Federici and Maddox, 1996). In addition, entomophthoraceous fungi have been reported from many insects, and many of them have been found to have a narrow host range (Harcourt *et al.*, 1974). Recently, recombinant DNA technology has been used to improve the efficacy of some of the mycoinsecticides. Additional copies of a gene encoding a regulated cuticle-degrading protease from an entomopathogenic fungus, *Metarhizium anisopliae*, were inserted into the same fungus so that the enzyme was constitutively overproduced in *Manduca sexta* infected with the recombinant fungus. The insects ate less, died quickly and the cadavers were black due to melanization. The resulting cadavers were poor substrates for sporulation ensuring minimal persistence of the genetically altered fungus (St Leger *et al.*, 1996).

Phylum Protozoa is now considered a sub-Kingdom, and the class Microspora has now been elevated to a phylum and includes the entomopathogenic microsporidia; this is important from an insect control stand point. *Nosema locustae* has been registered for controlling grasshoppers. Microsporidia are obligate intracellular pathogens and the larvae are infected when they ingest the spores. The spore extrudes a hollow polar filament through which the sporoplasm is injected into the midgut cells where they multiply vegetatively. Many forest insect pests have microsporidia and some of them have been shown to be transovarially transmitted (Canning, 1982). They are quite infective but lack the virulence required to compete with other control measures. Their large genome has been an impediment to developing recombinant DNA techniques to increase their virulence by inserting foreign genes.

The bacterium *Bacillus thuringiensis* (Bt) is the most widely used pathogen for insect control in North America. Bt is a common organism that is found in the soil, insects, frass, grain dust and leaves. It was first isolated in 1902 as the causative agent for 'Sotto disease' in silkworm in Japan. It is a Gram-positive, aerobic, spore-forming bacterium, which produces a characteristic parasporal crystal that contains the insecticidal toxins. The insecticidal parasporal proteins are activated by the midgut proteases and the activated toxin binds to a specific receptor on midgut cells of susceptible insects. The lysis and destruction of the midgut cells eventually leads to the death of the host. The genes that encode the insecticidal proteins are typically located on large transmissible plasmids that determine the host specificity of the isolate or serotype (Meadows, 1993). Bt has more than 50 subspecies and produces a variety of crystal (Cry) proteins (see Federici and Maddox (1996) for examples of commercially available microbial insecticides based on Bt). These Cry proteins are classified according to their activity spectrum and molecular mass. Cry I and Cry II are active against lepidoptera and are 135 and 65 kDa, respectively, in size. The Cry III protein of 65 kDa size is active against coleoptera such as the Colorado potato beetle, *Leptinostarsa decemlineata*. Cry IV proteins range in size from 65 to 135 kDa and are effective against mosquitoes and blackflies. The presence of these toxin genes in the plasmids made it relatively easy to isolate and purify them, and in the 1980s most of them were cloned. These toxins have been engineered into insect pathogenic viruses, which is described later, as well as into plants and trees, which is covered in Chapter 18.

17.9 Viral Pathogens of Insects

Insect viruses causing epizootics have been recognized for nearly a century and one of the earliest such occurrences described in North America was in 1913 by Glaser and Chapman on the wilt disease of the gypsy moth, *Lymantria dispar*, caused by a nuclear polyhedrosis virus (NPV) (David, 1975). Perhaps one of the most dramatic controls ever reported was the control of the European pine sawfly, *Neodiprion sertifer*, in Ontario, Canada by an NPV isolated from dead larvae sent from Sweden (Bird, 1953). Various types of viruses are pathogenic to insects but one group, the Baculoviridae, is uniquely pathogenic only to insects (Table 17.1) (Tinsley, 1979). By far the largest group of insects affected by viruses is the Lepidoptera. Baculoviruses and cytoplasmic polyhedrosis viruses form the two major groups of viruses that are infectious to insects (David, 1975; Martignoni and Iwai, 1981; Cunningham, 1995).

Table 17.1 Groups of pathogenic viruses found in insects

Family	Nucleic acid[a]	Particle symmetry	Biochemical and biophysical similarities	
			Vertebrate viruses	Plant viruses
Baculoviridae (Baculovirus groups A, B, C)	DNA	Rod (occluded)	None	None
Poxviridae	DNA	'Brick' (occluded)	Orthopoxvirus Avipoxvirus Capripoxvirus Leporipoxvirus Parapoxvirus	None
Reoviridae (cytoplasmic polyhedrosis viruses)	dsRNA	Isometric (occluded)	Reovirus Orbivirus	Plant Reoviruses
Iridoviridae	DNA	Isometric	African Swine Fever Frog Viruses 1-3 Lymphocystis virus	Fungal Algal
Parvoviridae (Densovirus)	ssDNA	Isometric	Parvovirus Adeno-associated group	None
Picornaviridae (Enterovirus; unclassified groups)	RNA	Isometric	Enterovirus	Small RNA viruses (single polypeptide)
Rhabdoviridae	RNA	Bullet/ bacciliform	Vesiculovirus Lyssavirus	Plant rhabdovirus

[a] dsRNA, double-stranded RNA; ssDNA, single-stranded DNA
From Maeda (1989) (reproduced with permission from Annual Reviews Inc.)

The host range of entomopathogenic viruses varies widely but generally it is quite narrow. Viruses such as the multicapsid nuclear polyhedrosis virus of the beet army worm, *Spodoptera exigua* (*Se*MNPV) are species specific whereas the virus of the alfalfa looper, *Autographa californica* (*Ac*MNPV) infects and replicates in at least 33 lepidopterous species from seven different families. However, most of the viruses have an intermediate host range and are best described as genus specific, as is illustrated by *Hz*MNPV that infects several species of *Heliothis*, including *H. zea* (Federici and Maddox, 1996). Many baculoviruses have been developed commercially or used in a practical manner for controlling specific pests (Table 17.2). The effective dosage is generally established in terms of the number of polyhedra or occlusion bodies (OB) required to affect 50 per cent of the treated insects (LD_{50}). Since the time taken to have an effect has an important bearing on the ability to protect the crop from damage, the LT_{50} or time taken to kill 50 per cent of the treated insects, usually in days, is also studied.

Larvae typically become infected by NPV when they ingest the polyhedra on the foliage. The polyhedra dissolve in the midgut and the virions are released. These

Table 17.2 Baculoviruses of commercial/practical use

Pest	Virus	Status
Cotton bollworm, Tobacco budworm, Corn earworm, Tomato fruitworm (*Heliothis* species): *Heliocoverpa* (=*Heliothis*) *zea, Heliothis virescens, H. armigera, H. paradoxa, H. peltigera, H. phloxiphaga, H. punctigera*	Single nucleocapsid NPV Registered as Viron H®, Biotrol-VH2, Eclar™	>1000000 ha treated in USA Dosage: 6×10^{11} occlusion bodies (OB) per hectare to 1.2×10^{12} OB/ha. Honeybees used to disseminate virus
Alfalfa looper, *Autographa californica.* Infects 43 species in 11 families. Used on Cabbage looper (*Trichoplusia ni*), Beet armyworm (*Spodoptera exigua*), Diamondback moth (*Plutella xylostella*), Douglas-fir tussock moth (*Orgyia pseudotsugata*)	Multiple nucleocapsid NPV. Commercial development as Gusano®, VPN 80	Being extensively tested. Dose: 2.5×10^{11} OH/ha to 2.5×10^{12} OB/ha
Velvetbean caterpillar, *Anticarsia gemmatalis*	At least five companies are producing commercial products of this MNPV	1 to 4×10^{6} ha annually. Dosage: 10^{4} OB/ha
Cassava caterpillar, *Erinnyis ello* in Brazil Cabbage looper, *Trichoplusia ni*	Granulosis virus (GV). Farmer level SNPV, MNPV, and GV. The MNPV is commercially produced as Biotrol VTN and Viron T	About 5×10^{3} ha annually
Celery looper, *Anagrapha falcifera*; also effective on *H. virescens, H. zea, T. ni,* Coddling moth (*Cydia pomonella*), European corn borer (*Ostrinia nubilalis*), *P. xylostella*	Patented by USDA and being commercialized by Sandoz	Kills host faster and at a lower doses than most other baculoviruses
Grape leaf skeletonizer, *Harrisina brillians*	GV	Early stage of development but very promising. Dosage: 5.8 g/ha

Table 17.2 Continued

Pest	Virus	Status
Beet armyworm, *Spodoptera exigua*	SeNPV; Commercialized as Spod-X®	Effective on a large number of lepidoptera infesting vegetable crops. Dosage: 2.5×10^{11} OB/ha to 1.25×10^{12} OB/ha
Cotton leafworm, *Spodoptera littoralis*	NPV; Spodopterin®	Extensively tested in Egypt, 5×10^{12} OB/ha
Fall armyworm, *Spodoptera frugiperda*	NPV; produced by National Soybean Research Center in Brazil	>5000 ha in Brazil
Soybean semilooper, *Trichoplusia orichalcea*. Also controls *Chrysodeixis chalcites*	NPV in Zimbabwe, domestically produced	$>15 \times 10^{6}$ ha treated
Cabbage moth, *Mamestra brassicae*. Effective on *H. zea*, *H. virescens*, *H. armigera*, *S. exigua*, and *Diperopsis watersi*	NPV; produced as Mamestrin®	Widely used in France
Coddling moth, *Cydia pomonella*	GV, produced as UCB 87; San 406; Decyde™, Carpovirusine®	Tested in Californian fruit orchards. Dosage: 2.5×10^{13} GV capsules/ha. Wide interest in commercialization of this GV
Indian meal moth, *Plodia interpunctella*	GV, patented by USDA	200 g of freeze dried product will treat 16 tons of dried fruit and nuts
Coconut rhinoceros beetle, *Oryctes rhinoceros* also controls *O. monoceros*	Non-occluded baculovirus discovered in Malaysia; infects adults as well	Palm industry in the countries of the South Pacific and Indian Ocean regions

Modified from Sohi, *et al.* (1981) © National Research Council of Canada, 1981

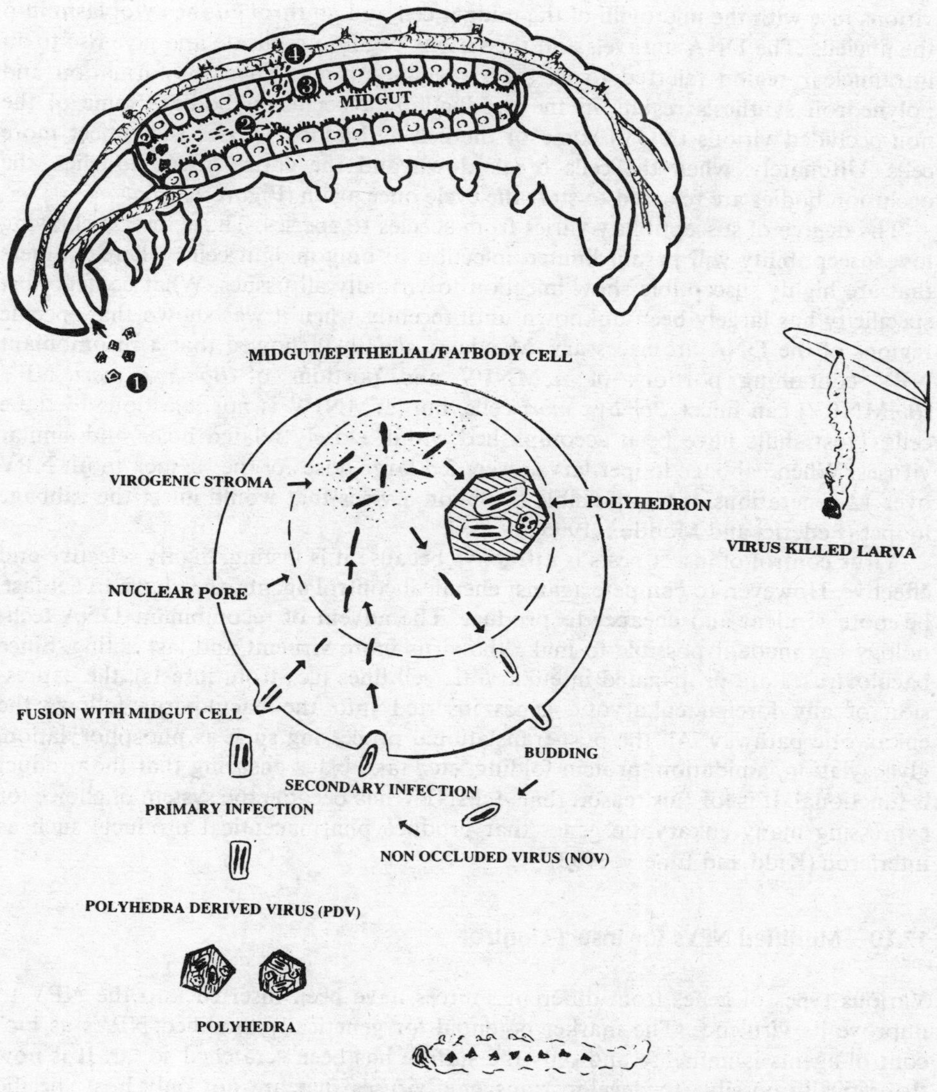

Figure 17.1 Life cycle of a nuclear polyhedrosis virus based on the research conducted at the Canadian Forest Service laboratory in Sault Ste Marie on the spruce budworm virus (*Cf*MNPV). Polyhedra are ingested by the larva feeding on contaminated foliage (1). The polyhedra are dissolved in the midgut by the high pH and the virions are released (2). The virions fuse with the midgut microvilli, enter the nucleus, replicate (virogenic stroma) and either form polyhedral occlusion bodies or bud non-occluded virus (NOV) for secondary infection (3). The virus entering the haemocoel from the midgut cells infects the tracheal epithelium and is spread via these cells to various tissues such as the fat body and epidermis (4). The polyhedra are contained in the larvae and are disseminated when the cadavers disintegrate

275

virions fuse with the microvilli of the midgut cell and go through the cytoplasm into the nucleus. The DNA unravels and the virions begin to replicate and give rise to an intranuclear region referred to as the virogenic stroma. Envelope formation and polyhedron synthesis results in the production of occlusion bodies. Some of the non-occluded virions (NOV) come of the cell as budded viruses that infect more cells. Ultimately, when the cells break down and the entire insect liquefies, the occlusion bodies are released to start the cycle once again (Figure 17.1).

The degree of susceptibility varies from species to species. Those insects showing low susceptibility will permit limited infection to only midgut cells, whereas others that are highly susceptible show infection to virtually all tissues. What confers host specificity has largely been unknown until recently when it was shown that specific regions of the DNA are necessary. Maeda *et al.* (1993) showed that a recombinant NPV containing portions of *Ac*MNPV and portions of *Bombyx mori* NPV (*Bm*MNPV) can infect *Bombyx mori* cells, but *Ac*MNPV is not infectious to these cells. Host shifts have been accomplished within closely related hosts and similar viruses. When cabbage looper larvae were fed large doses of the tussock moth NPV over 12 generations, it was possible to obtain a virus that would infect the cabbage looper (Federici and Maddox, 1996).

Virus control of insect pests is attractive because it is lasting, highly selective and effective. However, to compete against chemical control agents, they have to act fast, be more virulent and cheaper to produce. The advent of recombinant DNA technology has made it possible to make the virus more virulent and fast acting. Since baculoviruses are propagated in eukaryotic cell lines (i.e., from insects), the expression of any foreign eukaryotic genes inserted into the baculovirus follows the eukaryotic pathway. All the post-translational processing such as phosphorylation, glycosylation, amidation, protein folding, etc., take place ensuring that the product is functional. It is for this reason that *Ac*MNPV has become the system of choice for expressing many eukaryotic genes that produce pharmaceutical products such as interferon (Kidd and Emery, 1993).

17.10 Modified NPVs for Insect Control

Various types of genes from different sources have been inserted into the NPV to improve its virulence. The market potential for genetically modified NPVs as biocontrol agents is immense and only the surface has been scratched so far. It is now theoretically possible to develop transgenic viruses that are not only host specific but also as effective as conventional control agents.

17.10.1 Structure, Biology, and Ecology

Before embarking on genetic modification of the virus, it is essential that the biology of the virus is well understood. Fortunately, there is a fund of information in this area (Adams and McClintock, 1991; Bilimoria, 1991). The spruce budworm virus, *Choristoneura fumiferana*, multicapsid nuclear polyhedrosis virus (*Cf*MNPV) is one of the forestry-related viruses that has been well studied (Figure 17.2). The infectivity, host-range, LD_{50}, LT_{50} and ecology have to be studied to provide the baseline information for assaying the activities of the transformed viruses (Kaupp and Ebling, 1990).

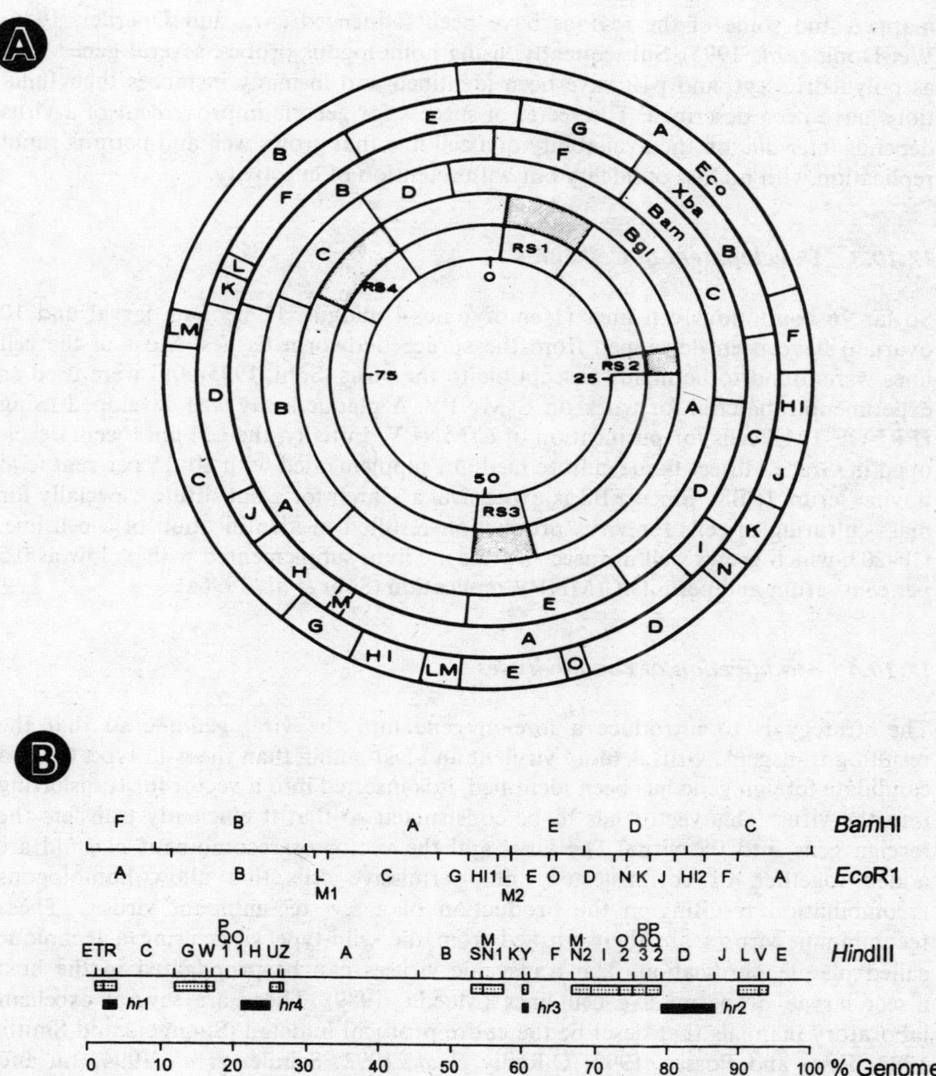

Figure 17.2 Physical map of *Cf*MNPV genome. (A) Circular map with the zero position at the junction of *Bam*H1 fragments B and F (Arif and Doerfler, 1984; reproduced with the permission of *EMBO Journal* and the Oxford University Press, UK). (B) Linearized version with the *Bam*H1 fragment F to the left (Wei-Dong *et al.*, 1995; reproduced with the permission of *Virology* and Academic Press)

17.10.2 Biochemical Characterization

NPVs from different insects have been characterized to various degrees. The relatively small size of the circular dsDNA has made it possible to elucidate the structure. The genomes of *Ac*MNPV and *Bm*MNPV have been completely sequenced (Ayres *et al.*, 1994; Maeda, personal communication). The *Cf*MNPV has been mapped and some of the regions have been sequenced (Arif and Doerfler, 1984; Wei-Dong *et al.*, 1995). Subsequently, using homologous probes, several genes such as polyhedrin, egt, and p10 have been identified and in many instances their functions have been described. The secret of success for genetic improvement of a virus depends *inter alia* on the availability of a cell line that grows well and permits rapid replication with no loss of fidelity but with retention of infectivity.

17.10.3 Development of Cell Lines

So far 26 continuous cell lines (1 embryonic, 4 midgut, 11 neonate larval and 10 ovarian) have been developed from the spruce budworm tissues. Most of the cell lines were found to be highly susceptible to the virus (Sohi, 1995) and were used as experimental material for work on *Cf*MNPV. A plaque assay was developed using IPRI-CF-124T cells for purification of *Cf*MNPV. Initially, the cell lines were developed in Grace's insect tissue culture medium supplemented with 10–15 per cent fetal bovine serum (FBS). Since FBS is expensive, a search for a substitute especially for mass-culturing of cells for virus production resulted in identification of a cell line, CF-203, which grows well in Insect-Xpress medium supplemented with as low as 0.5 per cent serum and permits *Cf*MNPV replication (Sohi *et al.*, 1996a).

17.10.4 Modification of Baculoviruses

The strategy is to introduce a foreign gene into the viral genome so that the resulting transgenic virus is more virulent and fast-acting than the wild type. Once a candidate foreign gene has been identified, it is inserted into a vector for transferring into the virus. This vector has to be constructed so that it efficiently transfers the foreign gene into the virus. The virus and the vector or recombinant plasmid are added together, or cotransfected, into permissive cells that allow homologous recombination resulting in the production of a few recombinant viruses. These recombinant viruses are then purified from the wild-type virus using a technique called plaque purification. The transgenic viruses can be propagated in the host insect larvae or permissive cell lines (Maeda, 1989). There are several excellent laboratory manuals that describe the entire protocol in detail (Summers and Smith, 1987; King and Possee, 1992; O'Reilly *et al.*, 1992; Schuler *et al.*, 1994; Lu and Miller, 1996).

17.10.5 Promoter Selection

One of the primary requirements for producing a successfully modified virus for pest control is the incorporation of the proper promoter that will direct the expression of the foreign gene at the appropriate time and place. For instance, a fast-acting neurotoxin has to be secreted quickly into haemolymph close to the nervous system. Several late gene viral promoters such as polyhedrin and p10 have been used successfully (King *et al.*, 1994). Large quantities of mRNA for genes expressed under these very late gene promoters will be produced, beginning 24 hours after infection. Polyhedrin promoter is the most commonly used promoter. The advantage of using polyhedrin promoter in a virus where the polyhedrin gene has been removed is that the modified virus is produced without the polyhedrin in a non-occluded form and therefore does not survive in the environment beyond a generation of the host insect. This feature is attractive to both environmentalists because it does not persist in the environment, and to commercial producers because the virus has to be produced every year for repeated application. The disadvantage of using such a system is that there is no reliable quantitative bioassay to assess the effectiveness of non-occluded virions. Currently, the assay is performed by either injecting the modified virus, which can be time consuming, or by feeding the preoccluded virus which is difficult to quantify. Many researchers have been persuaded to use the p10 promoter for the expression of foreign genes because of the difficulties encountered with non-occluded virus particles. The main disadvantage with the p10 promoter is that the modified virus will produce polyhedrin occlusion bodies that can persist and survive in the soil for a long time. Other promoters such as the combination of early and late promoters of baculoviruses and constitutive promoters such as the insect actin promoter that are active all the time have been tried, but none have enjoyed the success of the polyhedrin promoter (Johnson *et al.*, 1992; Tomalski and Miller, 1992; King *et al.*, 1994).

17.10.6 Candidate Foreign Genes

Table 17.3 lists the candidate genes that have been inserted into baculoviruses for improving their control potential; the most promising ones are those showing either the deletion of ecdysone glucosyl transferase (egt$^-$), or the expression of scorpion toxin, mite toxin or juvenile hormone esterase (JHE). The egt$^-$ virus has been one of the leading candidates for early registration because of its acceptance by the public. There is no foreign gene inserted into the egt$^-$ virus and by deleting this gene from the virus the moulting of the host larva is no longer prolonged, which results in less feeding damage. Other modified viruses such as the one expressing the scorpion toxin and the enzyme, JHE, are at various stages of evaluation (Bonning and Hammock, 1996). There are various developmental genes that have a regulatory function and may be tested in the future as candidates for insertion (Riddiford and Truman, 1993).

17.10.7 Future Considerations

Producing a genetically altered virus is at the early stages of development at present and there are several hurdles to be crossed before the modified virus can be successfully applied in the field. Some of the major problems that need to be solved are mass production, stability of the modified virus in the field, effect of modifications on the host range, effect of modified virus on non-target organisms, and formulation of the modified virus for field application. Mass producing a fast-acting transgenic virus in an insect host will be difficult because the virus will kill the insects before enough virus is produced. Strategies such as propagating the virus in cell cultures and using an inducible promoter than can be turned on at will by adding the inducers at the appropriate time will have to be employed. Mass production of recombinant viruses in SF-21 cells using large-scale fermenters appears promising (Gard, 1996). Most of the tests conducted on the modified viruses so far have indicated that the modifications have not affected the host range of the transformed

Table 17.3 Modified baculoviruses for insect control

Gene inserted/ deleted	Effect	Reference
Ecdysone glucosyl transferase(egt)	Insects infected with egt-deletion consumed 50% less food than the ones infected with wild-type virus. Insects infected with egt-deletion mutant died sooner than those infected with wild-type virus	O'Reilly and Miller (1989)
Diuretic hormone (DH)	Infection of *Bombyx mori* nuclear polyhedrosis virus expressing DH resulted in a decrease in haemolymph volume and the infected insects died one day earlier than controls	Maeda (1989)
Juvenile hormone esterase	The recombinant *Ac*MNPV expressing JHE caused reduced feeding in neonate *T. ni* larvae. The feeding inhibition was not observed in later larval instars	Hammock *et al.* (1990)
Bacillus thuringiensis (Bt) delta toxin	No significant improvement in LD_{50} compared with wild-type virus	Martens *et al.* (1990) Merryweather *et al.* (1990)
Androctonus australis insect specific neurotoxin (Aait)	*T. ni* larvae infected with recombinant virus expressing Aait, consumed 50% less diet and the LD_{50} was reduced slightly; there was a 25% decrease in survival time compared with controls	Stewart *et al.* (1991) Maeda (1989) McCutchen *et al.* (1991)
Pyemotes tritici (mite) toxin (MT)	*Ac*MNPV expressing MT caused significant reduction in survival time, feeding and weight gain in *T. ni*	Tomalski and Miller (1991) Tomalski and Miller (1992)

virus and most of the non-target beneficial insects are not affected (Gard, 1996). The occlusion bodies (OB) of baculoviruses are stable in the environment, especially in the soil, for several years (Evans and Harrap, 1982). To circumvent this problem, Wood and his colleagues at the Boyce Thomson Research Institute have been working on a procedure of using a polyhedrin minus (null) preoccluded virus preparation for field application. They have shown that such a polyhedrin null virus disappears rapidly from the soil (Wood, 1996).

A successful modified baculovirus is one that can outcompete a conventional chemical insecticide in its ability to kill faster and be produced at a competitive cost. Some of the promising new strategies include using multiple genes for not only adversely interfering with feeding but also derailing the normal development of the insect pest. The receptors and transcription factors that are involved in insect development will become one of the main target areas that will be used for adversely interfering with the normal developmental process. Integrating the rapidly developing field of biotechnology with insect molecular biology, endocrinology, physiology, ecology, cell culture and virology should make the dream of making an ideal recombinant virus a reality in the not too distant future.

17.11 Alternative Ways of Improving Baculoviruses

The traditional approach for improving baculoviruses is to modify the virus by either deleting an existing gene or adding a foreign gene into the viral genome.

However, there are alternative ways of improving the baculoviruses as pest control agents. Selecting the more effective strains from the mixture of strains that occur in nature and optimizing the combinations of the more active strains is one method. Another approach is to alter the host range of an effective virus such as *Ac*MNPV so that it can infect other important pests such as the gypsy moth and the spruce budworm.

17.11.1 Strain Selection and Mixed Strains

The baculovirus isolates from nature contain a mixture of genetically heterogeneous populations. For example *Cf*MNPV field isolates contain at least two different viruses, and one of the two is more infectious to spruce budworm larvae (Arif *et al.*, 1992; Sohi *et al.*, 1996b). Two variants were isolated from field-collected *Pf*MNPV (*P. flammea* NPV; Weitzman *et al.*, 1992). A strain of codling moth granulosis virus (GV) that was 5.6 times more resistant to artificial ultraviolet light in the laboratory and survived twice as long as the wild type in the field was selected for development (Brassel and Benz, 1979). Replication of *Ac*MNPV in the presence of 2-aminopurine resulted in a strain that had increased virulence (Wood *et al.*, 1981).

17.11.2 Host Range Alteration

*Ac*MNPV is one of the most virulent and well-characterized baculoviruses. Unfortunately though, *Ac*MNPV does not infect some of the important forest pests such as the gypsy moth and the spruce budworm. When gypsy moth, *Lymantria dispar*, IPLB-Ld652Y (Ld652Y) cells were infected with *Ac*MNPV alone the infection was aborted but, if they were coinfected with *Ld*MNPV, the *Ac*MNPV infection was enhanced (McClintock and Dougherty, 1987; Thiem *et al.*, 1996). An *Ld*MNPV gene host range factor 1 (*hrf*-1) that enabled *Ac*MNPV to infect Ld652Y cells was isolated, cloned and inserted into *Ac*MNPV. Modified *Ac*MNPV expressing the *hrf*-1 gene can successfully infect *L. dispar* (Thiem, personal communication). CF-203 cells inoculated with *Ac*MNPV showed a DNA ladder and morphological changes such as plasma membrane granulation, blebbing and nuclear fragmentation, which are characteristic of apoptosis (Palli *et al.*, 1996). The mRNA for the apoptosis suppressor gene p35 was detected 9 hours later in *Ac*MNPV-inoculated CF-203 cells than in SF-21 cells. Only a trace amount of mRNA for the *Ac*MNPV-inhibitor of apoptosis homologue (*Ac-iap*) gene and no mRNAs for the late genes, *Ac*MNPV-polyhedrin (*Ac-polh*) and *Ac*MNPV-p10 (*Ac-p10*), were detected in *Ac*MNPV-inoculated CD-203 cells. Inoculation of CF-203 cells with *Cf*MNPV at least 12 hours prior to inoculation with *Ac*MNPV prevented apoptosis-like cell death, and mRNAs for *Ac-iap*, *Ac-polh* and *Ac-p10* genes were expressed, resulting in successful virus replication and OB production. Work is currently in progress to identify the gene(s) in *Cf*MNPV that provide this protection and enhances *Ac*MNPV replication in CF-203 cells. Once this gene is identified, it should then be possible to engineer a recombinant *Ac*MNPV bearing this *Cf*MNPV gene similar to the recombinant *Ac*MNPV bearing the *Lymantria dispar* MNPV host range gene *hrf-1* (Thiem *et al.*, 1996). Such a recombinant *Ac*MNPV would most likely infect the spruce budworm and also will probably be more virulent than the wild-type *Cf*MNPV for this forest pest.

17.12   Other Applications of Biotechnology for Pest Control

The recombinant DNA technology used for altering the effectiveness of the virus can also be used on the insect. Incorporation of the genes into the host insect and its parasites and predators can be used in control strategies.

17.12.1   *Transgenic Insects*

Production of transgenic insects carrying an insecticidal gene under the control of an inducible promoter has been under investigation for several years. In this strategy the pest insect is transformed by incorporating a gene that is expressed under the control of an inducible promoter. The transformed insects are mass-produced and field-released to mate with natural populations. After a few generations when the population of the pest insect reaches damaging levels the inducers are sprayed. The insecticidal gene is thereby induced resulting in the death of the pest species. This strategy is similar to the sterile male release technique and can be used in an integrated pest management strategy. Due to the lack of a good procedure for germ line transformation for pest insects this approach has not been very successful so far. With the recent cloning and characterization of transposons from several key pest insects it may be possible to produce transgenic insects in the near future (Crampton and Eggleston, 1992).

17.12.2   *Improvement of Parasites and Parasitoids*

There have also been attempts to produce transgenic parasites and predators expressing genes that can enhance the survival of these organisms in the environment. Parasites and predators expressing insecticide resistance genes will work better in the integrated pest management strategies since they will survive insecticide treatments. This approach has had limited success due to the lack of availability of transformation protocols that can be used for producing transgenic parasites and predators. With the recent discovery of transposons in many insects this situation will soon change (Crampton and Eggleston, 1992).

17.13   Conclusions

Biological control agents in general are the best option for controlling pests in an environmentally friendly manner. Unfortunately however they do not always perform as efficiently as chemical control agents. Furthermore, it costs more to produce biological control agents than conventional chemical pesticides. The advent of recombinant DNA technology has changed the entire situation. It is now possible to enhance the control potential of biologicals by genetic manipulation. We can not only increase the activity to match chemical pesticides but also in many instances retain the host specificity. We are on the threshold of developing new, genetically modified organisms that will revolutionize the concept of pest control. We can now make 'designer pesticides' that will be able to control many pests with surgical precision with little or no impact on non-target organisms. New concepts and new techniques often come with their own set of problems. Establishing their safety, educating the users, and perception by the public are some of the aspects that will ultimately decide the fate of this new generation of genetically engineered transgenic biocontrol agents.

Acknowledgements

This study was supported by the Canadian Forest Service and the National Biotechnology Strategy fund. We thank Dr S. S. Sohi and Mrs K. Jamieson for critical comments on the manuscript, and Miss A. Ricci for assistance in the preparation of the manuscript. We thank all of our colleagues in the Biotechnology Project for use of their unpublished results.

References

ADAMS, J. R. and MCCLINTOCK, J. J. (1991) Nuclear polyhedrosis virus of insects. In: Adams, J. R. and Bonami, J. R., eds, *Atlas of Invertebrate Viruses*, Boca Raton: CRC Press, pp. 87–204.

ARIF, B. M. and DOERFLER, W. (1984) Identification and localization of reiterated sequences in the *Choristoneura fumiferana* MNPV genome. *EMBO J.* **3**, 525–529.

ARIF, B., JAMIESON, P., SOHI, S. S., KAUPP, W. and MACDONALD, A. (1992) Characterization of a helper dependent virus present in wild-type NPV of the eastern spruce budworm, *Choristoneura fumiferana*, Society for Invertebrate Pathology, 25th Annual Meeting, Heidelberg, Germany, August, *Abstracts of Meeting*, p. 68.

AYRES, M. D., HOWARD, S. C., KUZIO, J., LOPEZ-FERBER, M. and POSSEE, R. D. (1994) The complete sequence of *Autographa californica* nuclear polyhedrosis virus. *Virology* **202**, 586–605.

BAKKER, J., FOLKERSTMA, R. T., VAN DER VOORT, J. N. A. M. R., DE BOER, J. M. and GOMMERS, F. J. (1993) Changing concepts and molecular approaches in the management of virulence genes in potato cyst nematodes. *Annu. Rev. Phytopathol.* **31**, 169–190.

BILIMORIA, S. L. (1991) The biology of nuclear polyhedrosis viruses. In: Kurstak, E., ed., *Viruses of Invertebrates*, New York: Dekker, pp. 1–72.

BIRD, F. T. (1953) The use of virus disease in the biological control of the European pine sawfly, *Neodiprion sestifer* (Geoffr.). *Can. Entomol.* **85**, 437–446.

BONNING, B. C. and HAMMOCK, B. D. (1996) Development of recombinant baculoviruses for insect control. *Annu. Rev. Entomol.* **41**, 191–210.

BRASSEL, J. and BENZ, G. (1979) Selection of a strain of the granulosis virus of the codling moth with improved resistance against artificial ultraviolet radiation and sunlight. *J. Invert. Pathol.* **33**, 358–363.

CANNING, E. U. (1982) An evaluation of protozoal characteristics in relation to biological control of pests. *Parasitology* **84**, 119–129.

CARSON, R. (1962) *Silent Spring*, Boston: Houghton Mifflin.

CRAMPTON, J. M. and EGGLESTON, P. (eds) (1992) *Insect Molecular Science*, London: Academic Press.

CROFT, B. A. (1990) *Anthropod Biological Control Agents and Pesticides*, New York: Wiley.

CUNNINGHAM, J. C. (1995) Baculoviruses as microbial insecticides. In: Renveni, R., ed., *Novel Approaches to Integrated Pest Management*, Boca Raton: Lewis Publishers, pp. 261–292.

DANIELS, M. J., DOW, J. M. and OSBOURN, A. E. (1988) Molecular genetics of pathogenicity in phytopathogenic bacteria. *Annu. Rev Phytopathol.* **26**, 285–312.

DAVID, W. A. L. (1975) The status of viruses pathogenic for insects and mites. *Annu. Rev. Entomol.* **20**, 97–117.

ELLIS, J. G., LAWRENCE, G. J., PEACOCK, W. J. and PRYOR, A. J. (1988) Approaches to cloning plant genes conferring resistance to fungal pathogens. *Annu. Rev. Phytopathol.* **26**, 245–263.

EVANS, H. F. and HARRAP, K. A. (1982) Persistence of insect viruses. In: Mahy, B. W., Minson, A. C. and Darby, G. K., eds, *Virus Persistence*, Vol. 33, SGM Symposium, Cambridge: Cambridge University Press, pp. 57–96.

FEDERICI, B. A. and MADDOX, J. V. (1996) Host specificity in microbe-insect interactions – insect control by bacterial, fungal and viral pathogens. *BioScience* **46**, 410–421.

GARD, I. V. (1996) Development of recombinant baculoviruses as biopesticides. *Proceedings of the Symposium on Commercialization of Biopesticides, Applied Products and Transgenic Plants*, 22–24 January, Washington, DC, International Business Communications, Southborough, MA, USA.

GLASER, R. W. and CHAPMAN, J. W. (1913) The wilt disease of gypsy moth caterpillars, *J. Econ. Entomol.* **6**, 479–488.

GOEDEN, R. D. (1988) A capsule history of the biological control of weeds. *Biocontrol News and Information* **9**, 55–61.

HAMMOCK, B. D., BONNING, B. C., POSSEE, R. D., HANZLIK, T. N. and MAEDA, S. (1990) Expression and effects of the juvenile hormone esterase in a baculovirus vector. *Nature* **344**, 458–461.

HARCOURT, D., GUPPY, G. M. J. C., MACLEOD, D. M. and TYRRELL, D. (1974) The fungus *Entomophthora phytonomi* pathogenic to the alfalfa weevil, *Hypera postica* (Coleoptera: Curculionida). *Can. Entomol.* **109**, 1521–1532.

HASAN, S., DEL FOSSE, E. S., ARACIL, E. and LEWIS, R. C. (1992) Host-specificity of *Uromyces heliotropii*, a fungal agent for the biological control of common heliotrope (*Heliotropium europaeum*) in Australia. *Ann. Appl. Biol.* **121**, 697–705.

JOHNSON, R., MEIDINGER, R. G. and IATRON, K. (1992) A cellular promoter-based expression cassette for generating recombinant baculoviruses directing rapid expression of passenger genes in infected insects. *Virology* **190**, 815–823.

KAUPP, W. J. and EBLING, P. M. (1990) Response of third-, fourth-, fifth- and sixth-instar spruce budworm, *Choristoneura fumiferana* (Clem.), larvae to nuclear polyhedrosis virus. *Can. Entomol.* **122**, 1037–1038.

KIDD, I. M. and EMERY, V. C. (1993) The use of baculoviruses as expression vectors. *Appl. Biochem. Biotechnol.* **42**, 137–159.

KING, L. A. and POSSEE, R. D. (1992) *The Baculovirus Expression System – A Laboratory Guide*, London: Chapman and Hall.

KING, L. A., POSSEE, R. D., HUGHES, D. S., ATKINSON, A. E., PALMER, C. P., MARLOW, S. A., PICKERING, J. M., JOYCE, K. A., LAWRIE, A. M., MILLER, D. P. and BEADLE, D. J. (1994) Advances in insect virology. *Adv. Insect Physiol.* **25**, 1–73.

KUC, J. (1995) Phytoalexins, stress metabolism, and disease resistance in plants. *Annu. Rev. Phytopathol.* **33**, 275–297.

LIANG, X., ZHU, Y., MI, J. and CHEN, Z. (1994) Production of virus resistant and insect tolerant transgenic tobacco plants. *Plant Cell Rep.* **14**, 141–144.

LU, A. and MILLER, L. K. (1996) Generation of recombinant viruses by direct cloning. *Biotechniques* **21**, 63–68.

MAEDA, S. (1989) Expression of foreign genes in insects using baculovirus expression vectors. *Annu. Rev. Entomol.* **34**, 351–372.

MAEDA, S., KAMITA, S. G. and KONDO, A. (1993) Host range expansion of *Autographa californica* nuclear polyhedrosis virus (NPV) following recombination of a 0.6-kilobase-pair DNA fragment originating from *Bombyx mori. J. Virol.* **67**, 6234–6238.

MARTENS, J. W. M., HONEE, G., ZUIDEMA, D., VAN LEUT, J. W. M., VISSER, B. and VLOK, J. M. (1990) Insecticidal activity of a bacterial crystal protein expressing a recombinant baculovirus in insect cells. *Appl. Environ. Microbiol.* **56**, 2764–2770.

MARTIGNONI, M. E. and IWAI, P. J. (1981) A catalogue of viral diseases of insects, mites and ticks. In: Burges, H. D., ed., *Microbial Control of Pests and Plant Diseases, 1970–1980*, London: Academic Press, pp. 897–911.

MAZUR, B. J. and FALCO, S. C. (1989) The development of herbicide resistant crops. *Annu. Rev. Plant Physiol. Plant Mol. Biol.* **40**, 441–470.

McCLINTOCK, J. T. and DOUGHERTY, E. M. (1987) Superinfection of baculovirus infected gypsy moth cells with nuclear polyhedrosis viruses of *Autographa californica* and *Lymantria dispar. Virus Res.* **7**, 351–364.

McCOY, C. W., SAMSON, R. A. and BOUCIAS, D. G. (1988) Entomogenous fungi. In: Ignaffo, C. M. and Mandava, N. B., eds, *Handbook of Natural Pesticides, Vol. V., Microbial Pesticides. Part A: Entomogenous Protozoa and Fungi*, Boca Raton: CRC Press, pp. 151–236.

McCUTCHEN, B. F., CHOUDARY, P. V., CRENSHAW, R., MADDOX, K., KAMITA, S. G., PALEKAR, N., VOLRATH, S., FOWLER, E., HAMMOCK, B. D. and MAEDA, S. (1991) Development of a recombinant baculovirus expressing an insect-selective neurotoxin: potential for pest control. *Biotechnology* **9**, 848–852.

McEVOY, P. B. (1996) Host specificity and biological pest control. *BioScience* **46**, 401–405.

MEADOWS, M. P. (1993) *Bacillus thuringiensis* in the environment: ecology and risk assessment. In: Entwistle, P. F., Cory, J. S., Baily, M. J. and Higgs, S., eds, Bacillus thuringiensis. *An Environmental Biopesticide*, Chichester: John Wiley and Sons, pp. 193–320.

MERRYWEATHER, A. T., WEYER, U., HARRIS, M. P. G., HIRST, M., BOOTH, T. F. and POSSEE, R. D. (1990) Construction of genetically engineered baculovirus insecticides containing the *Bacillus thuringiensis* subspecies *kurstaki* delta endotoxin. *J. Gen. Virol.* **71**, 1535–1544.

O'REILLY, D. R. and MILLER, L. K. (1989) A baculovirus blocks insect molting by producing ecdysteroid UDP-glucosyl transferase. *Science* **245**, 1110–1112.

O'REILLY, D. R., MILLER, L. K. and KUCKOW, V. A. (1992) *Baculovirus Expression Systems – A Laboratory Manual*, New York: W. H. Freeman and Co.

PALLI, S. R., CAPUTO, G. F., BROWNRIGHT, A. J., LADD, T. R., COOK, B. J., PRIMAVERA, M., ARIF, B. M. and RETNAKARAN, A. (1996) *Cf*MNPV blocks AcMNPV-induced apoptosis in a continuous midgut cell line. *Virology* **222**, 203–213.

RIDDIFORD, L. M. and TRUMAN, J. W. (1993) Hormone receptors and the regulation of insect reproduction. *Am. Zool.* **33**, 340–347.

SCHULER, M. L., WOOD, H. L., GRANADOS, R. R. and HAMMER, D. A. (1994) *Baculovirus Expression Systems and Biopesticides*, New York: Wiley-Liss.

SOHI, S. S. (1995) Development of lepidopteran cell lines. In: Richardson, C. D., ed., *Baculovirus Expression Protocols, Methods in Molecular Biology*, Vol. 39, Totowa, NJ: Humana Press, pp. 397–412.

SOHI, S. S., PERCY, J., CUNNINGHAM, J. C. and ARIF, B. M. (1981) Replication and serial passage of a multicapsid nuclear polyhedrosis virus of *Orgyia pseudotsugata* (Lepidoptera: Lymantriidae) in continuous insect cell lines. *Can. J. Microbiol.* **28**, 1133–1139.

SOHI, S. S., CAPUTO, G. F., COOK, B. J. and PALLI, S. R. (1996a) Growth of a *Choristoneura fumiferana* midgut cell line in different culture media. *In Vitro Cell. Dev. Biol.* **32**, 39A.

SOHI, S. S., KAUPP, W. J., MacDONALD, J. A., COOK, B. J., EBLING, P. M., BROWN, K. W., CAPUTO, G. F., PALLI, S. R., ARIF, B. M. and RETNAKARAN,

A. (1996b) Isolation of a defective virus by passaging field-collected *Cf*MNPV in a homologous cell line. *In Vitro Cell. Dev. Biol.* (in press).

SPIERS, A. G. and HOPCROFT, D. H. (1988) Factors affecting *Chondrosteum purpureum* infection of *Salix. Eur. J. For. Path.* **257**, 257–278.

STEWART, L. M. D., HIRST, M., LOPEZ-FERBER, M., MERRYWEATHER, A. T., CAYLEY, P. J. and POSSEE, R. D. (1991) Construction of an improved baculovirus insecticide containing an insect-specific toxin gene. *Nature* **352**, 85–88.

ST LEGER, R. J., JOSHI, L., BIDOCHKA, M. J. and ROBERTS, D. W. (1996) Construction of an improved mycoinsecticide over-expressing a toxic protease. *Proc. Natl. Acad. Sci. USA* **93**, 6349–6354.

STRAND, M. R. and OBRYCKI, J. J. (1996) Host specificity of insect parasitoids and predators. *BioScience* **46**, 422–429.

SUMMERS, M. D. and SMITH, G. E. (1987) *A Manual of Methods for Baculovirus Vectors and Insect Cell Culture Procedures*, Texas: College Station.

SWEETMAN, H. L. (1958) *The Principles of Biological Control*, Dubuque, IA: Brown.

TE BEEST, T. O., YANG, X. B. and CISAR, C. R. (1992) The status of biological control of weeds with fungal pathogens. *Annu. Rev. Phytopathol.* **30**, 637–657.

THIEM, S. M., DU, X., QUENTIN, M. E. and BERNER, M. M. (1996) Identification of a baculovirus gene that promotes *Autographa californica* nuclear polyhedrosis virus replication in a non permissive cell line. *J. Virol.* **70**, 2221–2229.

TINSLEY, T. W. (1979) The potential of insect pathogenic viruses as pesticidal agents. *Am. Rev. Entomol.* **24**, 63–87.

TOMALSKI, M. D. and MILLER, L. K. (1991) Insect paralysis by baculovirus-mediated expression of a mite-neutoroxin gene. *Nature* **352**, 82–85.

TOMALSKI, M. D. and MILLER, L. K. (1992) Expression of the paralytic neurotoxin to improve baculoviruses as biopesticides. *Bio/Technology*, **10**, 545–549.

WAGNER, R. G. (1992) Toward integrated forest vegetation management. *J. Forestry* **92**, 26–30.

WALL, R. E., PRASAD, R. and SHAMOUN, S. R. (1992) The development and potential role of mycoherbicides for forestry. *For. Chronicle* **68**, 736–741.

WEI-DONG, X., ARIF, B. M., DOGAS, P. and KRELL, P. J. (1995) Identification and analysis of a putative origin of DNA replication in the *Choristoneura fumiferana* mutinucleocapsid nuclear polyhedrosis virus genome. *Virology* **209**, 409–419.

WEITZMAN, M. D., POSSEE, R. D. and KING, L. A. (1992) Characterization of two variants of *Panolis flammea* multiple nucleocapsid nuclear polyhedrosis virus. *J. Gen. Virol.* **7**, 1881–1886.

WOOD, H. A. (1996) Recombinant viral insecticides: overcoming production and environmental issues. *Proceedings of the symposium on Commercialization of Biopesticides, Applied Products and Transgenic Plants*, 22–24 January, Washington, DC, International Business Communications, Southborough, MA, USA.

WOOD, H. A., HUGHES, P. R., JOHNSON, L. B. and LANGRIDGE, W. H. R. (1981) Increased virulence of *Autographa californica* nuclear polyhedrosis virus by mutagenesis. *J. Invert. Pathol.* **38**, 236–241.

Transgenic Trees

ARMAND SÉGUIN, GILLES LAPOINTE AND PIERRE J. CHAREST

18.1 Introduction

Forest biotechnology includes a number of technologies such as cell culture, enzyme production to improve or obtain a product, and recombinant DNA technology. In this chapter, biotechnology for the production of transgenic trees with improved silvicultural characteristics will be covered. The various traits obtained by recombinant DNA methods, the tissue culture aspects and the gene transfer procedures to produce transgenic trees will be discussed. The main goal of this type of technology is to increase the productivity of managed forests while providing environmentally sound methods for the sustainable management of forest resources.

Tree genetic improvement is a tedious endeavour because of the long reproductive cycle of these plants. For instance, the most advanced programmes for conifer tree improvement, such as for radiata pine, are into their fourth generation. For deciduous tree species such as poplar, tree improvement is in the second generation of improved hybrids, and for eucalyptus, it is in the fourth generation of the breeding cycle. In contrast to crop plants, tree species are considered to be only semi-domesticated. To accelerate the tree improvement cycle, genetic engineering techniques offer the potential to transfer, in one tissue culture cycle, traits that are monogenic or oligogenic. This solves the problem of transferring entire sets of chromosomes in conventional breeding that can result in undesired genotypes. Also, it eliminates the need to produce seeds because of the tissue culture process used in most gene transfer procedures and allows for the direct regeneration of transgenic trees. Another advantage of genetic engineering is the ability to transfer genes from a different species or from a totally different organism.

Transgenic plant recovery is relatively recent and was first reported in the late 1970s using the plant pathogen *Agrobacterium tumefaciens* with tobacco (*Nicotiana tabacum*) (Chilton *et al.*, 1977). This bacterium has the property of transferring a fragment of DNA (T-DNA) present on a plasmid (Ti plasmid) into the genome of the infected plant cells. Following the demonstration of natural gene transfer, the first engineered genes were introduced into the T-DNA (Barton *et al.*, 1983) to produce transgenic plants. Also, functional selectable marker genes for the transformation event were used in laboratories in Europe (Herrera-Estrella *et al.*, 1983) and the United States (Fraley *et al.*, 1983) to produce transgenic plants. Since then, the list of plant species from which transgenic plants were regenerated has grown dramatically and over 100 different species have been used for gene transfer experiments. Initially, this was limited to model species included in the Solanaceae (*Nicotiana, Petunia, Solanum* and *Lycopersicon*), but now it also includes members of the Leguminosae, Cruciferae, and cereals (Charest and Michel, 1991; Christou, 1995; Walden and Wingender, 1995).

With trees, progress was slower due to a lack of scientists involved and to more difficulties in developing tissue culture protocols. For instance, somatic embryogenesis in conifers was used successfully to regenerate conifer trees for the first time in 1985 (Hakman *et al.*, 1985). In 1987, the first reports of transgenic tree regeneration were published using *Agrobacterium*-mediated transformation of poplars (Fillatti *et al.*, 1987; Pythoud *et al.*, 1987). The first evidence of transgenic conifer regeneration was published in 1993 where microprojectile-mediated DNA delivery was used to produce transgenic white spruce (Ellis *et al.*, 1993). The advances with both angiosperm and gymnosperm trees have been considerable but gene transfer procedures are routine only for certain hybrids of poplar (Jouanin *et al.*, 1993; Schuerman and Dandekar, 1993).

18.2 Steps in the Production of Transgenic Trees

Production of transgenic trees includes several independent steps that all result from different products of biotechnology that are based on recombinant DNA, microbiology, tissue culture and sometimes engineering. These steps can be divided as follows:

1 Gene isolation and characterization.
2 Transformation vector construction.
3 Tissue culture of recipient host tree.
4 Gene delivery into host cells.
5 Selection or screening for the transformed cells.
6 Transgenic tree regeneration.

18.3 Gene Isolation and Characterization

Progress in crop improvement through genetic engineering has been considerable over the last few years and many reviews and books have been published on this topic (Hooykaas and Schilperoort, 1992; Christou, 1993; Maliga *et al.*, 1995; Potrykus and Spangenberg, 1995; Vain *et al.*, 1995). These improvements were achieved by the transfer of single genes that were first isolated, characterized and manipulated. Several strategies can be utilized to isolate a gene of interest, such as differential screening of genes expressed in different tissues or in response to different stimuli, or by using homologous DNA sequences from genes of other organisms for DNA hybridization or for polymerase chain reaction amplification. We will not cover these technologies because of the complexity of the different steps and techniques involved. The reader should consult reference manuals such as *Current Pro-*

tocols in Molecular Biology (Ausubel *et al.*, 1987) or *Molecular Cloning* (Sambrook *et al.*, 1989). These strategies have been extremely successful and several traits of agricultural importance have been characterized at the gene level. We will cover single gene traits that are of silvicultural importance.

18.3.1 Insect Pest Resistance

Aerial spraying of bacterial preparations of *Bacillus thuringiensis* has been widely used in forestry for controlling insect pests such as the spruce budworm. The bacterial toxicity is due to a crystal toxin found in sporulating bacteria; the gene for this toxin has been isolated and characterized (Gill *et al.*, 1992; Knowles and Dow, 1993; Llewellyn *et al.*, 1994; Cannon, 1995). The gene called δ-endotoxin was manipulated to improve its efficacy in plants and has been used to produce transgenic plants resistant to insect pests. *Bacillus thuringiensis* δ-endotoxin genes effective against Lepidoptera, Coleoptera, Diptera and nematodes have been identified (Braun *et al.*, 1991; Raman and Altman, 1994). The introduction of this gene was possible in transgenic poplar (McCown *et al.*, 1991), spruce (Ellis *et al.*, 1993), and larch (Shin *et al.*, 1994), and conferred tolerance to target insects.

The other approach is to use gene products that are involved in inhibition of the action of digestive enzymes in insects and animals; these are called proteinase inhibitors (Duffey and Stout, 1996). These genes are often found naturally in some plants and are induced systematically by insect attack. Several were isolated and characterized (Ryan, 1990), and used to produce transgenic plants (Gatehouse *et al.*, 1993). For example, proteinase inhibitor genes were used to produce tolerance to insect pests in transgenic tobacco (Hilder *et al.*, 1987; Johnson *et al.*, 1989). In trees, a popular cDNA with a sequence similarity to Kunitz-type trypsin inhibitors has been isolated (Bradshaw *et al.*, 1989) and transformation experiments have shown that one member of this gene family is transcriptionally activated in response to mechanical wounding (Hollick and Gordon, 1993). Genetic engineering of poplar hybrids with proteinase inhibitor genes for improving pest resistance has been also investigated (Klopfenstein *et al.*, 1993; Leple *et al.*, 1995).

Special consideration has to be given to the environmental aspects of genetic engineering of pest resistance due to the potential development of resistant pests following long-term exposure to the resistance mechanism in transgenic trees (Raffa, 1989; Strauss *et al.*, 1995a). For example, constitutive expression of a single gene encoding for pest resistance in a plantation of transgenic trees is not desirable. A preferable approach is to express the introduced genes only after insect attack or in specific tissues. Furthermore, strategies of introducing more than one gene to provide resistance/tolerance to an insect pest or to use mosaics of transgenic and non-transformed trees are more environmentally sound.

18.3.2 Fungal and Bacterial Disease Resistance

Both fungi and bacteria can cause severe damage to trees either at the seedling or mature stage. Genetic engineering of fungal disease resistance was achieved in crop plants by transferring genes coding for antifungal proteins from bacteria or other plants. These proteins have the capacity to inhibit the growth of fungi *in vitro*

(Cornelissen and Melchers, 1993). From another point of view, characterization of the molecular signals involved in pathogen recognition and of the molecular events that specify the expression of resistance may lead to novel strategies for plant disease control (Lamb *et al.*, 1992; Lindsay *et al.*, 1993; Staskawicz *et al.*, 1995).

Bacterial disease resistance could be engineered into crop plants by using genes from insects that are part of their immune system, such as genes encoding for cecropins and magainins (Broekaert *et al.*, 1995; Rao, 1995). An alternative strategy is to transfer genes encoding detoxification enzymes for bacterial pathogen toxins (Anzai *et al.*, 1989).

18.3.3 Herbicide Resistance

Herbicide resistance in trees could be a useful characteristic in nurseries and in intensively managed plantations where weeds have to be controlled to favour the establishment and growth of young trees. Three different strategies were used in transgenic crop plants for engineering herbicide resistance: introduction of a mutant gene isolated from a plant coding for an enzyme affecting the herbicide binding site (e.g. resistance to glyphosate and sulphonylureas); introduction of a chimeric gene for overproduction of the herbicide target plant enzyme (e.g. glyphosate); and transfer of a microorganism gene coding for an enzyme that detoxifies a herbicide (e.g. phosphinotricin, bromoxynil). Transgenic poplars engineered to be resistant to glyphosate (Fillatti *et al.*, 1987; Donahue *et al.*, 1994), phosphinotricin (De Block, 1990; Jouanin *et al.*, 1993) and sulphonylureas (Miranda Brasileiro *et al.*, 1992) have been produced. For conifers, transgenic larch expressing the *aroA* gene for glyphosate resistance has been also obtained (Shin *et al.*, 1994).

18.3.4 Flower Sterility

There are two reasons behind the engineering of flower sterility in trees related to productivity and to gene flow containment. On the productivity side, considerable energy is spent by trees to produce flowers and it is hypothesized that prevention of flowering would increase biomass production. The other potential outcome of the release of transgenic trees is associated with their long life cycle and their semi-domestication. To be environmentally acceptable, use of transgenic trees in reforestation programmes could require that genetically transformed trees be unable to reproduce sexually (Strauss *et al.*, 1995a). Engineering of sterility can be achieved by cell ablation of reproductive structures done through expression of cytotoxic genes (Barnase or DTA) or RNAses in flower cells (Goldberg *et al.*, 1993, 1995). Male sterility was achieved using this last approach to destroy pollen (Mariani *et al.*, 1991).

A different approach is to modify the expression of tree genes involved in flower formation such as homeotic genes (Weigel and Meyerowitz, 1994; Weigel, 1995) in order to prevent normal gamete development. This approach is being pursued by several groups around the world and has been demonstrated in transgenic *Arabidopsis* and tobacco (Rutledge, unpublished). This work is being done in transgenic poplar (Strauss *et al.*, 1995b). On the other hand, transgenic aspens in which the flower-meristem-identity gene LEAFY of *Arabidopsis thaliana* is constitutively expressed showed accelerating flower development (Weigel and Nilsson, 1995).

18.3.5 *Phytoremediation*

Bioremediation technologies have become increasingly important as a means of cleaning up sites contaminated with undesirable chemicals. These methods take advantage of natural biological processes divided in two broad categories: immobilization, sequestration, or inhibition of the toxic material in order to avoid uptake by other living organisms; and *in vivo* detoxification by chemical modification (oxidative pathway). So far, most bioremediation schemes use microorganisms to metabolize or chemically to convert hazardous or toxic substances. An alternative strategy could be based on the hypothesis of using plants as pumping and filtering systems. The potential of some herbaceous species such as *Thlaspi* spp. for bioremediation of soils contaminated with heavy metals was evaluated (Baker and McGrath, 1991) and the physiological processes of metal tolerance in plants were investigated in a number of plant species including trees (Robinson, 1990; Macnair, 1993). Enhanced metal tolerance in tobacco plants through genetic engineering was investigated using mammalian metallothioneins and increased resistance to cadmium chloride was obtained in a few cases (Lefebvre *et al.*, 1987; Maiti *et al.*, 1988; Misra and Gedamu, 1989; Brandle *et al.*, 1993; Pan *et al.*, 1994).

The perennial nature of woody plants as well as their extensive root systems will make them ideal candidates for phytoremediation. Moreover, root surfaces of trees act as a microenvironment for the rhizosphere microorganisms where root exudates support the growth of a wide diversity of microorganisms including bacteria, mycorrhizal fungi, nitrogen fixing actinomycetes and blue-green algae. Genetic engineering of poplar for phytoremediation (Stomp *et al.*, 1994) was achieved by transferring bacterial genes encoding enzymes involved in the detoxification of chlorophenols such as trichloroethylene (TCE). The added benefits of using trees for bioremediation are the prevention of soil erosion and of contaminant dispersion.

18.3.6 *Lignin Modification*

Removal of lignin from wood fibres during pulping is expensive and pollutive. Consequently, trees with decreased lignin content would represent a significant improvement. Lignin is the most abundant organic compound in the biosphere after cellulose; in trees, it represents 15–35 per cent of their dry weight. Lignin is formed from phenylalanine through the phenylpropanoid biosynthetic pathway (Whetten and Sederoff, 1995). In addition to lignin, branching of the phenylpropanoid pathway supports production of many other aromatic compounds involved in defence against environmental aggressions (ultraviolet light and/or pathogens) (Dixon and Paiva, 1995) as well as in the development of specific structures like reproductive organs (Koes *et al.*, 1994). To engineer reduced lignin content, it will be important to target the enzymes involved in the final step of lignin synthesis and deposition to avoid weakening these defence mechanisms.

During plant development, lignin is embedded in the secondary cell walls of lignifying tissues (mainly xylem), providing hardness and structural support to the cell walls. Biochemical investigations of enzymes involved in lignin synthesis provided information on the key steps in lignification (Boudet *et al.*, 1995). Several reports were published on the production of angiosperm transgenic plants with modified lignin biosynthetic enzyme expression and in some instances modified lignin content (Whetten and Sederoff, 1995; Campbell and Sederoff, 1996). Transgenic poplars expressing different regions of the COMT (caffeic acid, 5-hydroxyferulic acid O-methyltransferase) and CAD (cinnamyl alcohol dehydrogenase) genes in sense and antisense orientation were obtained and analyzed. Transgenic lines with low expression for COMT and CAD genes were regenerated with reduced corresponding enzyme activities (Boerjan *et al.*, 1997). For the down-regulated CAD plants, the enzymatic inhibition was correlated with a red coloration of the wood, with modified lignin characteristics and structures but with the identical lignin quantity. For down-regulated COMT transgenic poplar, a 95 per cent reduction of the COMT activity did not result in a reduction of the amount of lignin but lignin composition was modified. Altered expression of the anionic peroxidase (*prxA1*) in poplar by genetic transformation resulted in increased total peroxidase activity (Kajita *et al.*, 1994) but lignin characteristics and structures were not evaluated.

18.3.7 *Frost and Drought Tolerance*

Tolerance to extreme environmental conditions can be modified in plants through genetic engineering and could prove critical for tree species given the threats of climate change and global warming. Frost tolerance has been genetically engineered in plants by either introducing an anti-freeze gene from fish (Georges *et al.*, 1990) or by manipulating the fatty acid metabolism (Murata *et al.*, 1992). Drought tolerance is at the conceptual stage now and plants such as the resurrection plant (*Craterostima plantagineum*) that can resist high desiccation were studied at the molecular level to identify possible gene products involved in this resistance (Bartels *et al.*, 1992). Several approaches have been considered such as modification of the sugar or the amino acid biosynthetic pathways (Bohnert *et al.*, 1995).

18.4 *In Vitro* Tissue Culture

Gene transfer requires the availability of tree cells that are competent both to receive and integrate the foreign DNA, and to regenerate complete trees. In most transformation procedures with plants, tissue or cell cultures were used as recipient material for gene transfer. However, in an increasing number of cases, *in planta* transformation can be achieved by infiltrating *Agrobacterium* into developing meristems, thus allowing transfer of the foreign genes into these tissues and subsequent transmission of the foreign genes into the progeny of the transformed plant (Bechtold *et al.*, 1993; Chang *et al.*, 1994). For trees, only *in vitro* tissue and cell cultures were used successfully to regenerate transgenic plants.

Tissue culture refers to the maintenance and propagation of plant tissues (material as small as a single cell) in an axenic (microbiologically sterile conditions) controlled environment. It is possible to induce cell proliferation *in vitro* from a variety of tree explants such as leaf and petiole pieces for poplars, and zygotic embryos and young buds for conifers. With regard to the regeneration of entire trees, two processes can be used: organogenesis and somatic embryogenesis. Organogenesis is the regeneration of plants through organ formation on an explant

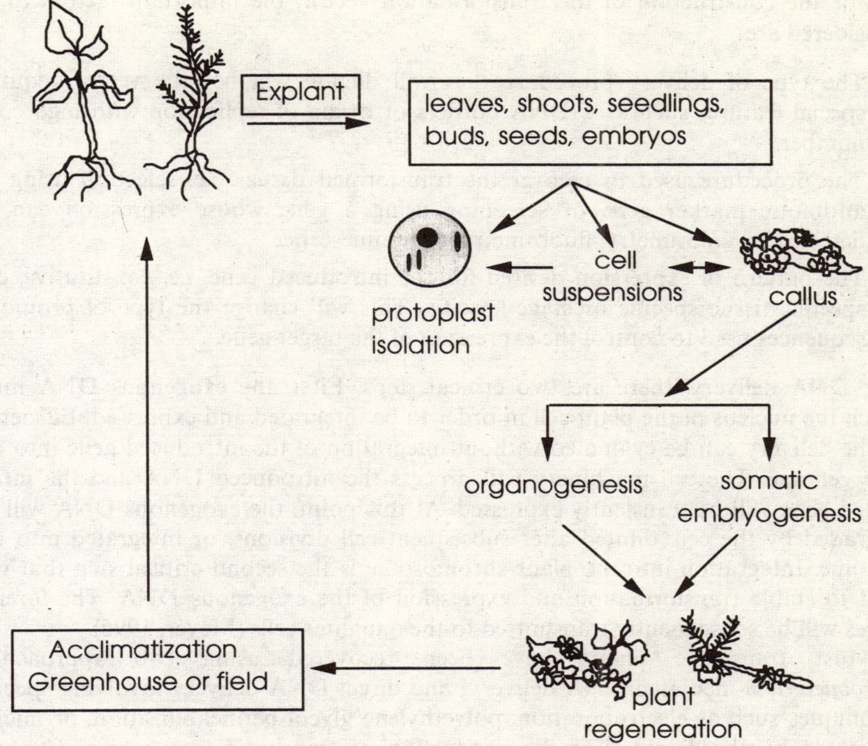

Figure 18.1 Various steps in tree tissue culture. A whole plant can be regenerated from a single cell (a process called totipotency). This system is used to produce genetic clones of a specific plant. Practically any part of a plant can be used for regeneration; however, less differentiated tissue will be easier to initiate *in vitro*.

or from cell masses; somatic embryogenesis is done through the formation of embryo-like structures. Organogenesis has been the method of choice for species such as poplar and eucalyptus, and embryogenesis has been used very successfully with conifers (Tautorus *et al.*, 1991, Lelu *et al.*, 1993; Attree and Fowke, 1995). Both processes provide the means to propagate large numbers of élite trees clonally for research and reforestation. The ability to regenerate plants is a prerequisite for gene transfer and the tissue culture process is critical for integration into a tree improvement scheme. An overview of the various steps in plant tissue culture is presented in Figure 18.1.

18.5 Genetic Transformation Procedures

Once isolated genes are available, they can be modified to ensure their proper expression in the target organism and introduced into the proper cloning vehicle for the DNA delivery method being used. After gene delivery into host cells, transformed cells have to be either screened or selected for the integration of the foreign DNA into their host genome. This section will cover the aspects of transformation vector construction and gene delivery into host cells.

For the construction of the transformation vector, the important factors to be considered are:

1 The type of delivery procedure that will dictate whether the vector requires special features such as T-DNA borders or origin of replication with high copy number.

2 The procedure used to recover the transformed tissues, i.e. selection using an antibiotic marker gene or screening using a gene whose expression can be detected by colorimetry, fluorometry, or luminescence.

3 The pattern of expression desired for the introduced gene, i.e. constitutive, cell specific, tissue specific or stage specific. This will change the type of promoter sequences used to control the expression of the target gene.

For DNA delivery, there are two critical steps. First, the exogenous DNA must reach the nucleus of the plant cell in order to be integrated and expressed. Efficiency of the delivery can be evaluated without integration of the introduced gene into the tree genome. The cell machinery will process the introduced DNA and the introduced gene will be transiently expressed. At this point, the exogenous DNA will be degraded by the cell, diluted after subsequent cell divisions, or integrated into the genome. Integration into the plant chromosome is the second critical step that will lead to stable transformation and expression of the exogenous DNA. The foreign genes will be subsequently transmitted to the daughter cells (Meyer, 1995).

Most transgenic plants have been recovered using two approaches: *Agrobacterium*-mediated DNA delivery; and direct DNA delivery involving specific techniques such as electroporation, polyethylene glycol permeabilization, or microprojectile bombardment. For the production of transgenic trees, *Agrobacterium*-mediated transformation has been used mainly for deciduous (angiosperm) trees and microprojectile bombardment for conifer (gymnosperm) trees. An overview of the approaches used in the genetic transformation of trees is presented in Figure 18.2.

18.5.1 Agrobacterium-*mediated DNA delivery*

This technology is based on the natural gene transfer system inherent to the bacterial pathogens *Agrobacterium tumefaciens* and *Agrobacterium rhizogenes* which are the causative agents of the crown gall and hairy root diseases respectively (Hooykaas and Schilperoort, 1992; Walden and Wingender, 1995). Both species carry large plasmids called Ti for tumour inducing or Ri for root inducing. A fraction of these plasmids, designated T-DNA, is transmitted by the bacteria into single plant cells, usually at a wounded site. Thereafter, the T-DNA will be translocated to the nucleus where it will be stably integrated randomly into the genome. Expression of the genes present in this T-DNA results in the synthesis of gene products involved in hormone biosynthesis that direct phenotypic changes such as tumour or hairy root formation. For genetic engineering purposes, these genes were removed and replaced by marker genes or other genes coding for the traits of interest. As mentioned earlier, deciduous trees such as poplars are efficiently transformed by wild-type *Agrobacterium* but are more difficult to engineer with *Agrobacterium* containing disarmed (without hormone genes) T-DNA (Table 18.1).

Table 18.1   Genetic transformation of tree species

Genus	Method	Transferred genes	Reference
Allocasuarina	*Agrobacterium rhizogenes*	T-DNA	Phelep *et al.* (1991)
Azadiracta	*Agrobacterium tumefaciens*	ocs/nptII	Naina *et al.* (1989)
Eucalyptus	*A. rhizogenes*	T-DNA	MacRae and Van Staden (1993)
Juglans	*A. tumefaciens*	nptII	McGranahan *et al.* (1988)
	A. tumefaciens	nptII/uidA	McGranahan *et al.* (1990)
	A. tumefaciens	nptII/uidA	Miranda Brasileiro *et al.* (1991)
	A. tumefaciens	nptII/uidA/aschs	Jouanin *et al.* (1993)
Larix	*A. rhizogenes*	T-DNA	Huang *et al.* (1991)
Liriodendron	Microprojection	nptII/uidA	Wilde *et al.* (1992)
Picea	Microprojection	uidA/nptII/cryIA	Ellis *et al.* (1993)
	Microprojection	uidA/nptII	Robertson *et al.* (1992)
	Microprojection	uidA/nptII	Charest *et al.* (1996)
Populus	*A. tumefaciens*	aroA/nptII	Fillatti *et al.* (1987)
	A. rhizogenes	bar/nptII	De Block (1990)
	A. tumefaciens	nptII/crs1-1	Miranda Brasileiro *et al.* (1992)
	A. tumefaciens	nptII/uidA	Miranda Brasileiro *et al.* (1991)
	Microprojection	nptII/uidA/cryIAa	McCown *et al.* (1991)
	A. tumefaciens	nptII/pin2-cat	Klopfenstein *et al.* (1991)
	A. tumefaciens	nptII/uidA	Leple *et al.* (1992)
	A. tumefaciens	nptII, hptII/luxab	Nilsson *et al.* (1992)
	A. rhizogenes	nptII/bar	Jouanin *et al.* (1993)
	A. rhizogenes	nptII	Pythoud *et al.* (1987)
	A. tumefaciens	aroA/nptII	Donahue *et al.* (1994)
	A. tumefaciens	uidA	Confalonieri *et al.* (1995)
Robinia	*A. rhizogenes*	nptII	Han *et al.* (1993)

18.5.2   *Direct DNA Transfer*

Direct DNA transfer methods are based on DNA delivery into the plant cells without any intermediary. Some methods, such as electroporation that uses an electric pulse or chemical poration with polyethylene glycol (PEG) or polyvinyl alcohol to allow the entry of DNA into the cells, increase the permeability of plant cell walls and membranes. These methods can be used with whole tissues, cells or protoplasts (cells from which the walls have been removed enzymatically). Transgenic plants of important crops such as rice and corn have been obtained following protoplast electroporation. However this approach was, tested with deciduous and coniferous trees and only yielded transgenic poplar (Chupeau *et al.*, 1994).

The other commonly used method is microprojectile-mediated DNA delivery and uses a strong force to accelerate tungsten or gold microprojectiles (1–10 μm in size) coated with DNA which will penetrate cell walls and membranes (Klein *et al.*, 1988; Gordon-Kamm *et al.*, 1990; Christou, 1995). After penetration into the host cell the DNA is released, resulting in expression of the genes present. The major advantage of this technology is that it can be used to introduce DNA into a wide range of species and tissues (Séguin *et al.*, 1995). It has been used in tree species to introduce DNA into various differentiated tissues and organs (Charest *et al.*, 1993) and allow

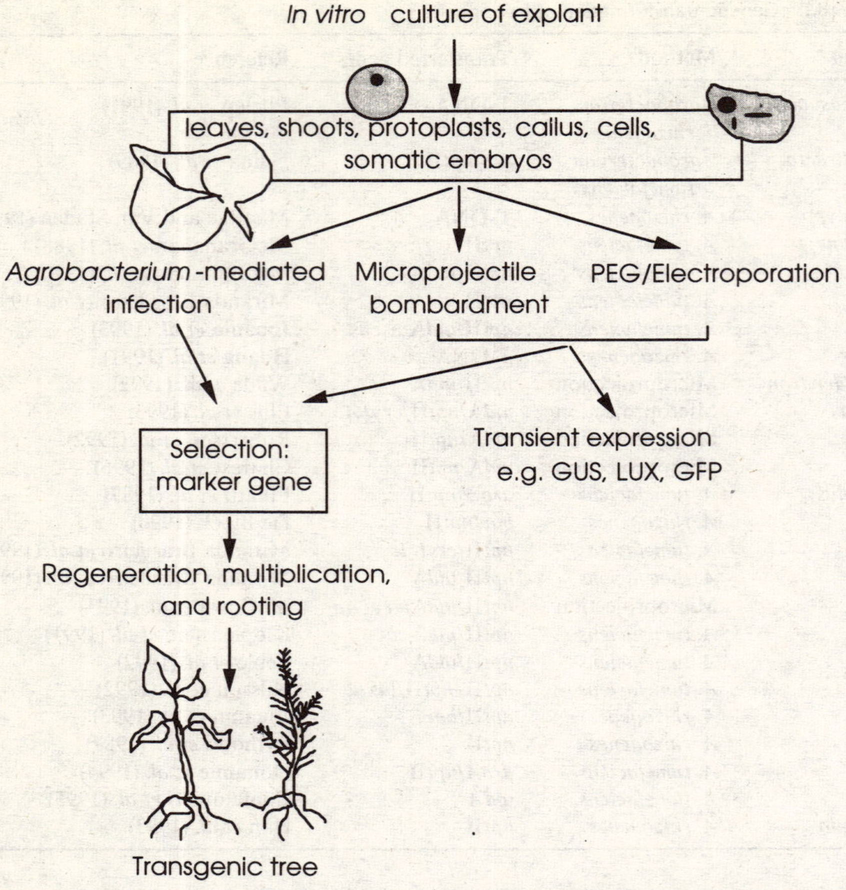

Figure 18.2   Approaches to genetic transformation of trees. PEG, polyethyleneglycol; GUS, β-glucuronidase; LUX, bacterial luciferase; GFP, green fluorescent protein

rapid testing of various gene constructions (Ellis *et al.*, 1991; Duchesne and Charest, 1992; Newton *et al.*, 1992).

Furthermore, microprojectile bombardment is the approach that initially yielded the recovery of transgenic conifers. Transgenic white spruce, black spruce, tamarack and radiata pine were regenerated (Ellis *et al.*, 1993; Walter *et al.*, 1994; Charest *et al.*, 1996; Klimaszewska *et al.*, 1997).

Other direct DNA uptake methods include precise transfer of DNA into a cell through a microscopic needle (microinjection) or transfer of DNA through the use of microscopic silicon carbide fibres that perforate cell walls and membranes. Both methods have been successfully used to produce transgenic herbaceous plants but not tree species (Potrykus and Spangenberg, 1995).

18.6   Environmental Aspects of Deploying Transgenic Trees

Field tests of transgenic trees have been established in several countries with poplar and spruce. These tests aim to evaluate the long-term stability of the genes introduced and to investigate the potential environmental effects of the introduced genes. Essentially, the aspects being studied at this stage are:

1 Development of weediness.

2 Gene flow to natural tree populations.

3 Impact on target and non-target pests.

4 Impact on food chain.

5 Impact on associated organisms (beneficial ones such as mycorrhizae).

Some of these points were covered earlier in the section on genes of silvicultural importance. These potential effects have to be evaluated before any transgenic tree is used on a large scale and they have to be placed in a larger perspective with the deployment strategy, which can either minimize or increase the risks to the environment (Rogers and Parkes, 1995).

18.7   Integration of Genetic Engineering in the Tree Improvement Cycle

Tree improvement is a relatively long process limited by the rate, of growth of the species involved and by the quantity of improved seeds that can be obtained through each breeding cycle. It is also a costly process because it requires the establishment of seed orchards, complex breeding plots and often extensive programmes of accelerated flowering. Genetic engineering makes it possible to shortcut this process by allowing the transfer of single gene traits and by using tissue culture procedures that permit the production of an unlimited number of propagules. This biotechnology is a critical component for the development of intensively managed tree plantations that would resemble agriculture more than forestry. Furthermore, biotechnological methods for pest management have to be integrated into this type of managed plantation to protect the large investment that would be made.

Tree genetic engineering is a logical complement to tree breeding with the difference that it is more precise and more focused. Traits such as lignin content and pest tolerance that can be manipulated through the introduction of single genes could allow the production of designer trees for either specific industrial use or for situations where no other alternative is available. In the latter circumstance, it may be that a specific tree species could not be grown in a given area because of the presence of a highly devastating disease. Introduction of disease tolerance through genetic engineering could thus be the only solution to this problem. As more becomes known about the molecular biology of tree species, more opportunities will be identified for improving tree growth.

References

ANZAI, H., YONEYAMA, K. and YAMAGUCHI, I. (1989) Transgenic tobacco resistant to a bacterial disease by the detoxification of a pathogenic toxin. *Mol. Gen. Genet.* **219**, 492–494.

ATTREE, S. M. and FOWKE, L. C. (1995) Conifer somatic embryogenesis, embryo development, maturation drying and plant formation. In: Gamborg, O. L. and Phillips, G. C., eds, *Plant Cell, Tissue and Organ Culture*, Berlin, Heidelberger, Germany: Springer-Verlag, pp. 103–113.

AUSUBEL, F. M., BRENT, R., KINGSTON, R. E., MOORE, D. D., SEIDMAN, J. G., SMITH, J. A. and STRUHL, K. (1987) *Current Protocols in Molecular Biology*, New York: Wiley.

BAKER, A. J. M. and MCGRATH, S. P. (1991) *In situ* decontamination of heavy metal polluted soils using crops of metal-accumulating plants – a feasibility study. In: Hichee, R. E. and Olfenbuttel, R. F., eds, *In Situ Bioreclamation*, Stoneham, MA: Butterworth-Heinemann, pp. 400–405.

BARTELS, D., HANKE, C., SCHNEIDER, K., MICHEL, D. and SALAMINI, F. (1992) A desiccation-related Elip-like gene from the resurrection plant *Craterostigma plantagineum* is regulated by light and ABA. *EMBO J.* **11**, 2771–2778.

BARTON, K. A., BINNS, A. N., MATZKE, A. J. M. and CHILTON, M. D. (1983) Regeneration of intact tobacco plants containing full length copies of genetically engineered T-DNA, and transmission of T-DNA to R1 progeny. *Cell* **32**, 1033–1043.

BECHTOLD, N., ELLIS, J. and PELLETIER, G. (1993) *In planta Agrobacterium*-mediated gene transfer by infiltration of adult *Arabidopsis thaliana* plants. *C R Acad. Sci. Paris, Life Sci.* **316**, 1194–1199.

BOERJAN, W., BAUCHER, M., CHABBERT, B., PETIT-CONIL, M., LEPLE, J. C., PILATE, G., CORNU, D., MONTIES, B., INZE, D., VAN DOORSSELAERE, J., JOUANIN, L. and VAN MONTAGU, M. (1997) Genetic modification of lignin biosynthesis in quaking aspen (*Populus tremuloides*) and poplar (*Populus tremula* × *Populus alba*). Klopfenstein, N. and Chu, Y. W., eds, Lincoln, NE: USDA, National Agroforestry Centre (in press).

BOHNERT, H. J., NELSON, D. E. and JENSEN, R. G. (1995) Adaptations to environmental stresses. *Plant Cell* **7**, 1099–1111.

BOUDET, A. M., LAPIERRE, C. and GRIMA-PETTENATI, J. (1995) Biochemistry and molecular biology of lignification. *New Phytol.* **129**, 203–236.

BRADSHAW, H. JR, HOLLICK, J., PARSONS, T., CLARKE, H. and GORDON, M. (1989) Systematically wound-responsive genes in poplar trees encode proteins similar to sweet potato sporamins and legume Kunitz trypsin inhibitors. *Plant Mol. Biol.* **14**, 51–59.

BRANDLE, J. E., LABBE, H., HATTORI, J. and MIKI, B. L. (1993) Field performance and heavy metal concentrations of transgenic flue-cured tobacco expressing a mammalian metallothionein-beta-glucuronidase gene fusion. *Genome* **36**, 255–260.

BRAUN, C. J., JILKA, J. M., HEMENWAY, C. L. and TUMER, N. E. (1991) Interactions between plants, pathogens and insects: possibilities for engineering resistance. *Curr. Opin. Biotechnol.* **2**, 193–198.

BROEKAERT, W. F., TERRAS, F. R. G., CAMMUE, B. P. A. and OSBORN, R. W. (1995) Plant defensins: novel antimicrobial peptides as components of the host defense system. *Plant Physiol.* **108**, 1353–1358.

CAMPBELL, M. M. and SEDEROFF, R. R. (1996) Variation in lignin content and composition. *Plant Physiol.* **110**, 3–13.

CANNON, R. J. C. (1995) *Bacillus thuringiensis* in pest control. In: Hokkanen, H. M. T. and Lynch, J. M., eds, *Biological Control: Benefits and Risks*, Cambridge: Cambridge University Press, pp. 190–200.

CHANG, S. S., PARK, S. K., KIM, B. C., KANG, B. J., KIM, D. U. and NAM, H. G. (1994) Stable genetic transformation of *Arabidopsis thaliana* by *Agrobacterium* inoculation *in planta*. *Plant J.* **5**, 551–558.

CHAREST, P. J. and MICHEL, M. F. (1991) *Basics of Plant Genetic Engineering and its Potential Applications to Tree Species*, Report No. PI-X104; Canadian Forest Service, Petawawa National Forest Institute.

CHAREST, P. J., CALERO, N., LACHANCE, D., MITSUMUNE, M. and YOO, B. Y. (1993) The use of microprojectile DNA delivery to bypass the long life cycle of tree species in gene expression studies. In: Menon, J., ed., *Current Topics in Botanical Research*, India: Council of Scientific Research Integration, pp. 151–163.

CHAREST, P. J., DEVANTIER, Y. and LACHANCE, D. (1996) Stable genetic transformation of *Picea mariana* (black spruce) via particle bombardment. *In Vitro Cell. Dev. Plant* **32**, 91–99.

CHILTON, M. D., DRUMMOND, M. H., MERLO, D. J. S. D., MONTOYA, A. L., GORDON, M. P. and NESTER, E. W. (1977) Stable incorporation of plasmid DNA into higher plant cells: the molecular basis of crown gall tumorigenesis. *Cell* **11**, 263–271.

CHRISTOU, P. (1993) Philosophy and practice of variety-independent gene transfer into recalcitrant crops. *In Vitro Cell Dev. Biol. Plant* **29P**, 119–124.

CHRISTOU, P. (1995) Strategies for variety-independent genetic transformation of important cereals, legumes and woody species utilizing particle bombardment. *Euphytica* **85**, 13–27.

CHUPEAU, M. C., PAUTOT, V. and CHUPEAU, Y. (1994) Recovery of transgenic trees after electroporation of poplar protoplasts. *Transgen. Res.* **3**, 13–19.

CONFALONIERI, M., BALESTRAZZI, A., BISOFFI, S. and CELLA, R. (1995) Factors affecting *Agrobacterium tumefaciens*-mediated transformation in several black poplar clones. *Plant Cell Tiss. Organ Cult.* **43**, 215–222.

CORNELISSEN, B. J. C. and MELCHERS, L. S. (1993) Strategies for control of fungal diseases with transgenic plants. *Plant Physiol.* **101**, 709–712.

DE BLOCK, M. (1990) Factors influencing the tissue culture and the *Agrobacterium tumefaciens*-mediated transformation of hybrid aspen and poplar clones. *Plant Physiol.* **93**, 1110–1116.

DIXON, R. A. and PAIVA, N. L. (1995) Stress-induced phenylpropanoid metabolism. *Plant Cell* **7**, 1085–1097.

DONAHUE, R. A., DAVIS, T. D., MICHLER, C. H., RIEMENSCHNEIDER, D. E., CARTER, D. R., MARQUARDT, P. E., SANKHLA, N., SANKHLA, D., HAISSIG, B. E. and ISEBRANDS, J. G. (1994) Growth, photosynthesis and herbicide tolerance of genetically modified hybrid poplar. *Can. J. For. Res.* **24**, 2377–2383.

DUCHESNE, L. C. and CHAREST, P. J. (1992) Effect of promoter sequence on transient expression of the *β*-glucuronidase gene in embryogenic calli of *Larix × eurolepsis* and *Picea mariana* following microprojection. *Can. J. Bot.* **70**, 175–180.

DUFFEY, S. S. and STOUT, M. J. (1996) Antinutritive and toxic components of plant defense against insects. *Arch. Insect Biochem. Physiol.* **32**, 3–37.

ELLIS, D. D., MCCABE, D., RUSSEL, D., MARTINELL, B. and MCCOWN, B. H. (1991) Expression of inductible angiosperm promoters in a gymnosperm, *Picea glauca* (white spruce). *Plant Mol. Biol.* **17**, 19–27.

ELLIS, D. D., MCCABE, D. E., MCINNIS, S., RAMACHANDRAN, R., RUSSELL, D. R., WALLACE, K. M., MARTINELL, B. J., ROBERTS, D. R., RAFFA, K. F. and MCCOWN, B. H. (1993) Stable transformation of *Picea glauca* by particle acceleration. *Bio/Technology* **11**, 84–89.

FILLATTI, J. J., SELLMER, J., MCCOWN, B., HAISSIG, B. and COMAI, L. (1987) *Agrobacterium*-mediated transformation and regeneration of *Populus*. *Mol. Gen. Genet.* **206**, 192–199.

FRALEY, R. T., ROGERS, S. G., HORSCH, R. B., SANDERS, P. R., FLICK, J. S., ADAMS, S. P., BITTNER, M. L., BRAND, L. A., FINK, C. L., FRY, J. S., GALLUPPI, G. R., GOLDBERG, S. B., HOFFMANN, N. L. and WOO, S. C. (1983) Expression of bacterial genes in plant cells. *Proc. Natl. Acad. Sci. USA* **80**, 4803–4807.

GATEHOUSE, A. M. R., SHI, Y., POWELL, K. S., BROUGH, C., HILDER, V. A., HAMILTON, W. D. O., NEWELL, C. A., MERRYWEATHER, A., BOULTER, D. and GATEHOUSE, J. A. (1993) Approaches to insect resistance using transgenic plants. *Phil. Trans. R. Soc. Lond. B.* **342**, 279–286.

GEORGES, F., SALEEM, M. and CUTLER, A. J. (1990) Design and cloning of a synthetic gene for the flounder antifreeze protein and its expression in plant cells. *Gene* **91**, 159–165.

GILL, S. S., COWLES, E. A. and PIETRANTONIO, P. V. (1992) The mode of action of *Bacillus thuringiensis* endotoxins. *Annu. Rev. Entomol.* **37**, 615–636.

GOLDBERG, R. B., BEALS, T. P. and SANDERS, P. M. (1993) Another development: basic principles and practical applications. *Plant Cell* **5**, 1217–1229.

GOLDBERG, R. B., SANDERS, P. M. and BEALS, T. P. (1995) A novel cell-ablation strategy for studying plant development. *Phil. Trans. R. Soc. Lond. B.* **350**, 5–17.

GORDON-KAMM, W. J., SPENCER, T. M., MANGANO, M. L., ADAMS, T. R., DAINES, R. J., START, W. G., O'BRIEN, J. V., CHAMBERS, S. A., ADAMS, W. R. J., WILLETTS, N. G., RICE, T. B., MACKEY, C. J., KRUEGER, R. W., KAUSCH, A. P. and LEMAUX, P. G. (1990) Transformation of maize cells and regeneration of fertile transgenic plants. *Plant Cell* **2**, 603–618.

HAKMAN, I., FOWKE, L. C., VON ARNOLD, S. and ERIKSSON, T. (1985) The development of somatic embryos in tissue cultures initiated from immature embryos of *Picea abies* (Norway spruce). *Plant Sci.* **38**, 53–59.

HAN, K. H., KEATHLEY, D. E., DAVIS, J. M. and GORDON, M. P. (1993) Regeneration of a transgenic woody legume (*Robinia pseudoacacia* L., black locust) and morphological alterations induced by *Agrobacterium rhizogenes*-mediated transformation. *Plant Sci.* **88**, 149–157.

HERRERA-ESTRELLA, L., DE BLOCK, M., MESSENS, E., HERNALSTEENS, J. P., VAN MONTAGU, M. and SCHELL, J. (1983) Chimeric genes as dominant selectable markers in plant cells. *EMBO J.* **2**, 987–995.

HILDER, V., GATEHOUSE, A., SHEERMAN, S., BARKER, R. and BOULTER, D. (1987) A novel mechanism of insect resistance engineered into tobacco. *Nature* **330**, 160–163.

HOLLICK, J. B. and GORDON, M. P. (1993) A poplar tree proteinase inhibitor-like gene promoter is responsive to wounding in transgenic tobacco. *Plant Mol. Biol.* **22**, 561–572.

HOOYKAAS, P. J. J. and SCHILPEROORT, R. A. (1992) *Agrobacterium* and plant genetic engineering. *Plant Mol. Biol.* **19**, 15–38.

HUANG, Y., DINER, A. M. and KARNOSKY, D. F. (1991) *Agrobacterium rhizogenes*-mediated genetic transformation and regeneration of a conifer: *Larix decidua*. *In Vitro Cell. Dev. Biol.* **27P**, 201–207.

JOHNSON, R., NARVAEZ, J., AN, G. and RYAN, C. (1989) Expression of proteinase inhibitors I and II in transgenic tobacco plants: effects on natural defense against *Manduca sexta* larvae. *Proc. Natl Acad. Sci. USA* **89**, 9871–9875.

JOUANIN, L., BRASILEIRO, A. C. M., LEPLE, J. C., PILATE, G. and CORNU, D. (1993) Genetic transformation: a short review of methods and their applications, results and perspectives for forest trees. *Ann. Sci. For.* **50**, 325–336.

KAJITA, S., OSAKABE, K., KATAYAMA, Y., KAWAI, S., MATSUMOTO, Y., HATA, K. and MOROHOSHI, N. (1994) *Agrobacterium*-mediated transformation of poplar using a disarmed binary vector and the overexpression of a specific member of a family of poplar peroxidase genes in transgenic poplar cell. *Plant Sci.* **103**, 231–239.

KLEIN, T. M., HARPER, E. C., SVAB, Z., SANFORD, J. C., FROMM, M. E. and MALIGA, P. (1988) Stable genetic transformation of intact *Nicotiana* cells by the particle bombardment process. *Proc. Natl Acad. Sci. USA* **85**, 8502–8505.

KLIMASZEWSKA, K., DEVANTIER, Y., LACHANCE, D., LELU, M. A. and CHAREST, P. J. (1997) *Larix laricina* (tamarack): somatic embryogenesis and genetic transformation. *Can. J. For. Res.* **27**, 538–550.

KLOPFENSTEIN, N. B., SHI, N. Q., KERNAN, A., McNABB, H. S., HALL, R. B., HART, E. R. and THORNBURG, R. W. (1991) Transgenic *Populus* hybrid expresses a wound-inducible potato proteinase inhibitor II–CAT gene fusion. *Can. J. For. Res.* **21**, 1321–1328.

KLOPFENSTEIN, N. B., MCNABB, H. S., HART, E. R., HALL, R. B., HANNA, R. D., HEUCHELIN, S. A., ALLEN, K. K., SHI, N. Q. and THORNBURG, R. W. (1993) Transformation of populus hybrids to study and improve pest resistance. *Silvae Genet.* **42**, 86–90.

KNOWLES, B. H. and DOW, J. A. T. (1993) The crystal delta-endotoxins of *Bacillus thuringiensis*: models for their mechanism of action on the insect gut. *BioEssays* **15**, 469–476.

KOES, R. E., QUATTROCHIO, F. and MOL, J. N. M. (1994) The flavonoid biosynthetic pathway in plants: function and evolution. *BioEssays* **16**, 123–132.

LAMB, C. J., RYALS, J. A., WARD, E. R. and DIXON, R. A. (1992) Emerging strategies for enhancing crop resistance to microbial pathogens. *Bio/Technology* **10**, 1436–1445.

LEFEBVRE, D. D., MIKI, B. L. and LALIBERTÉ, J. F. (1987) Mammalian metallothionein functions in plants. *Bio/Technology* **5**, 1053–1056.

LELU, M. A., KLIMASZEWSKA, K. K., JONES, C., WARD, C., VON ADERKAS, P. and CHAREST, P. J. (1993) *A Laboratory Guide to Somatic Embryogenesis in Spruce and Larch*, Report No. PI-X-111F, Petawawa National Forestry Institute.

LEPLE, J. C., BRASILEIRO, A. C. M., MICHEL, M. F., DELMOTTE, F. and JOUANIN, L. (1992) Transgenic poplars: expression of chimeric genes using four different constructs. *Plant Cell Rep.* **11**, 137–141.

LEPLE, J. C., BONADEBOTTINO, M., AUGUSTIN, S., PILATE, G., LETAN, V. D., DELPLANQUE, A., CORNU, D. and JOUANIN, L. (1995) Toxicity to *Chrysomela tremulae* (Coleoptera: Chrysomelidae) of transgenic poplars expressing a cysteine proteinase inhibitor. *Mol. Breeding* **1**, 319–328.

LINDSAY, W. P., LAMB, C. J. and DIXON, R. A. (1993) Microbial recognition and activation of plant defense systems. *Trends Microbiol.* **1**, 181–187.

LLEWELLYN, D., COUSINS, Y., MATHEWS, A., HARTWECK, L. and LYON, B. (1994) Expression of *Bacillus thuringiensis* insecticidal protein genes in transgenic crop plants. *Agr. Ecosyst. Environ.* **49**, 85–93.

MACNAIR, M. R. (1993) The genetics of metal tolerance in vascular plants. *New Phytol.* **124**, 541–559.

MACRAE, S. and VAN STADEN, J. (1993) *Agrobacterium rhizogenes*-mediated transformation to improve rooting ability of eucalypts. *Tree Physiol.* **12**, 411–418.

MAITI, I. B., HUNT, A. G. and WAGNER, G. J. (1988) Seed-transmissible expression of mammalian metallothionein in transgenic tobacco. *Biochem. Biophys. Res. Commun.* **150**, 640–647.

MALIGA, P., KLESSING, D. F., CASHMORE, A. R., GRUISSEM, W. and VARNER, J. E. (1995) *Methods in Plant Molecular Biology; A Laboratory Course Manual*, Cold Spring Harbor, New York: Cold Spring Harbor Laboratory Press.

MARIANI, C., GOLDBERG, R. B. and LEEMANS, J. (1991) Engineered male sterility in plants. *Symp. Soc. Exp. Biol.* **45**, 271–279.

MCCOWN, B. H., MCCABE, D. E., RUSSELL, D. R., ROBISON, D. J. and BARTON, K. A. (1991) Stable transformation of *Populus* and incorporation of pest resistance by electric discharge particle acceleration. *Plant Cell Rep.* **9**, 590–594.

MCGRANAHAN, G. H., LESLIE, C. A., URATSU, S. L., MARTIN, L. A. and DANDEKAR, A. M. (1988) *Agrobacterium*-mediated transformation of walnut somatic embryos and regeneration of transgenic plants. *Bio/Technology* **6**, 800–804.

MCGRANAHAN, G. H., LESLIE, C. A., URATSU, S. L. and DANDEKAR, A. M. (1990) Improved efficiency of the walnut somatic embryo gene transfer system. *Plant Cell Rep.* **8**, 512–516.

MEYER, P. (1995) Understanding and controlling transgene expression. *Trends Biotechnol.* **13**, 332–337.

MIRANDA BRASILEIRO, A. C., LEPLE, J. C., MUZZIN, J., OUNNOUGHI, D., MICHEL, M.-F. and JOUANIN, L. (1991) An alternative approach for gene transfer in trees using wild-type *Agrobacterium* strains. *Plant. Mol. Biol.* **17**, 441–452.

MIRANDA BRASILEIRO, A. C., TOURNEUR, C., LEPLE, J. C., COMBES, V. and JOUANIN, L. (1992) Expression of the mutant *Arabidopsis thaliana* acetolactate synthase confers chlorsulfuron resistance to poplar. *Transgen. Res.* **1**, 133–141.

MISRA, S. and GEDAMU, L. (1989) Heavy metal tolerant transgenic *Brassica napus* L. and *Nicotiana tabacum* L. plants. *Theor. Appl. Genet.* **78**, 161–168.

MURATA, N., ISHIZAKI-NISHIZAWA, O., HIGASHI, S., HAYASHI, H., TASAKA, Y. and NISHIDA, I. (1992) Genetically engineered alteration in chilling sensitivity of plants. *Science* **356**, 710–713.

NAINA, N. S., GUPTA, P. K. and MASCARENHAS, A. F. (1989) Genetic transformation and regeneration of transgenic neem (*Azadirachta indica*) plants using *Agrobacterium tumefaciens. Curr. Sci.* **58**, 184–187.

NEWTON, R. J., YIBRAH, H. S., DONG, N., CLAPHAM, D. H. and VON ARNOLD, S. (1992) Expression of an abscisic acid responsive promoter in *Picea abies* (L.) Karst. following bombardment from an electric discharge particle accelerator. *Plant Cell Rep.* **11**, 188–191.

NILSSON, O., TORSEN, A., SITBON, F., LITTLE, C. H. A., CHALUPA, V., SANDBERG, G. and OLSSON, O. (1992) Spatial pattern of cauliflower mosaic 35S promoter-luciferase expression in transgenic hybrid aspen trees monitored by enzymatic assay and non-destructive imaging. *Transgen. Res.* **1**, 209–220.

PAN, A., YANG, M., TIE, F., LI, L., CHEN, Z. and RU, B. (1994) Expression of mouse metallothionein-I gene confers cadmium resistance in transgenic tobacco plants. *Plant Mol. Biol.* **24**, 341–351.

PHELEP, M., PETIT, A., MARTIN, L., DUHOUX, E. and TEMPE, J. (1991) Transformation and regeneration of a nitrogen-fixing tree, *Allocasuarina verticillata* Lam. *Bio/Technology* **9**, 461–466.

POTRYKUS, I. and SPANGENBERG, G. (1995) *Gene Transfer to Plants*, Berlin: Springer-Verlag.

PYTHOUD, F., SINKAR, V. P., NESTER, E. W. and GORDON, M. P. (1987) Increased virulence of *Agrobacterium rhizogenes* conferred by the VIR region of pTi BO542: application to genetic engineering of poplar. *Bio/Technology* **5**, 1323–1327.

RAFFA, K. F. (1989) Genetic engineering of trees to enhance resistance to insects. *BioScience* **39**, 524–534.

RAMAN, K. V. and ALTMAN, D. W. (1994) Biotechnology initiative to achieve plant pest and disease resistance. *Crop Prot.* **13**, 591–596.

RAO, A. G. (1995) Antimicrobial peptides. *Mol. Plant Microb. Inter.* **8**, 6–13.

ROBERTSON, D., WEISSINGER, A. K., ACKLEY, R., GLOVER, S. and SEDEROFF, R. R. (1992) Genetic transformation of Norway spruce (*Picea abies* (L) Karst) using somatic embryo explants by microprojectile bombardment. *Plant Mol. Biol.* **19**, 925–935.

ROBINSON, N. J. (1990) Metal-binding polypeptides in plants. In: Shaw, J. A., ed., *Heavy Metal Tolerance in Plants: Evolutionary Aspects*, Boca Raton, FL: CRC Press.

ROGERS, H. J. and PARKES, H. C. (1995) Transgenic plants and the environment. *J. Exp. Bot.* **46**, 467–488.

RYAN, C. (1990) Protease inhibitors in plants: genes for improving defenses against insect and pathogens. *Annu. Rev. Phytopathol.* **28**, 425–449.

SAMBROOK, J., FRITSCH, E. F. and MANIATIS, T. (1989) *Molecular Cloning: A Laboratory Manual*, 2nd edition, Cold Spring Harbor, NY: Cold Spring Harbor Laboratory.

SCHUERMAN, P. L. and DANDEKAR, A. M. (1993) Transformation of temperate woody crops – progress and potentials. *Sci. Hort.* **55**, 101–124.

SÉGUIN, A., LACHANCE, D. and CHAREST, P. J. (1995) Transient gene expression and stable genetic transformation into conifer tissues by microprojectile bombardment. In: Lindsey, K., ed., *Plant Molecular Biology Manual*, Vol. B13, Dordrecht: Kluwer, pp. 1–46.

SHIN, D. I., PODILA, G. K., HUANG, Y., KARNOSKY, D. F. and HUANG, Y. H.

(1994) Transgenic larch expressing genes for herbicide and insect resistance. *Can. J. For. Res.* **24**, 2059–2067.

STASKAWICZ, B. J., AUSUBEL, F. M., BAKER, B. J., ELLIS, J. G. and JONES, J. D. G. (1995) Molecular genetics of plant disease resistance. *Science* **268**, 661–667.

STOMP, A. M., HAN, K. H., WILBERT, S. M., GORDON, M. P. and CUNNINGHAM, S. D. (1994) Genetic strategies for enhancing phytoremediation. *Recomb. DNA Technol. II* **721**, 481–491.

STRAUSS, S. H., ROTTMANN, W. H., BRUNNER, A. M. and SHEPPARD, L. A. (1995a) Genetic engineering of reproductive sterility in forest trees. *Mol. Breeding* **1**, 5–26.

STRAUSS, S. H., SHEPPARD, L. A., ROTTMANN, W. H., BRUNNER, A. M. and MEILAN, R. (1995b) Expression of floral homeotic genes in *Populus* and use for engineering reproductive sterility. In: *IUFRO Meeting on Somatic Cell Genetics and Molecular Genetics of Trees*, Gent, Belgium: Kluwer.

TAUTORUS, T. E., FOWKE, L. C. and DUNSTAN, D. I. (1991) Somatic embryogenesis in conifers. *Can. J. Bot.* **69**, 1873–1899.

VAIN, P., DE BUYSER, J., BUITRANG, V., HAICOIR, R. and HENRY, Y. (1995) Foreign gene delivery into monocotyledonous species. *Biotechnol. Adv.* **13**, 653–671.

WALDEN, R. and WINGENDER, R. (1995) Gene-transfer and plant-regeneration techniques. *Trends Biotechnol.* **13**, 324–331.

WALTER, C., APPLEBY, R. D., GARDNER, R., GRACE, L., HARGREAVES, C., STEWART, C. and SMITH, D. L. (1994) Biolistic transformation of *Pinus radiata* embryogenic tissues. In: *The Fourth Annual Queenstown Molecular Biology Meeting*, Queenstown, New Zealand.

WEIGEL, D. (1995) The genetics of flower development: from floral induction to ovule morphogenesis. *Annu. Rev. Genet.* **29**, 19–39.

WEIGEL, D. and MEYEROWITZ, E. M. (1994) The ABCs of floral homeotic genes. *Cell* **78**, 203–209.

WEIGEL, D. and NILSSON, O. (1995) A developmental switch sufficient for flower initiation in diverse plants. *Nature* **377**, 495–500.

WHETTEN, R. and SEDEROFF, R. (1995) Lignin biosynthesis. *Plant Cell* **7**, 1001–1013.

WILDE, H. D., MEAGHER, R. B. and MERKLE, S. A. (1992) Expression of foreign genes in transgenic yellow-poplar plants. *Plant Physiol.* **98**, 114–120.

Use of Molecular Methods for the Detection and Identification of Wood Decay Fungi

JOHN W. PALFREYMAN

19.1 Introduction

The use of molecular methods for organism and/or metabolite detection is a fundamental feature of many aspects of applied biology and biotechnology most notably in medicine but increasingly in other fields such as plant pathology and forensic science. Molecular tools have only recently been considered for the analysis of organisms which cause biodeterioration of wood. Reasons for this include the difficulty, until recently, of producing molecular probes to the main organisms involved, i.e. fungi; the relatively low value of primary timber products; and the lack of a perceived need to identify decay organisms as treatment systems were not organism specific. The drive to understand the decay process more fully and develop wood preservatives with less general toxicity and greater organism specificity, coupled with an increased understanding of the ecology of wood decay fungi led to the initiation of molecular studies of wood decay fungi and, more recently, to the reporting of many simple and sensitive test systems for these organisms. Currently the use of such systems is generally limited to the realms of research scientists though it seems certain that more commercial applications will develop in the next few years.

This chapter will: discuss some of the types of methodology which molecular biology offers to the field of wood biodeterioration; detail early work undertaken on the molecular identification of the basidiomycete *Serpula lacrymans*; discuss various types of immunoassay system now available for other important basidiomycetes; introduce the range of DNA based systems now available; and discuss some applications of DNA technology to wood biodeterioration.

19.2 Methodologies

Detection and identification of wood decay fungi is possible using a range of different techniques ranging from classical identification via morphology, isolation and study of mating patterns through to DNA sequencing. Currently the most widely utilized systems depend on protein analysis, antibody binding or DNA sequence recognition. Other systems such as lipid analysis, enzyme profiles and volatile production are also possible though are less widely used at present. All systems offer the possibility of the detection of wood decay organisms early on in the decay process, i.e. during incipient decay; this is a critical period if remedial treatment is to be undertaken before wood structure is compromised. The difficulties associated with recognizing incipient decay have been detailed by Schultz and Nicholas (1987).

19.2.1 Protein Analysis

Foremost among the protein based technologies available for fungal identification and detection are various refinements of electrophoresis which allow the separation of individual protein components of complex systems. After separation proteins can be detected by various systems which include the use of protein binding dyes/stains such as coommassie blue, or the more sensitive silver stains or specific protein probes such as polyclonal or monoclonal antibodies (Western or immunoblotting). Sub-classes of proteins such as glycoproteins can be detected using general carbohydrate binding stains or more specific stains such as enzyme bound lectins. None of these systems depend on protein activity being maintained during the separation stage and they can all be linked with the sodium dodecyl sulphate polyacrylamide gel electrophoresis system (SDS-PAGE) introduced by Laemmli (1970). Such protein denaturing electrophoresis systems cannot be used if the protein identification after separation depends on its retained activity, i.e. the protein remaining in a non-denatured (native) configuration. For example identification of enzyme activity after separation cannot be coupled with SDS-PAGE; the alternative isoelectric focusing gel electrophoresis is used since the separated proteins remain in their native folded state in this technique. The combination of SDS-PAGE with isoelectric focusing allowed the development of two-dimensional separation techniques (O'Farrell, 1975) with the possibility of identifying many thousands of individual protein species from cell extracts (Strahler *et al.*, 1989).

The output from most of these techniques is a protein fingerprint either as a series of bands on a gel or, in the case of 2-D analysis, a series of spots. The relative positions of bands/spots can be used not just to identify organisms but also to reveal/confirm phylogenetic relationships as reported, for example, for the tree pathogen *Heterobasidion annosum* (Karlsson and Stenlid, 1991).

19.2.2 Antibody Based Systems (Immunotechnology)

The mammalian immunological system has developed around the exquisite ability of specific proteins (antibodies) to detect particular molecular shapes (antigens). It is possible for antibodies to be produced which can specifically interact with any shape that an organism recognizes as foreign, and these antibodies can be easily produced for *in vitro* studies. Antibodies produced *in vivo* (polyclonal) or *in vitro* (monoclonal) can be used in a whole range of detection systems such as Western blotting, immunocytochemistry, enzyme linked immunoassays, in mechanistic studies on antigen function and in purification systems. Polyclonal antiserum produced against whole organisms has not been widely used in fungal analysis because of the difficulty of producing specific reagents; however the introduction of exoantigen technology by Kaufman and Standard (1987) and its application to monoclonal antibody production by Dewey *et al.* (1989) has resulted in a reliable method for reagent production. (Exoantigens are essentially proteins secreted by, or loosely attached to the external surface of fungi, often their actual identity remains unclear.) An alternative approach to the development of specific immunological reagents has been the production of monoclonal antibodies to specific, highly purified enzymes, e.g. the β-1,4-xylanase of *Postia placenta* as reported by Clausen *et al.* (1993)

19.2.3 *DNA Based Techniques*

The application of DNA based techniques for the detection and identification of fungi has been restricted until recently by the lack of information available about fungal DNA sequences. This situation has been completely transformed in the last 5 years by the introduction of techniques dependent on the polymerase chain reaction (PCR). Using PCR it is possible to amplify specific DNA sequences even in the absence of any sequence knowledge. The system relies on the binding of short, synthetically produced DNA molecules (called primers) to native, for example fungal, DNA and the amplification of the DNA in the intervening space between binding sites. Sufficient amplified DNA can be produced to allow it to be easily detected after separation on gel electrophoresis systems. Assuming that more than one intervening sequence can be amplified from the DNA of any one organism, a fingerprint, analogous to a protein fingerprint, can be produced for the organism.

Initial PCR techniques still required some sequence information to be available as primers need to be complementary with sequences on the genomic DNA. For fungi it was possible to use well conserved sequences found in ribosomal DNA as the basis for primers; however the introduction of the RAPD–PCR (random amplified polymorphic DNA–PCR) system, which utilizes random primer sequences (Caetano-Annoles *et al.*, 1991) has overcome the necessity for any genomic DNA information. RAPD–PCR depends on the likelihood of short arbitrary DNA primers binding to genomic DNA in such a way as to allow amplification. Essentially it is necessary that bindings sites are not too distant and that there are not too many of them. Generally 10 base primers are used and ideally amplification of perhaps 5–15 intervening sequences from genomic DNA is required, thus allowing a reasonable number of bands to be produced on a fingerprint gel. When developing RAPD–PCR identification systems for a new organism it is normal to try out a range (up to 20) of different random primers to find the one which produces the most useful information. For further information about PCR technologies a range of texts are available (e.g. Innis *et al.*, 1990; McPherson *et al.*, 1995). A description of some of the techniques currently used in fungal systems is given in a later section of this chapter.

Fundamental distinctions exist between detection and identification systems based on DNA analysis and protein analysis. DNA based systems do not rely on expression; the material detected is always present in an organism. By contrast proteins may not be expressed throughout the life cycle of an organism, may be modified at different stages, and their absence cannot be used as proof for absence of a specific organism. DNA systems also have limitations however since DNA sequences may well vary both within and between species and the use of DNA based systems to demonstrate identity or non-identity of samples is not as straight-forward as appears at first sight. Both types of system offer the possibility of detecting organisms, or their products, in the presence of contaminating material such as the appropriate fungal growth medium, for example wood.

19.3 Applications of Molecular Technology to the Wood Biodeteriogen, *Serpula lacrymans*

The basidiomycete fungus *S. lacrymans*, is a major biodeteriogen of timber in houses in various temperate regions of the world. The fungus is notable for its ability to cause major devastation to building timbers, its apparent ability to transport water over large distances through cord-like structures (rhizomorphs), its environmental sensitivity notably to mildly elevated temperatures and humidity levels, and, despite its prevalence in badly designed and maintained buildings, its apparent absence from the natural environment in areas where it causes most devastation to buildings. Molecular techniques have started to be applied to investigate some of these features of the organism.

Molecular identification of *S. lacrymans* was first reported by Vigrow *et al.* (1989) and by Schmidt and Kebernik (1989). Both groups used SDS–PAGE for identification and both groups commented on the high degree of similarity between *S. lacrymans* isolates originating from diverse areas of the world, and the ability of SDS–PAGE to discriminate between *S. lacrymans* and the closely related organism *S. himantioides*. Further reports of protein analysis of *S. lacrymans* using lectin markers and polyclonal antisera (in Western blots) have confirmed these observations (Vigrow *et al.*, 1991a,b; Palfreyman and Vigrow, 1994; White *et al.*, 1996a) although one 'rogue' isolate of *S. lacrymans*, BF-050, has been identified; this may be an intermediate between *S. lacrymans* and the closely related *S. himantioides* (Palfreyman *et al.*, 1995).

RAPD–PCR has also been used to identify relationships between isolates of *S. lacrymans* (Palfreyman *et al.*, 1995; Theodore *et al.*, 1995). The latter authors report that isolates of *S. lacrymans* which appear similar by SDS–PAGE can be differentiated by RAPD–PCR, indicating that the latter technique may have more use in epidemiological and phylogenetic studies. Specific isolates from Germany, Poland and Australia could be differentiated from a second Australian strain and one from Japan. The recent confirmation by SDS–PAGE of the isolation of *S. lacrymans* from the wild, i.e. not from within buildings (White *et al.*, 1996a) suggests that interesting studies on the natural origin of the fungus and its spread around the world will probably be susceptible to PCR analysis.

The detection of wood decay fungi within the natural substrate can be achieved in the laboratory by the use of SDS–PAGE on extracts from artificially infected wood blocks (Vigrow, 1992); however once field material is used interference from wood products masks banding patterns and prevents identification of fungi. The use of antibody based systems for detection has proved of more value and initial studies by Goodell and Jellison (1986), Palfreyman *et al.* (1987) and Breuil *et al.* (1988) on *Postia placenta*, *Neolentinus lepideus* and *Ophiostoma* sp. C28, respectively, indicated the potential of these systems. Detection of *S. lacrymans* using an antiserum was first reported by Vigrow *et al.* (1991c) who showed that even using a very non-specific reagent (i.e. an antiserum that could react with a wide range of different fungi) identification of *S. lacrymans* was possible if the correct immunological system was used; in this case Western blotting. A much more specific reagent was reported by Toft (1993), viz. an antiserum to an unidentified 23 kDa antigen present in *S. lacrymans* but not in *S. himantioides*. Unfortunately neither of these antiserum based reagents allowed detection of fungus in field samples but this has now been overcome by the advent of more specific monoclonal based reagents (Glancy and Palfreyman, 1993; Burge *et al.*, 1994). The extent to which such systems can become of practical use in identification of *S. lacrymans* remains to be demonstrated.

As noted previously a possible disadvantage of any detection/identification system based on proteins is that differential expression of specific proteins may occur. For example Vigrow *et al.* (1991a) showed that 'old' and 'juvenile' regions of an agar culture plate of *S. lacrymans* demonstrated different antigens. Such results could be either an indication of differential enzyme expression in different colony regions, as reported for proteinases by Venables and Watkinson (1989), or related to structural changes occurring since old mycelium often demonstrates cord formation (e.g. Vigrow *et al.*, 1991a). Other changes in protein expression during the life cycle of *S. lacrymans* are associated with environmental stress either by changes in the physical or biological environment of the organism. Thus analysis of the heat shock response of *S. lacrymans* (i.e. the reaction of the organism to stress) reveals changes in expression of specific genes (Palfreyman *et al.*, 1995; Sienkiewicz *et al.*, 1997) and differences in enzyme expression have been shown during interactions of *S. lacrymans* with potential control fungi such as certain *Trichoderma* spp. (Score *et al.*, 1997). These data indicate that the use of unidentified antigens to produce monoclonal antibodies, as undertaken by Glancy and Palfreyman (1993) and Burge *et al.* (1994) may be less reliable as the basis of a detection system than the use of a specific enzyme such as the xylanase of *P. placenta* (Clausen *et al.*, 1993). The recent demonstration (Low, Palfreyman and White, unpublished observations) that air movement can affect the morphology and growth patterns of *S. lacrymans* indicates yet another instance where protein expression will be variable.

This discussion is generally limited to detection and identification systems based on macromolecules. However indications that fungi such as *S. lacrymans* can be identified from their volatile emissions suggest another potentially useful system based on much smaller markers (Esser and Tas, 1992; Bjurman and Kristensson, 1992). Compounds such as ethyl benzene, 1-methoxy-2-propanol, furfural, acetone and ethanol have all been reported in air samples from *S. lacrymans* cultures. None of these compounds is likely to be *S. lacrymans* specific but fingerprints of the range of substances should aid identification. The major obstacle to be overcome before such methods can be routinely used relates again to differential expression of volatiles. It seems highly likely that substrates, competing fungi, environmental conditions, etc. will all impact on the actual volatile production by a fungus at any particular moment. However the well-known ability of both experienced remedial treaters and appropriately trained dogs to detect the rot caused by *S. lacrymans* indicates that this will be a valuable line of research.

19.4 Uses of Immunotechnology

Immunoassay systems are now widely used to detect pathogenic fungi both in medical and plant systems. Their application in areas related to forest products technology are much more limited, particularly in situations related to harvested wood. Assays for *P. placenta*, a major brown-rot organism, and *Ophiostoma piceae* have however been described, as have assays for reagents designed to prevent colonization of wood by fungi.

19.4.1   Postia placenta *and Other Brown-Rot Fungi*

Postia placenta, also known as *Poria placenta*, can be regarded as a 'typical' brown-rot fungus and is used by many groups as a representative test organism. Goodell *et al.* (1988) reported on the serological detection of *P. placenta* using a rather non-specific antiserum and subsequently demonstrated the inhibitory effects of wood extractives on immunoassay systems (Jellison and Goodell, 1989) which, while they can no doubt be overcome by suitable extraction procedures, need always to be considered if assays are behaving in an anomalous fashion. Monoclonal antibodies derived against both *P. placenta* and purified Mn(II)-peroxidase were subsequently used by Daniel *et al.* (1991) to identify sites of action of enzymes in actively decayed wood. Mn(II)-peroxidase was found in the lignin-rich cell corner area of the middle lamella whereas antibodies developed against the brown-rot fungus localized to cellulose rich regions of the wood cell and were only noted in poorly lignified areas of the middle lamella. The ability, as would be expected, of degrading enzymes to be found at some distance from the fungal hyphae explains the results reported by Galbraith and Palfreyman (1993) who showed positive detection of the heart-rot fungus *H. annosum* with a highly specific monoclonal antibody in regions somewhat distant from the actual fungus. Studies on the mechanisms of action of brown-rot fungi have been helped by immunoelectron microscopy where penetration of labelled antibodies into partially degraded timber was observed by Srebotnik and Messner (1991) and helped confirm that the initial phases of degradation must be undertaken by relatively small molecules.

Other mechanisms of the wood decay process have been studied by Jellison *et al.* (1991) based on the ability to produce antibodies to a wide range of types of molecule. Thus siderophores have been visualized in *Gloeophyllum trabeum* colonized wood, and measured in immunoassay systems in both wood extracts and liquid cultures.

Quantification of brown-rot fungi, or their metabolites, in wood has been reported by Burge *et al.* (1994) (*S. lacrymans*), Kim *et al.* (1991) (*P. placenta*), and Clausen *et al.* (1991) (a range of six common brown-rot fungi, including *P. placenta*, *Gloeophyllum trabeum*, *Antrodia carbonica*, *Lentinus lepideus*, *Serpula incrassata* and *Coniophora puteana*). While there can be little doubt that quantification of specific metabolites is relatively easily achieved, quantification of whole organisms is more problematic due, among other reasons, to differential expression of target antigens. Vigrow (1992) demonstrated for *S. lacrymans* that at very high weight losses in infected blocks relatively lower immunoassay results were found. As well as expression changes, autolysis and reutilization of resources could explain this result. Linking immunoassay results to biomass is thus likely to prove difficult. However there is no doubt that immunoassay can detect colonizing organisms even at early stages of decay and positive immunoassay results have been reported in wood blocks giving 3.6 per cent weight loss in *A. carbonica* and *L. lepideus* (Clausen *et al.*, 1991) and *P. placenta* (Clausen *et al.*, 1993) and 1.6 per cent weight loss in *S. lacrymans*

(Vigrow *et al.*, 1991c). Since weight losses of these amounts are generally considered to be within the experimental error of decayed block experiments it is reasonable to assume that fungal colonization in the absence of weight loss can be detected by immunoassay.

19.4.2 Types of Immunoassay

The majority of assays used in reports on wood decay fungi have been based on simple enzyme linked reagents. As expected, assay type can have important effects on both sensitivity and specificity (Kim *et al.*, 1991), and Vigrow *et al.* (1991c) reported on the differential sensitivity of an *S. lacrymans* antiserum when used in a dot blot assay and a Western blot. A novel, dyed particle immunoassay has been described by Clausen (1994) which may be particularly appropriate to the detection of incipient decay caused by brown-rot fungi. This assay uses a xylanase specific monoclonal antibody immobilized to an area (the capture zone) of a hydrophobic polyester cloth. Polyclonal-labelled latex particles are applied to one end of the strip which is dipped into extracts of fungally infected wood blocks. Movement of the latex particles is inhibited by the presence of fungal antigens which effectively cross-link the monoclonal and polyclonal antibodies. Movement, or lack of movement, of the latex particles can be used as an assessment of colonization. The assay is reported to be able to detect fungus in blocks with a weight loss of <2 per cent. The simplicity of this assay, together with the lack of need for any specialized detection equipment, suggest that it could become a widely used field system.

19.4.3 Detection of Sapstain Fungi

Using a polyclonal antiserum of probably low specificity, Breuil *et al.* (1988) demonstrated that detection of the sapstain organism *Ophiostoma* sp. C28 was possible before discoloration of wood became apparent. As would be expected for an immunological technique, sensitivity was very high and less than 1 μg of fungal biomass could be detected in 1 mg dry weight wood blocks. A monoclonal antibody produced by immunization with a cell wall protein extract from the sapstain fungus *Ophiostoma piceae* subsequently proved useful in early detection of the fungus (Banerjee *et al.*, 1994) as well as in distinguishing the fungus from a wide range of potential biocontrol organisms. It is hoped that this antibody will allow study both of the colonization of wood by *O. piceae* and the mechanisms by which the fungus gains access to woody nutrients.

19.4.4 Detection of Biocides

Because of the diversity of antibodies it is realistic to assume that it is possible to develop them against almost any molecular shape. However, size does limit the ability to produce reagents and molecules of <1000 Da are not normally immunogenic – if injected into an animal they will not initiate the production of specific antibodies. This can be overcome quite simply by linking small molecules to larger carrier proteins (for example, bovine serum albumin) and using the complex as an immunogen. In this way it is possible to visualize the production of immunological reagents designed to detect many of the chemicals used in wood preservation. The only limitation to such a system would be that only organic based preservatives could be used as effective immunogens. Reports of a sensitive assay to hexaconazole (5-(2,4-dichlorophenyl)-5-hydroxy-6-(1H-1,2,4-triazol-1-yl)hexanoic acid) demonstrate the potential for biocide detection (Chen *et al.*, 1996). This assay can detect hexaconazole at levels as low as 0.1 ng/ml and though cross-reaction was found with some related compounds appropriate specificity was reported. Although this assay is specifically designed to detect hexaconazole as an environmental pollutant there is little doubt that it could be used as a simple test to confirm the appropriate treatment of wood. The same group (Chen *et al.*, 1995) have also developed an immunoassay for didecyldimethylammonium chloride (DDAC) and both assays have been used to confirm the active ingredients of biocide formulations. The advent of simple biocide assays based on immunological reagents opens up the possibility of easy monitoring of biocides in treated wood which will aid in the production of better preservation treatment systems as well as resolving issues related to efficacy of prior treatment. Both the DDAC and the hexaconazole assays offer a simplicity of use, the ability to function in the presence of a wide range of non-target molecules, an ability to supply accurate measurements in the absence of complex extraction procedures and appropriate sensitivity. It is highly likely that immunoassays to other biocides used in the forest products industry will appear in the next few years.

Finally, from the same research group, comes an assay developed for a typical resin acid, dehydroabietic acid (DHA) (Li *et al.*, 1994). This assay may have a role in the monitoring of effluent from pulp mills and has a sensitivity of 1.9 ppb. The potential of immunoassay systems in the forest products sector is only now starting to be realized and we can expect assays to be developed for many other molecules used, or produced, by the industry.

19.5   DNA Systems

Application of DNA systems to identification and detection of wood decay fungi is still in its infancy. However, given the power of the techniques developed in the last 5 years, there can be little doubt of the impact they will make. As well as RAPD–PCR already discussed, PCR using specific primers, RFLP analysis (restriction fragment length polymorphisms), mitochondrial DNA analysis, sequence information based on ribosomal DNA (rDNA), rDNA analysis and various other techniques have been applied to other types of fungi and applications to wood biodeteriogens will occur in the coming years. RAPD–PCR and rDNA analysis have already proved useful in characterizing both *S. lacrymans* and various potential biodeteriogen control fungi as well as the edible shiitake mushroom belonging to *Lentinula* spp.

19.5.1   PCR Using Specific Primers

In the absence of specific sequence information for particular biodeteriogenic organisms, highly conserved sequences, found in all fungi, can be used as the recognition sequences for PCR primers. The most widely utilized sequences are those bounding the internal transcribed spacer (ITS) regions of ribosomal DNA (rDNA sequences). Primers which can be used in ITS amplification have been described by White *et al.* (1990). The ITS regions of rDNA evolve relatively rapidly whereas their flanking regions, which are transcribed to form the small and large rRNA molecules, are highly conserved. Variation in the ITSs can be revealed, after amplification, by restriction enzyme digestion to reveal any polymorphisms (RFLPs – sequence differences) between fungal isolates. While identification of fungi by this method may not be appropriate it is particularly useful for determining relationships between closely related organisms (phylogenetic trees) as well as detecting suspected microorganisms in particular ecosystems. The ITS/RFLP methodology, like all DNA based techniques, is not affected by the morphological form of a fungus. New developments of this technology are appearing all the time, many coming from the medical field. For example Walsh *et al.* (1995) undertook PCR amplification of a conserved region of the 18s rRNA region of *Candida* spp. followed by electrophoresis of the amplicon (amplified region) to reveal so-called 'single-strand length conformational polymorphisms'. Differences in sequence, down to one base pair, are revealed by changes in mobility on gels due to alterations in DNA tertiary structure.

19.5.2   RFLP Analysis

Given known DNA sequences or specific DNA probes (i.e. DNA molecules capable of binding to genomic DNA), RFLP analysis is possible without prior PCR amplification. In this technique DNA is isolated from pure cultures, digested with a range of restriction enzymes, separated on agarose or polyacrylamide gels, then hybridized with a radioactive or enzyme labelled DNA probe. Differences in banding patterns, caused by the presence or absence of specific restriction enzyme recognition sequences, reveal relationships between organisms. Again this technique is most useful for confirming the identity of organisms and/or analyzing phylogenetic relationships among species/isolates.

19.5.3   Mitochondrial DNA Analysis

Although ITS sequences evolve rapidly due to lack of conservation pressure and nuclear rDNA sequences are relatively highly conserved, mitochondrial rDNA genes evolve at an intermediate level and are therefore useful at the ordinal or family level. While universal primers capable of amplifying sequences from all fungi cannot be based on mitochondrial rDNA, primers which react differently, for example with ascomyctes and some basidiomycetes, have been reported (White *et al.*, 1990).

19.5.4   PCR Followed by Sequence Analysis

The use of universal rDNA primers allows the production of species-, perhaps isolate-, specific DNA probes if amplification is followed by sequence analysis.

Such highly specific probes could be of particular use in the analysis of fungal communities where large numbers of organisms may be present in relatively small locales. Simple PCR, either by RAPD–PCR or using universal primers, is unlikely to produce unambiguous fingerprinting data. Analysis of community structures is of interest in ecological studies of wood decay (e.g. White *et al.*, 1996b) and might also be important in assessing the role and performance of potential biocontrol agents. An example of the use of this technique in complex ectomycorrhizal communities, and comparison with other potential technologies, is given by Egger (1995). Of course, due to polymorphisms within DNA sequences, information may, in many instances where detection or identification is required, be too precise for the practical mycologist. Sequence specific probes are likely to remain in the realm of the research scientist for the present.

19.5.5 *Quantitative PCR*

Compared with protein based technologies and particularly immunoassay systems, PCR based techniques do not offer the basis for simple quantification. However, recent developments have been made in this area and a number of systems have been reported (e.g. Kang *et al.*, 1995; Siebert and Kellogg, 1995). Kang *et al.* (1995) described PCR linked to temperature gradient gel electrophoresis (TGGE), an electrophoresis system capable of distinguishing amplified DNA molecules which differ in single nucleotide residues. Target DNA is mixed with standard DNA which is almost identical to the relevant amplicon. DNA is amplified and homo- and heteroduplex molecules develop which migrate at different rates in TGGE. The ratio between hetero- and homo-duplexes will be related to the original ratio of target and standard DNA. The Siebert and Kellogg (1995) technique depends on the development of a PCR–MIMIC. Here standard DNA molecules containing unique restriction sites are used as competitor sequences in the PCR reaction. Neither of these techniques matches the simplicity of, for example, enzyme based immunoassay systems and advances in PCR based quantification methods can be expected in the next few years.

19.5.6 *Applications*

Specific applications of DNA technology for the detection and identification of wood decay fungi are limited. Two areas which are demonstrating the potential are now discussed.

Typing Trichoderma *isolates using PCR fingerprinting*

Kuhls *et al.* (1995) reported on the use of PCR for the comparison of ex-type strains of *Trichoderma* spp. To validate the technique the group considered identical strains deposited in various culture collections around the world. PCR fingerprinting was undertaken using a standard protocol and with primers – (GACA)$_4$, (GTG)$_5$ and an M13 phage core sequence. Strains/species of *Trichoderma* studied included *T. longibrachiatum*, *T. pseudokoningii*, *T. harzianum* and *T. reesei*. Different strains of, for example, *T. longibrachiatum* showed differences of an average 20 per cent in banding patterns. By contrast ex-type strains showed identical PCR profiles. This was true even when they had been held, and cultivated, in separate culture collections over many years. Though some minor differences in band intensity were seen after subculturing, this seemed to be a result of differing DNA quality or concentration. Similarities in banding profiles of *T. parceramosum* and *T. todica* allowed the authors to conclude that these species are conspecific, a result supported by the identify of their ITS sequences and their morphology.

Schlick *et al.* (1994) used the methodology of Kuhls *et al.* (1995) to identify strains of *T. harzianum* of potential use in biocontrol. The importance of being able unambiguously to identify such strains, particularly when, as in the case of Schlick *et al.* (1994), they have been patented, is apparent. Even closely related strains, for example gamma-ray induced mutants of *T. harzianum*, could be distinguished by PCR fingerprinting. In the longer term, proof of efficacy or otherwise of organisms, by demonstrating their longevity in a biocontrol situation, will no doubt be reported after identification by PCR fingerprinting.

Shiitake Mushroom Studies

At least three types of analysis have been undertaken to demonstrate relationships between isolates and species of the edible shiitake mushroom (*Lentinula*, Tricholomataceae): isozyme analysis; rDNA ITS sequences; and mitochondrial DNA (mtDNA) sequences (Hibbett *et al.*, 1995). Phylogenetic trees for isolates from Japan, Thailand, Borneo, New Guinea, Tasmania and New Zealand were reported by Hibbett *et al.* (1995). While the stimulus for this work was academic, in that disagreements over species limits for *Lentinula* have been discussed for many years, there are practical consequences of the work. Most breeding cultivars of shiitake were shown to be derived from Japanese, Chinese or Korean strains. The wild population of shiitake present outside north-east Asia represents a rich genetic pool which should be useful for future breeding programmes. The ability to identify specifically new commercial strains by DNA technology will help in the patenting process. In addition, Hibbett *et al.* (1995) discuss the potential threat to indigenous shiitake populations by exogenous genotypes coming from mushroom farms. The result of this could be a loss in genetic diversity just at the time when expansion of shiitake cultivation seems assured. Again good identification systems will at least allow events to be monitored. At an academic level the studies of Hibbett *et al.* (1995) reveal that the ancestral area for shiitake is probably the South Pacific (the most diverse area studied), that various unidentified isolates were in fact *L. edodes sensu stricto*, and that nuclear ITSs and mtDNA may have different evolutionary histories. It seems certain that DNA studies of this type will also resolve disputes regarding the taxonomy of *Lentinula*, which currently includes at least five species as defined by morphology – *L. guarapiensis, L. boryana*, (the type species) *L. lateritia, L. novaezelandieae*, and the commercially important *L. edodes*.

19.5.7   Other Nucleic Acid Based Systems

A wide range of other techniques are now available which may impact on detection and identification of wood biodeteriogens in the coming years. Examples are: use of PCR fingerprinting of microsatellite DNA (important in typing yeast and *Candida* strains; Lieckfeldt *et al.*, 1993); electrophoretic analysis of tRNA preparations (can be used to type and identify *Candida* species; Santos *et al.*, 1994); PCR-coupled ligase chain reaction (used to distinguish between isolates of *Erwinia stewartii* differing by single base changes; Wilson *et al.*, 1994); karyotype analysis, which can now be undertaken electrophoretically; and PCR/sequencing, which can be carried out simultaneously as described by Deng *et al.* (1993) for various types of microorganism.

References

BANERJEE, S., LITTLE, J., CHAN, M., LUCK, B. T., BREUIL, C. and BROWN, D. L. (1994) Production and characterisation of monoclonal antibodies to the sap staining fungus *Ophiostoma piceae*. *Can. J. Microbiol.* **40**, 35–44.

BJURMAN, J. and KRISTENSSON, J. (1992) *Analysis of Volatile Emissions as an Aid in the Diagnosis of Dry-Rot*, Document No. IRG/WP/2393-92, International Research Group on Wood Preservation.

BREUIL, C., SEIFERT, K. A., YAMADA, J., ROSSIGNOL, L. and SADDLER, J. N. (1988) Quantitative estimation of fungal colonisation of wood using an enzyme-linked immunosorbent assay. *Can. J. For. Res.* **18**, 374–377.

BURGE, M. N., MSUYA, J. C., CAMERON, M. and STIMSON, W. M. (1994) A monoclonal antibody for the detection of *Serpula lacrymans*. *Mycol. Res.* **98**, 356–362.

CAETANO-ANNOLES, G., BASSAM, B. J. and GRESHOFF, P. M. (1991) DNA amplification fingerprinting using very short arbitrary primers. *Biotechnology* **9**, 553–557.

CHEN, T., DWYREGYGAX, C., SMITH, R. S. and BREUIL, C. (1995) Enzyme linked immunosorbent assay for didecyldimethylammonium chloride, a fungicide used by the forest products industry. *J. Agric. Food Chem.* **43**, 1400–1406.

CHEN, T., DWYREGYGAX, C., HADFIELD, S. T., WILLETTS, C. and BREUIL, C. (1996) Development of an enzyme-linked-immunoabsorbent assay for a broad spectrum triazole fungicide – hexaconazole. *J. Agric. Food Chem.* **44**, 1352–1356.

CLAUSEN, C. A. (1994) Dyed particle capture immunoassay for detection of incipient brown-rot decay. *J. Immunoassay* **15**, 305–316.

CLAUSEN, C. A., GREEN, F. and HIGHLEY, T. L. (1991) Early detection of brown-rot decay in Southern-Yellow pine using immunodiagnostic procedures. *Wood Sci. Technol.* **26**, 1–8.

CLAUSEN, C. A., GREEN, F. and HIGHLEY, T. L. (1993) Characterisation of monoclonal antibodies to wood derived beta-1,4-xylanase of *Postia placenta* and their application to the detection of incipient decay. *Wood Sci. Technol.* **27**, 219–228.

DANIEL, G., JELLISON, J., GOODELL, B., PASZCZYNSKI, A. and CRAWFORD, R. (1991) Use of monoclonal antibodies to detect Mn(II)-peroxidase in birch wood degraded by *Phanerochaete chrysoporium*. *Appl. Microbiol. Biotechnol.* **35**, 674–680.

DENG, S. J., FORSTER, R. J., HIRUKI, C. and TEATHER, R. M. (1993) Simultaneous amplification and sequencing of genomic DNA (SAS). *J. Microbiol. Methods* **17**, 103–113.

DEWEY, F. M., MACDONALD, M. M. and PHILLIPS, S. I. (1989) Development of monoclonal-antibody-ELISA, dot-blot and dip stick immunoassays for *Humicola lanuginosa* in rice. *J. Gen. Microbiol.* **28**, 719–725.

EGGER, K. N. (1995) Molecular analysis of ectomycorrhizal fungal communities. *Can. J. Bot.* **73**, S1415–S1422.

ESSER, P. M. and TAS, A. C. (1992) *Detection of Dry Rot by Air Analysis*, Document No. IRG/WP/2399-92, International Research Group on Wood Preservation.

GALBRAITH, D. and PALFREYMAN, J. W. (1993) Detection of *Heterobasidion annosum* using monoclonal antibodies. In: Shot, A., Dewey, F. M. and Oliver, R., eds, *Modern Assays for Plant Pathogenic Fungi*, Oxford: CAB International, pp. 105–110.

GLANCY, H. and PALFREYMAN, J. W. (1993) *The Development and Use of Monoclonal Antibodies to the Dry Rot Fungus* Serpula lacrymans, Document No. IRG/WP/93-10004, International Research Group on Wood Preservation.

GOODELL, B. S. and JELLISON, J. (1986) *Detection of a Brown-Rot Fungus using Serological Assays*, Document No. IRG/WP/1305, International Research Group on Wood Preservation.

GOODELL, B. S., JELLISON, J. and HOSLI, J. P. (1988) Serological detection of wood decay fungi. *For. Prod. J.* **38**, 59–62

HIBBETT, D. S., FUKUMASA-NAKAI, Y., TSUNEDA, A. and DONOGHUE, M. J. (1995) Phylogenetic diversity in shiitake inferred from ribosomal DNA sequences. *Mycologia* **87**, 618–638.

INNIS, M. A., GELFAND, D. H., SNINSKY, J. J. and WHITE, T. J. (1990). *PCR Protocols: A Guide to Methods and Applications*, San Diego: Academic Press.

JELLISON, J. and GOODELL, B. (1989). Inhibitory effects of undecayed wood and the detection of *Postia placenta* using the enzyme-linked immunosorbent assay. *Wood Sci. Technol.* **23**, 13–20.

JELLISON, J., CHANDHOKE, V., GOODELL, B. and FEKETE, F. A. (1991) The isolation and immunolocalisation of iron-binding compounds produced by *Gloeophyllum trabeum*. *Appl. Microbiol. Biotechnol.* **35**, 805–809.

KANG, J., KUHN, J. E., SCHAFER, P., IMMELMANN, A. and HENCO, K. (1995) Quantification of DNA and RNA by PCR. In: Rickwood, D. and Hames, D. D., eds, *PCR 2: A Practical Approach*, Oxford: IRL Press, pp. 119–132.

KARLSSON, J. O. and STENLID, P. G. (1991). Pectic isozyme profiles of intersterility groups in *Heterobasidion annosum*. *Mycol. Res.* **95**, 537–542.

KAUFMAN, L. and STANDARD, P. G. (1987). Specific and rapid identification of medically important fungi by exoantigen detection. *Annu. Rev. Microbiol.* **41**, 209–225.

KIM, Y. S., JELLISON, J., GOODELL, B., TRACY, V. and CHANHOKE, V. (1991) The use of ELISA for the detection of white-rot and brown-rot fungi. *Holzforschung* **45**, 403–406.

KUHLS, K., LIECKFELDT, E. and BORNER, T. (1995) PCR-fingerprinting used for comparison of ex type strains of *Trichoderma* species deposited in different culture collections. *Microbiol. Res.* **150**, 363–371.

LAEMMLI, U. K. (1970) Cleavage of structural proteins during the assembly of the head of the bacteriophage T4. *Nature* **227**, 680–685.

LI, K., CHESTER, M., KUTNEY, J. P., SADDLER, J. N. and BREUIL, C. (1994) Production of polyclonal antibodies for the detection of dehydroabietic acid in pulp-mill effluents. *Anal. Lett.* **27**, 1671–1688.

LIECKFELDT, E., MEYER, W. and BORNER, T. (1993) Rapid identification and differentiation of yeasts by DNA and PCR fingerprinting. *J. Basic Microbiol.* **33**, 413–426.

McPHERSON, M. J., HAMES, B. D. and TAYLOR, G. R. (1995) *PCR 2: A Practical Approach*, Oxford: IRL Press.

O'FARRELL, P. H. (1975) High resolution two-dimensional electrophoresis of proteins. *J. Biol. Chem.* **256**, 4007–4021.

PALFREYMAN, J. W. and VIGROW, A. (1994) Molecular analysis of certain isolates of *Serpula lacrymans*. *FEMS Microbiol. Lett.* **117**, 281–286.

PALFREYMAN, J. W., BRUCE, A., BUTTON, D., GLANCY, H., VIGROW, A. and KING, B. (1987) Immunological methods for the detection and characterisation of wood decay basidiomycetes. In: Houghton, D. R., Smith, R. M. and Eggins, H. O. W., eds, *Biodeterioration 7*, London: Elsevier, pp. 709–713.

PALFREYMAN, J. W., WHITE, N. A., BUULTJENS, T. E. J. and GLANCY, H. (1995) The impact of current research on the treatment of infestations by the dry rot fungus *Serpula lacrymans*. *Int. Biodet. Biodeg.* **95**, 369–395.

SANTOS, M. A. S., ELADLOUNI, C., COX, A. D., LUZ, J. M., KEITH, G. and TUITE, M. F. (1994). Transfer-RNA profiling, a new method for the identification of pathogenic *Candida* species. *Yeast* **10**, 625–636.

SCHLICK, A., KUHLS, K., MEYER, W., LIECKFELD, E., BORNER, T. and MESSNER, K. (1994) *Application of DNA Fingerprinting Methods to Identify Biocontrol Strains of Fungi Imperfecti*, Document No. 94-10068, International Research Group on Wood Preservation.

SCHMIDT, O. and KEBERNICK, U. (1989) Characterization and identification of the dry rot fungus *Serpula lacrymans* by polyacrylamide gel electrophoresis. *Holzforschung* **43**, 195–198.

SCHULTZ, T. P. and NICHOLAS, D. D. (1987) Fourier transform infrared spectrometry. Detection of incipient brown rot decay of wood. *Int. Anal.* **1**, 35–39.

SCORE, A. J., PALFREYMAN, J. W. and WHITE, N. A. (1997) Extracellular phenoloxidase and peroxidase enzyme production during interspecific fungal interactions. *Int. Biodet. Biodeg.* **39**, 225–233.

SIEBERT, P. D. and KELLOGG, D. E. (1995). PCR MIMIC's: competitive DNA fragments for use in quantitative PCR. In: Rickwood, D. and Hames, B. D., eds, *PCR 2: A Practical Approach*, Oxford: IRL Press, pp. 135–148.

SIENKIEWICZ, N., BUULTJENS, T. E. J., WHITE, N. A. and PALFREYMAN, J. W. (1997) *Serpula lacrymans* and the heat shock response. *Int. Biodet. Biodeg.* **39**, 217–224.

SREBOTNIK, E. and MESSNER, K. (1991) Immunoelectron microscopic study of the porosity of brown-rot degraded pine wood. *Holzforschung* **45**, 95–101.

STRAHLER, J. R., KUICK, R. and HANASH, S. M. (1989) Two-dimensional electrophoresis. In: Creighton, T. E., ed., *Protein Structure. A Practical Approach*, Oxford: IRL Press.

THEODORE, M. L., STEVENSON, T. W., JOHNSON, G. C., THORNTON, J. D. and LAWRIE, (1995) Comparison of *Serpula lacrymans* isolates using RAPD-PCR. *Mycol. Res.* **99**, 447–450.

TOFT, L. (1993) Immunological identification *in vitro* of the dry rot fungus *Serpula lacrymans*. *Mycol. Res.* **92**, 273–277.

VENABLES, C. E. and WATKINSON, S. C. (1989) Production and localisation of proteinases in colonies of timber-decaying basidiomycete fungi. *J. Gen. Microbiol.* **135**, 1369–1374.

VIGROW, A. (1992) Molecular analysis of the dry rot fungus *Serpula lacrymans*. PhD Thesis, CNAA Dundee Institute of Technology.

VIGROW, A., BUTTON, D., PALFREYMAN, J. W., KING, B. and HEGARTY, B. (1989) *Molecular Studies on Isolates of* Serpula lacrymans, Document No. IRG/WP/1421, International Research Group on Wood Preservation.

VIGROW, A., KING, B. and PALFREYMAN, J. W. (1991a) Studies of *Serpula lacrymans* mycelial antigens by Western blotting techniques. *Mycol. Res.* **95**, 1423–1428.

VIGROW, A., PALFREYMAN, J. W. and KING, B. (1991b) On the identity of certain isolates of *Serpula lacrymans*. *Holzforschung* **45**, 153–154.

VIGROW, A., GLANCY, H., PALFREYMAN, J. W. and KING, B. (1991c) *The Antigenic Nature of* S. lacrymans, Document No. IRG/WP/1492, International Research Group on Wood Preservation.

WALSH, T. J., FRANCESCONI, A., KASIA, M. and CHANOCK, S. J. (1995) PCR and single-stranded conformational polymorphism for recognition of medically important opportunistic fungi. *J. Clin. Microbiol.* **33**, 3216–3220.

WHITE, T. J., BRUNS, T., LEE, S. and TAYLOR, J. (1990) Amplification and direct sequencing of fungal ribosomal RNA genes for phylogenetics. In: *PCR Protocols: A Guide to Methods and Applications*. San Diego: Academic Press, pp. 315–324.

WHITE, N. A., LOW, G. A., SINGH, J., STAINES, H. and PALFREYMAN, J. W. (1997) Isolation and environmental study of 'wild' *Serpula lacrymans* and *Serpula himantioides* from the Himalayan forests. *Mycol. Res.* **101**, 580–584.

White, N. A., Palfreyman, J. W. and Staines, H. J. (1996b) Fungal colonisation of the keelson timbers of a nineteenth century wooden frigate. *Holzforschung* **30**, 117–131.

Wilson, W. J., Wiedmann, M., Dillard, H. R. and Batt, C. A. (1994) Identification of *Erwinia stewartii* by a ligase chain reaction assay. *Appl. Environ. Microbiol.* **60**, 278–284.

Index